THE THEORY OF PROBABILITY

BY

B. V. GNEDENKO

TRANSLATED FROM THE RUSSIAN BY
B. D. SECKLER

Martino Publishing
Mansfield Centre, CT
2014

Martino Publishing
P.O. Box 373,
Mansfield Centre, CT 06250 USA

ISBN   978-1-61427-710-1

© 2014 Martino Publishing

All rights reserved. No new contribution to this publication may be reproduced, stored in a retrieval system, or transmitted, in any form or by any means, electronic, mechanical, photocopying, recording, or otherwise, without the prior permission of the Publisher.

Cover design by T. Matarazzo

Printed in the United States of America On 100% Acid-Free Paper

THE THEORY OF PROBABILITY

BY

B. V. GNEDENKO

TRANSLATED FROM THE RUSSIAN BY
B. D. SECKLER

CHELSEA PUBLISHING COMPANY
NEW YORK, N. Y.

SECOND EDITION

© 1962, CHELSEA PUBLISHING COMPANY
COPYRIGHT 1962, BY CHELSEA PUBLISHING COMPANY
COPYRIGHT © 1962, BY CHELSEA PUBLISHING COMPANY
LIBRARY OF CONGRESS CATALOGUE CARD NUMBER 61-13496

© 1963, CHELSEA PUBLISHING COMPANY

THE PRESENT WORK IS AN ENGLISH TRANSLATION, BY
PROF. B. D. SECKLER, OF THE RUSSIAN-LANGUAGE WORK

KURS TEORII VEROYATNOSTEI
BY PROF. B. V. GNEDENKO (Б. В. Гнеденко)

PRINTED IN THE UNITED STATES OF AMERICA

TRANSLATOR'S PREFACE

THIS BOOK IS ESSENTIALLY a close translation of an amplified version of the second Russian edition of Gnedenko's *Kurs Teorii Veroyatnostei*. While the translation was in preparation, Prof. Gnedenko was gracious enough to send the translator additional manuscript for incorporation into the text. Thus, the present work represents an even more valuable addition to the English scientific literature.

The book still retains its basic form, the eleven chapters of the present work corresponding exactly to the eleven of the original. However, for those who may have references to particular sections of the Russian edition or the German translation thereof, §§ 20-49 and §§ 50-64 are now §§ 21-50 and §§ 52-66, respectively. The major changes are as follows. The chapter on Markov chains has been considerably enlarged, § 17 and § 18 having been combined and augmented with some new material, and § 18 and § 20 being entirely new. Kolmogorov's inequality and his strong law of large numbers now appear as corollaries to an inequality of Hájek and Rényi. The sufficiency part of the proof of the Bochner-Khintchine Theorem is new, as is the entire section on the Poisson process. A number of smaller changes have been made within the various sections and in the exercises, but there is no need to mention them in detail.

I have made several amendments and corrections and have eliminated the typographical errors noted. In this connection, I should like to thank Prof. Warren Hirsch of New York University and Prof. Howard G. Tucker of Stanford University for their suggested changes in two of the theorems of the book, namely, the addition to the proof of the last theorem on the law of large numbers (beginning with the Borel-Cantelli Lemma) and the partial revision of Glivenko's Theorem, respectively. I have incorporated both into the present translation.

In the second edition of this translation, the answers to the Exercises have been included and all errata found since the first edition have been corrected.

B. D. SECKLER

PREFACE TO THE SECOND EDITION

THE PRESENT EDITION differs considerably from the first. In preparing it, I have endeavored to take into account as fully as possible the comments and suggestions contained in the reviews of the first edition as well as those communicated to me orally and in correspondence. Perhaps the most significant change has been the addition of exercises to the first nine chapters. Furthermore, I have eliminated the mathematical appendixes at the end of the book. The first appendix has been incorporated into the text in the first, fourth, and fifth chapters. A few other changes as well have been made in these chapters. The tenth chapter has been added to considerably, the additions being mainly concerned with the extension of the material on the theory of stationary stochastic processes. The last chapter, which is devoted to mathematical statistics, has also been very much modified. There are some new sections and, at the same time, some of the material of the first edition has been deleted. Thus, for example, the cumbersome proof of Kolmogorov's Theorem on the limit distribution of the maximum deviation of the empirical distribution from the true one has been omitted; also omitted is the section on sequential analysis. Finally, the misprints and errors noted in the first edition have been corrected.

I take this opportunity to thank cordially those of my colleagues who have given me their frank opinion on the book and whose criticisms have been responsible for the correction of its shortcomings. I especially wish to thank Prof. Yuri Linnik for his continuing interest in the present book, for his numerous comments on the first edition, and for his criticism of the manuscript of the second edition.

I realize that, even in its present form, the book is not free of defects, and I therefore ask the reader to please inform me of any shortcomings he may find in this second edition and also let me know of any suggestions he may have as regards the content and the arrangement of the material. I also wish to thank those who are kind enough to send me interesting problems for use in subsequent editions.

B. GNEDENKO

FROM THE PREFACE TO THE FIRST EDITION

THE PRESENT BOOK is divided into two parts: elementary (Chapters I-VI) and special (Chapters VII-XI). The last five chapters can be used as the basis for special courses on the theory of sums of random variables, the theory of stochastic processes, and the elements of statistics.

In this book, the theory of probability is studied as a mathematical discipline exclusively, and the acquisition of specific scientific or engineering results is therefore not an end in itself. All the examples cited in the text of the book are merely intended to elucidate the general principles of the theory and to indicate the connection between these principles and the problems of the natural sciences. Of course, at the same time, these examples indicate the possible areas of applicability of the general theoretical results as well as develop the reader's ability to apply these results to concrete problems. Such a course of study enables the reader to develop a distinctive intuition for probability theory that will enable him to foresee a probability-theoretic result in general terms, even before analytical techniques are brought into play. Further, we observe that probability theory cannot be studied—especially at first—without the systematic solution of problems.

The first four sections of the first chapter constitute a slightly modified version of some unpublished manuscripts of Andrei Kolmogorov.

It gives me great pleasure to thank my esteemed teachers, Professors Andrei Kolmogorov and Alexander Khintchine for the assistance and advice they have given me in discussions with them on the main questions of the theory of probability.

B. GNEDENKO

TABLE OF CONTENTS

Translator's Preface	1
Preface to the Second Edition	2
From the Preface to the First Edition	3
Introduction	9

I. The Concept of Probability 15

§ 1.	Certain, impossible, and random events	15
§ 2.	Different approaches to the definition of probability	18
§ 3.	The field of events	21
§ 4.	The classical definition of probability	26
§ 5.	Examples	30
§ 6.	Geometrical probability	37
§ 7.	The statistical definition of probability	44
§ 8.	Axiomatic construction of the theory of probability	51
§ 9.	Conditional probability and the simplest basic formulas	58
§ 10.	Examples	67
	Exercises	76

II. Sequences of Independent Trials 79

§ 11.	The probability $P_n(m_1, m_2, \ldots, m_k)$	80
§ 12.	The Local Limit theorem	84
§ 13.	The Integral Limit Theorem	93
§ 14.	Applications of the Integral Theorem of DeMoivre-Laplace	107
§ 15.	Poisson's Theorem	112
§ 16.	Illustration of the scheme of independent trials	118
	Exercises	122

III. MARKOV CHAINS 125

§ 17. Definition of a Markov Chain. Transition matrix..... 125
§ 18. Classification of possible states...................... 130
§ 19. A theorem on limiting probabilities.................. 132
§ 20. Generalization of the DeMoivre-Laplace Theorem to a sequence of chain-dependent trials.................. 136
EXERCISES 144

IV. RANDOM VARIABLES AND DISTRIBUTION FUNCTIONS...... 145

§ 21. Fundamental properties of distribution functions..... 145
§ 22. Continuous and discrete distributions............... 152
§ 23. Multi-dimensional distribution functions............ 157
§ 24. Functions of random variables..................... 166
§ 25. The Stieltjes integral.............................. 180
EXERCISES 185

V. NUMERICAL CHARACTERISTICS OF RANDOM VARIABLES.. 189

§ 26. Mathematical expectation 189
§ 27. Variance 195
§ 28. Theorems on expectation and variance.............. 202
§ 29. The definition of mathematical expectation in Kolmogorov's axiomatic treatment....................... 210
§ 30. Moments 213
EXERCISES 219

VI. THE LAW OF LARGE NUMBERS 222

§ 31. Mass phenomena and the Law of Large Numbers...... 222
§ 32. Tchebychev's form of the Law of Large Numbers...... 225
§ 33. A necessary and sufficient condition for the Law of Large Numbers 233
§ 34. The Strong Law of Large Numbers.................. 237
EXERCISES 247

VII. CHARACTERISTIC FUNCTIONS 250

§ 35. The definition and simplest properties of characteristic functions 250
§ 36. The inversion formula and the Uniqueness Theorem... 256
§ 37. Helley's theorems 263
§ 38. Limit theorems of characteristic functions........... 268
§ 39. Positive-semidefinite functions 273
§ 40. Characteristic functions for multi-dimensional random variables 277
EXERCISES 283

VIII. THE CLASSICAL LIMIT THEOREM 286

§ 41. Statement of the problem......................... 286
§ 42. Liapounov's Theorem 290
§ 43. The Local Limit Theorem......................... 295
EXERCISES 301

IX. THE THEORY OF INFINITELY DIVISIBLE DISTRIBUTION LAWS 303

§ 44. Infinitely divisible laws and their fundamental properties 304
§ 45. Canonical representation of infinitely divisible laws... 307
§ 46. A limit theorem for infinitely divisible laws........... 312
§ 47. Limit theorems for sums: Formulation of the problem. 315
§ 48. Limit theorems for sums......................... 316
§ 49. Conditions for convergence to the Normal and Poisson laws 320
EXERCISES 323

X. THE THEORY OF STOCHASTIC PROCESSES 325

§ 50. Introductory remarks 325
§ 51. The Poisson process.............................. 330
§ 52. Conditional distribution functions and Bayes' formula 338
§ 53. The generalized Markov equation.................. 342
§ 54. Continuous stochastic processes. Kolmogorov's equations 344

§ 55. Purely discontinuous stochastic processes. The Kolmogorov-Feller equations.................. 353
§ 56. Homogeneous stochastic processes with independent increments 360
§ 57. The concept of a stationary stochastic process. Khintchine's theorem on the correlation coefficient.......... 366
§ 58. The notion of a stochastic integral. Spectral decomposition of stationary processes...................... 374
§ 59. The Birkhoff-Khintchine Ergodic Theorem........... 378

XI. **ELEMENTS OF STATISTICS** 384

§ 60. Some problems of mathematical statistics............ 384
§ 61. Variational series and empirical distribution functions. 387
§ 62. Glivenko's Theorem and Kolmogorov's compatibility criterion 389
§ 63. Comparison of two distribution functions............ 395
§ 64. The concept of critical region. Type I and Type II errors. Comparison of two statistical hypotheses...... 401
§ 65. The classical procedure for estimating the distribution parameters 409
§ 66. Confidence limits 419

TABLES 429
BIBLIOGRAPHY 445
INDEX 455
ANSWERS TO THE EXERCISES 465

INTRODUCTION

THE PURPOSE OF THIS BOOK is to present the foundations of the theory of probability, the mathematical discipline that studies the laws governing random phenomena.

The origin of probability theory dates from the middle of the 17th century and is associated with the names of Huyghens, Fermat, Pascal, and James Bernoulli. Such important concepts as probability and mathematical expectation were gradually crystallized in the correspondence carried on between Pascal and Fermat, occasioned by problems on gambling that did not come within the scope of the mathematics of that time. It should, of course, be realized quite clearly that the outstanding scholars in dealing with problems on gambling also foresaw the fundamental role of the science that makes a study of random phenomena. They were convinced that random events of a mass nature could give rise to clear-cut laws. However, owing to the low level of development of the natural sciences at that time, games of chance continued for a long time to provide the only concrete basis for the development of the concepts and methods of probability theory. This also left its mark on the formal mathematical tools that were used to solve the problems which had arisen in probability theory: they boiled down to elementary arithmetical and combinatorial procedures exclusively. The subsequent development of probability theory as well as the wide attraction its results and methods of investigation have had for the natural sciences and, above all, for physics have shown that the classical concepts and classical methods have not lost their value even at the present time.

The great demands made by the natural sciences (the theory of errors, the problems of the theory of ballistics, the problems of statistics, primarily population statistics) made it necessary to develop the theory of probability further and to bring in more advanced analytical tools. An especially important part in developing the analytical methods of probability theory was played by DeMoivre, Laplace, Gauss, and Poisson. From the point of view of formal analysis, we may add to this the work of Lobachevsky—one of the inventors of non-euclidean geometry—devoted to the theory of errors in measurements on a sphere and motivated by his desire to establish the Geometry of the universe.

From the middle of the 19th century up to approximately the twenties of this century, the development of probability theory was associated in great measure with the names of the Russian scholars Tchebychev [Chebyshev], Markov, and Liapounov [Lyapunov]. The way for this progress was paved by the activity of Victor Bunyakovsky, who extensively promoted research on the application of probability theory to statistics and, more particularly, the insurance business and demography. His book on probability was the first one written in Russia, and it proved to have a great effect there on the development of interest in this branch of mathematics. The contribution of lasting significance made by Tchebychev, Markov, and Liapounov in probability theory was their introduction and extensive use of the concept of a random variable. We shall meet with the results of Tchebychev on the law of large numbers, with ''Markov chains,'' and with the limit theorems of Liapounov in the appropriate sections of the present book.

The modern development of probability theory is characterized by a general rise of interest in the theory itself as well as a broadening of the range of its practical application. Many scientists in the United States, France, China, Italy, Great Britain, Poland, Hungary, and other countries of the world are enriching it with important results. The Soviet school of probability theory continues to occupy a prominent place in this vigorous scientific effort. Among the representatives of Soviet science we must mention first the names of Serge N. Bernstein, Andrei Kolmogorov, and Alexander Khintchine [Khinchin]. Because of the natural order of events, we shall be familiarizing the reader during the course of our presentation with the ideas and results of contemporary mathematicians that have altered the make-up of the subject. Thus, in the very first chapter, reference will be made to the basic work of Bernstein and Kolmogorov on the foundations of probability theory. In the first decade of this century, Emile Borel introduced some ideas connecting the theory of probability with the measure-theoretic aspects of the functions of a real variable. Somewhat later, in the 1920's, Khintchine, Kolmogorov, Eugene Slutsky, Paul Lévy, Anton Lomnitsky and others considerably developed these ideas, which proved to be very fruitful to the growth of the subject. We note, in particular, that in this very way a definitive solution was successfully obtained for classical problems which had been formulated as far back as Tchebychev. Fundamental advances in this area are associated with the names of Lindeberg, Bernstein, Kolmogorov, Khintchine, Lévy, William Feller, and a number of others. The notions of measure theory and, subsequently, functional analysis have lead to a considerable extension of the content of probability theory. Dating back to the thirties is the creation of the theory of stochastic (probabilistic, random) processes, which has now become a chief area of research in probability theory. This theory serves as a beautiful example of the organic synthesis of mathematical and scientific thought, in which the mathematician, in mastering the physical essence of the main problem of some science, finds a suitable mathematical language in which to express it. Apparently, Poincaré had already mentioned the idea of creating such a theory, but the first rough drafts of a realization may be found in the papers of Bachelier, Fokker, and Planck. However, the construction of a mathematically rigorous foundation of the theory of stochastic processes is associated with the names of Kolmogorov and Khintchine. It may be noted that the solution of the classical problems of probability theory has turned out to be closely connected with the theory of stochastic processes. The elements of this important new branch of the theory will be presented in the tenth chapter. Finally, we mention the new subdivision of mathematical statistics called *non-parametric methods of statistics*. In §§ 60-63, some idea of the content of this new branch of the subject will be given.

For the past decade, the part which the theory of probability has played in modern science has grown immeasurably. After the molecular theory of the structure of matter had received universal recognition, the wide use of probability theory both in physics and chemistry became inevitable. From the standpoint of molecular physics, every substance is composed of an enormous number of very small particles which are in constant motion and which in the process of moving about interact with one another. Little is known about the nature of these particles, the interaction that takes place between them, the character of their motion, etc. This knowledge basically consists of the fact that every substance is composed of very many particles, and that in a homogeneous body their properties are very similar. Of course, under these circumstances standard mathematical research methods have become useless in the investigation of physical theories. Thus, for example, the techniques of differential equations are incapable of yielding any serious results under these conditions. In reality, neither the structure of, nor the laws of interaction of material particles have been studied to any sufficient extent, and under such circumstances the application of the techniques of differential equations is therefore a rather arbitrary procedure. But even if this difficulty did not exist, the very fact that the motion of a large number of particles must be investigated is itself an obstacle which cannot be overcome by means of the usual equations of mechanics.

Besides, such an approach is also methodologically unsound. The problem here is not, in fact, that of investigating the motion of individual particles but that of determining the laws to which aggregates of a large number of interacting particles are subject. However, the laws that arise in consequence of the mass character of the participating components have their own peculiarities and do not amount to a simple summation of the individual motions. Furthermore, within certain limits, these laws are found to be independent of the individual properties of the particles that give rise to them. Of course, in order to investigate these new laws, one also has to find new and appropriate mathematical research methods. And what are the primary requirements that we wish these methods to satisfy? Clearly, in the first place, they should take into account the fact that the phenomenon is of a mass nature; thus, the presence of a large number of interacting particles should not offer any obstacle to the use of these methods, but rather should make it easier to study the laws that are being evolved. A lack of further knowledge concerning the nature and structure of the particles as well as the character of their interaction should also not restrict the effectiveness of their use. The methods of the theory of probability satisfy these requirements best of all.

In order for these remarks not to be misunderstood, we once again emphasize the following. In saying that the techniques of probability theory are best adapted to study molecular phenomena, we in no way wish to imply that the philosophical premises underlying the usage of probability in science are those of "insufficient reason." The basic rule is that *distinctive new laws* arise whenever "mass" phenomena are studied. In studying phenomena which depend on the action of a great number of molecules, an accounting of all the properties of each molecule is unnecessary. In fact, when studying natural phenomena, one must avoid taking *unessential* details into account. For, the consideration of all details and all existing relationships, including those which are unessential to the phenomenon in question, merely leads to one result: the phenomenon is obscured and a mastery of it is delayed because the situation has been so artificially complicated.

INTRODUCTION 13

As to how well a phenomenon has been schematized or how well the mathematical tools used in investigating it have been selected, can be judged by how well the theory agrees with experiment and with practice. The development of the sciences, especially physics, shows that the techniques of probability theory have proved to be very well suited to the study of many natural phenomena.

The connection just mentioned that exists between probability theory and the requirements of modern physics provides the best explanation as to why the theory of probability has been one of the most rapidly developing branches of mathematics in the last ten years. New theoretical results are revealing new possibilities for the use of the methods of probability theory in the sciences. The comprehensive study of natural phenomena spurs on a search in probability theory for new laws of chance. The theory of probability is not shirking the questions raised by the other sciences and is keeping abreast of the general development of science. This, of course, does not mean that the theory of probability is merely an auxiliary means for solving various practical problems. On the contrary, in the last thirty years the theory of probability has emphatically become a well-ordered mathematical discipline with its own problems and methods of proof. But at the same time, the most important questions of the subject have proved to be concerned with the solution of various scientific problems.

At the very start, we defined the theory of probability as the science that studies random phenomena. We shall postpone explaining the meaning of the concept of "random phenomenon (event)" until the first chapter, and we shall confine ourselves at this point to a few observations. If to the ordinary way of thinking and in everyday usage a random event is considered to be an extreme rarity going counter to the established order of things and to the natural development of events, then in the theory of probability we repudiate such ideas. Random events, as they are understood in probability theory, possess a number of characteristic features: in particular, they all occur in mass phenomena. By a mass phenomenon we mean one that takes place for aggregates of a large number of objects that have equal or almost equal status, the phenomenon being determined by this mass character itself and being only slightly dependent on the nature of the component objects.

We note that the entire development of probability theory shows evidence of how its concepts and ideas were crystallized in a severe struggle between materialistic and idealistic conceptions. Because of their idealistic conceptions, a number of mathematicians and statisticians (Karl Pearson, Paul Nekrasov, Richard von Mises, etc.) were openly opposed to the elemental materialistic views of James Bernoulli, Laplace, Lobachevsky, Tchebychev, Markov and many other outstanding scientists of the past. This struggle is continuing even at present. In the case of the von Mises definition of probability, we shall see that opposed to both the views of Soviet scientists, which have been developed along Marxist-Leninistic philosophical lines, and those of numerous materialists of all countries of the world are idealistic formulations which from time to time have been painstakingly camouflaged with the words *experiment, practice,* and *natural science.*

The theory of probability, like other branches of mathematics, has evolved out of the needs of practical application; in its abstract form, it reflects the laws inherent in random events of a mass nature. These laws play an exceptionally important part in physics and other natural sciences, in military matters, in the most diversified of engineering disciplines, economics, etc. Recently, in connection with the extensive growth of enterprises carrying on mass production, the results of probability theory have not only been used in the sorting of articles spoiled in manufacture but, what is more important, in organizing the very process of production (industrial quality control).

As already mentioned, the tie that exists between the theory of probability and practical requirements has been the basic cause of its vigorous development in the last thirty years. Many of its subdivisions have evolved just in connection with the answering of practical questions. It is opportune to recall here the striking words of Tchebychev, the founder of our native school of probability theory: "The rapprochement of theory with practice yields very beneficial results, and it is not practice alone that gains thereby; the sciences themselves develop under its influence: new subjects are opened up for investigation and new aspects of subjects already known are discovered . . . While theory benefits greatly from new applications or further extensions of an old method, theory benefits still more by the discovery of new methods, and in this case, Science finds a reliable guide in Practice."

CHAPTER I

THE CONCEPT OF PROBABILITY

§ 1. Certain, Impossible, and Random Events

On the basis of observation and experiment science arrives at the formulation of the natural laws that govern the phenomena it studies. The simplest and most widely used scheme of such laws is the following:

1. *Whenever a certain set of conditions $\mathfrak{S}$ is realized, the event A occurs.*

Thus, for example, if water under 760 mm. atmospheric pressure is heated to over 100° Centigrade (the set of conditions $\mathfrak{S}$), it is transformed into steam (the event A). Or another example: In any chemical reaction whatsoever in which no interchange of substance takes place with the surrounding medium (the set of conditions $\mathfrak{S}$), the total amount of substance remains unchanged (the event A). This last statement is called the law of conservation of matter. The reader will require no assistance in giving other examples of similar laws in physics, chemistry, biology, and other sciences.

An event whose occurrence is inevitable whenever the set of conditions $\mathfrak{S}$ is realized is called *certain* (or *sure*). If the event A can never occur when the set of conditions $\mathfrak{S}$ is realized, it is called *impossible*. An event that may or may not occur when the set of conditions $\mathfrak{S}$ is realized, is called *random*.

From these definitions, it is clear that when we speak of the certainty, impossibility, or randomness of any event, we shall always mean that it is certain, impossible, or random with respect to some definite set of conditions $\mathfrak{S}$.

Proposition 1 above states that the event A is certain whenever the set of conditions $\mathfrak{S}$ is realized. The assertion that some event is impossible whenever a given set of conditions is realized does not yield anything essentially new, since it can readily be reduced to a statement of type 1:

the impossibility of the event A is equivalent to the certainty of the opposite, or complementary, event $\bar{A}$, consisting in the non-occurrence of A.

The mere assertion of the randomness of an event is of very limited interest as regards the information it gives us; it amounts only to an indication that the set of conditions $\mathfrak{S}$ does not completely reflect all of the necessary and sufficient conditions for the event A to occur. Such an indication should not be regarded as utterly devoid of content, since it could serve as stimulus for the further study of the conditions under which the event A occurs, but by itself, nonetheless, it does not yield any affirmative knowledge.

However, a wide range of phenomena exists for which, whenever the set of conditions $\mathfrak{S}$ is realized *repeatedly,* the proportion of occurrences of the event A only seldom deviates significantly from some average value, and this number can thus serve as a characteristic index of the *mass phenomenon* (the repeated realization of the set of conditions $\mathfrak{S}$) with respect to the event A.

For such phenomena it is possible not only simply to state that the event is random but also to estimate in quantitative terms the chance of its occurrence. This estimate is expressed by propositions of the form:

2. *The probability that the event A will occur whenever the set of conditions $\mathfrak{S}$ is realized is equal to p.*

The laws of this second kind are called *probabilistic,* or *stochastic,* laws. Probabilistic laws play an important role in the most diversified fields of science. For example, no methods exist for predicting whether or not a given atom of radium will disintegrate in a given interval of time, but on the basis of experimental data it is possible to determine the probability of this disintegration: an atom of radium disintegrates in a period of t years with the probability

$$p = 1 - e^{-0.000436 t}.$$

The set of conditions $\mathfrak{S}$ in this example consists in the atom of radium being subjected to no unusual external influences during the years it is under consideration, such as bombardment by fast particles, all other conditions under which it exists being immaterial: the medium in which the atom is embedded, the temperature of the medium, etc. are of no importance; the event A consists in the fact that this atom disintegrates in a given period of t years.

The idea, which now seems perfectly natural to us, that the probability of a random event A under given conditions can be estimated quanti-

§ 1. CERTAIN, IMPOSSIBLE, AND RANDOM EVENTS

tatively by means of a certain number

$$p = \mathsf{P}(A)$$

was first systematically developed in the 17th century in the works of Fermat (1601-65), Pascal (1623-62), Huyghens (1629-95) and especially J. Bernoulli (1654-1705). Their investigations laid the foundations of the theory of probability. Since that time the theory of probability has developed continuously as a mathematical discipline, being constantly enriched with new and important results. Its applicability to the study of real occurrences of very diverse nature also continually receives new and brilliant confirmation.

Undoubtedly, the concept of mathematical probability deserves a thorough philosophical study. The basic specific philosophical question raised by the very existence of probability theory and by its successful application to real occurrences is the following: *Under what conditions does the quantitative estimate of the probability of a random event A by means of a definite number* $\mathsf{P}(A)$—*called the mathematical probability of the event A—have an objective meaning, and what is that meaning?* A clear understanding of the interrelation between the philosophical categories of randomness and necessity is an inescapable prerequisite for the successful analysis of the concept of mathematical probability, but this analysis cannot be complete without an answer to our question as to what the conditions are under which randomness admits of a quantitative estimate in the form of a number—a probability.

Every investigator who concerns himself with the application of probability theory to physics, biology, ballistics, economics, or any other specific science essentially starts out in his work with the conviction that *probabilistic judgments express certain objective properties of the phenomena that are studied.* To say that the occurrence of an event A under a certain set of conditions $\mathfrak{S}$ has a probability p is to assert that between the set of conditions $\mathfrak{S}$ and the event A there is a well-defined—although quite distinctive, but not on that account any the less objective—relation that exists independently of the observer. The philosophical problem is that of clarifying the nature of this relation. It is only the difficulty of this problem that has made possible the paradoxical situation in which, even among scholars who do not take the idealistic position on general philosophical questions, one may find a tendency to dismiss the problem rather than to seek its positive solution, by asserting that probabilistic judgments have to do only with the state of the observer (measuring the degree of his certainty that the event A will take place).

All of our varied experience with the application of probability theory to the most diverse subjects teaches that the very problem of quantitatively estimating the probability of some event makes reasonable objective sense only under certain completely definite conditions.

The definition given above that an event A is random under a set of conditions $\mathfrak{S}$ is a purely negative one: an event is random if it is not inevitable and if it is not impossible. However, it by no means follows from the randomness of an event in this purely negative sense that it is meaningful to talk about its probability as though it were a definite, even if unknown, number. In other words, not only the statement "Under the set of conditions $\mathfrak{S}$ the event A has the probability $\mathsf{P}(A)$" but also the mere statement that the probability in question *exists* is a highly meaningful one that requires substantiation in every individual case or, when taken as an hypothesis, calls for subsequent verification.

For example, a physicist finding a new radioactive element will assume *a priori* that there exists a certain probability that an undisturbed atom of this element (i.e., which is not subjected to external influences of extremely high intensity) will decay in an interval of time t which is expressible as a function of t by

$$p = 1 - e^{-at},$$

and he will set about determining the coefficient a which characterizes the rate of decay of the new radioactive element. The question might be raised as to how the probability of decay depends on external conditions —for example, on the intensity of cosmic radiation; here, the investigator would proceed from the assumption that to each *sufficiently definite* set of external conditions there corresponds some definite value of a.

This is precisely the situation in all the other applications of the theory of probability that have proved successful in practice. Therefore, the problem of philosophically clarifying the real essence of the concept of "mathematical probability" can be made hopeless in advance if we demand a definition applicable to any event A under any set of conditions $\mathfrak{S}$.

§ 2. Different Approaches to the Definition of Probability

The number of different definitions of mathematical probability that have been proposed by various authors is very large. We shall not stop at this point to examine all of the logical subtleties of these numerous definitions. Every scientific definition of a fundamental concept such as

§ 2. DIFFERENT APPROACHES TO THE DEFINITION OF PROBABILITY

the concept of probability is merely the refinement and logical processing of a series of very simple observations and practical methods that have proved their value by successful use over a long period of time. The interest in a logically irreproachable "foundation" of probability theory is a later historical development than the ability to determine the probabilities of various events and to calculate with these probabilities and make use of the results of these calculations in practical problems and in scientific investigations. Therefore, at the root of most attempts to give a general scientific definition of the concept of probability one can easily recognize one aspect or another of the specific cognitive process which leads in each individual case to the practical definition of the probability of some particular event, whether it be the probability of throwing at least one six in four rolls of a die, the probability of radioactive decay, or the probability of hitting a target. Some definitions take as their starting point unessential, secondary aspects of such specific processes—these definitions are quite fruitless. Others emphasize some one side of the matter or some practical methods for finding the probability which are not applicable to every case—such definitions, in spite of their narrowness, must be examined more carefully.

From the point of view just outlined, the majority of definitions can be subdivided into three groups:

1. Definitions of mathematical probability as a quantitative measure of the "degree of certainty" of the observer.

2. Definitions that reduce the concept of probability to the more primitive notion of "equal likelihood" (the so-called "classical" definition).

3. Definitions that take as their point of departure the "relative frequency" of occurrence of the event in a large number of trials ("statistical" definition).

The definitions of the second and third groups will be discussed in §§ 4 and 7. We now devote the remainder of the present section to a critique of the first group. If mathematical probability is the quantitative measure of the degree of certainty of the observer, then the theory of probability is something not unlike a branch of psychology. The final outcome of consistently using such a purely subjectivistic interpretation of probability is inevitably subjective idealism. Indeed, if we assume that the evaluation of probability only concerns the state of the observer, then all conclusions based on probabilistic judgments (judgments of the form 2.) are deprived of the objective meaning that they have independent of the observer. Meanwhile, science has established many positive results on the basis of probabilistic judgments of type 2. which do not differ in significance from results obtained without recourse to probability. For example, in physics all "macroscopic" properties of gasses are deduced from assumptions concerning the probabilistic nature of the behavior of the individual molecules. If we are to ascribe an objective meaning independent of the observer to these deductions, then the initial probabilistic hypotheses concerning the course of the "macroscopic" molecular processes must be something more than just a statement of the psychological state into which we are thrown when we think about the motion of molecules.

To those who take the point of view that the external world has a reality independent of ourselves and is in principle cognizable, and who take into account the fact that probabilistic judgments can be used successfully to obtain knowledge about the world, it should be perfectly clear that a purely subjective definition of mathematical probability is quite untenable. With this statement we could have concluded our discussion of the first group of definitions if it were not for the support they find in the original, common usage of the word "probability." The fact is, that in everyday speech such expressions as "it is probable," "it is very probable," "it is improbable," etc. merely convey the attitude of the speaker to the question of the truth or falsity of some *single* judgment. Therefore, it is necessary to lay stress on one fact which we have not especially paid attention to until now. When, in § 1, we straight-way concentrated our attention on probabilistic laws of the form 2., comparing them with rigorous causal laws such as 1., we proceeded in complete accordance with the whole of the successful scientific application of mathematical probability, but from the very outset we deviated slightly from the usual "pre-scientific" meaning of the word "probability": whereas in all practical scientific application of probability theory "probability" means the probability that some event A will occur upon the realization of a certain set of conditions $\mathfrak{S}$ which *in principle is reproducible an infinite number of times* (only in such a setting does the statement

$$p = \mathsf{P}(A)$$

express a law which has objective meaning), in everyday speech it is customary to speak of some *well-defined* judgment as having greater or lesser probability. For example, concerning the judgments:

(a) Every even natural number greater than two can be expressed as the sum of two primes $(4=2+2,\ 6=3+3,\ 8=5+3,$ etc.);

(b) It will snow in New York City on April 7, 1986;

one could state the following: Nothing more is known about (a) at present, but many believe it is very probable; as to judgment (b), one must assume that a definitive answer will be given on April 7, 1986 only. However, since it rarely snows in New York City in April, it is necessary at present to consider (b) as being improbable.

To such statements concerning the probability of individual facts or, in general, specific judgments (even if of a general character) there may in fact be ascribed subjective meaning only: such statements merely reflect the attitude of the speaker to the given question. Indeed, in saying that a specific judgment has a greater or a lesser probability, we do not as a rule have the least intention of questioning that the principle of the excluded middle is applicable to it. No one, for example, doubts that each of the propositions (a) and (b) is in fact either *true* or *false*. Even if the so-called intuitionists were to express such doubts as regards proposition (a), then, at any rate for the average intellect, having occasion to speak of the greater or lesser probability of this proposition would in no way be related to doubts as to whether the principle of the excluded middle were applicable to it. If at some time proposition (a) should be proved or disproved, then all prior estimates of its probability would become meaningless. In exactly the same way, when April 7, 1986 arrives, it will be easy to find out whether judgment (b) is true or not: if it snows on that day, then the opinion that this event is improbable becomes meaningless.

A thorough analysis of the complete range of psychic states of doubt lying between the categorical acceptance and categorical denial of a single opinion, no matter how interesting from a psychological point of view would lead us far afield from our basic problem of explaining the meaning of probabilistic laws that have objective scientific value.

§ 3. The Field of Events

In the preceding section, we saw that the definition of mathematical probability as a quantitative measure of the "degree of certainty" of the observer did not capture the essence of the concept of probability. We therefore return to the question of what the origin is of objective probabilistic laws. The classical and the statistical definitions claim to give simple and straightforward answers to this question. We shall see later on that both of these definitions do reflect essential aspects of the true meaning of the concept of probability, although each by itself is inadequate. A full understanding of the nature of probability thus calls for a synthesis of these definitions. In the next few sections, we shall be concerned exclusively with the classical definition of probability, which springs from the notion of *equal likelihood* as an objective property of the various possible outcomes of the phenomena studied, a property based on their actual symmetry. In what follows, we shall deal with this interpretation of equal likelihood only. However, the definition of probability in terms of the notion of "equal likelihood" taken in the purely subjective sense of equal "likelihood" to the observer, belongs to the group of definitions of probability in terms of the "degree of certainty" of the observer which we have already excluded from consideration.

Before proceeding to the classical definition of the concept of probability, we shall make several preliminary remarks. Let us consider a fixed set of conditions $\mathfrak{S}$ and some family S of events $A, B, C, \ldots$[1] each of which must either occur or not occur[2] whenever the set of conditions $\mathfrak{S}$ is realized. Certain relations can exist between the events of the family S which we shall continually make use of and which, therefore, we shall begin by studying.

1) If for every realization of the set of conditions $\mathfrak{S}$ for which the event A occurs the event B also occurs, then we shall say that A *implies*[3] B and we shall denote this fact by writing $A \subset B$, or $B \supset A$.

2) If A implies B and at the same time B implies A, i.e., if for every realization of the set of conditions $\mathfrak{S}$ either A and B both take place or both do not take place, then we shall say that the events A and B are *equivalent* and this fact will be denoted by writing $A = B$.

3) The event consisting in the simultaneous occurrence of A and B will be called the *product*, or intersection, of the events A and B and will be denoted by AB.

4) The event consisting in the occurrence of at least one of the events A or B will be called the *sum*, or union, of the events A and B and will be denoted by $A + B$.

---

[1] In the sequel, events will be denoted by capital italic letters $A, B, C, \ldots$.
[2] Instead of "occur," one also says "appear," "take place," or "ensue."
[3] Instead of "A implies B," we also say "A is contained in B."

§ 3. THE FIELD OF EVENTS

5) The event consisting in the occurrence of A and the non-occurrence of B will be called the *difference* of the events A and B and will be denoted by $A - B$.

For example, let the set of conditions $\mathfrak{S}$ consist in the selection at random of a point inside the square shown in Fig. 1 but not on the cir-

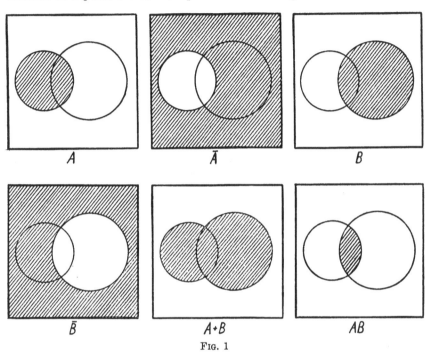

FIG. 1

cumference of either of the two circles shown in that figure. Let A and B be the following two events:

A: The point selected lies inside the left-hand circle.

B: The point selected lies inside the right-hand circle.

Then A, $\bar{A}$, B, $\bar{B}$, $A + B$, and AB are the events consisting in the selected point falling inside the regions that have been shaded in the correspondingly labeled diagrams of Fig. 1.

Let us consider another example. We shall assume that the set of conditions $\mathfrak{S}$ consists in a single throw of a die[4] onto a table. Let A, B, C,

---

[4] A die is a cube made of ivory, wood, or the like, on whose respective faces the numbers 1 through 6 are marked by means of pips.

and D denote, respectively, the event that a six, a three, an even number, and a multiple of three turn up. Then the events A, B, C, and D are connected by the following relations:

$$A \subset C, \quad A \subset D, \quad B \subset D, \quad A + B = D, \quad CD = A.$$

The definition of the sum and product of two events can be generalized to any number of events:

$$A + B + \ldots + N$$

denotes the event consisting of the occurrence of at least one of the events $A, B, \ldots, N$ and

$$AB \ldots N$$

denotes the event consisting of the simultaneous occurrence of all of the events $A, B, \ldots, N$.

6) An event is called *certain* if it must inevitably occur (whenever the set of conditions $\mathfrak{S}$ is realized). For example, if a pair of dice is thrown, then it is certain that the sum of the numbers thrown will not be smaller than two.

An event is called *impossible* if it is certain not to occur (for any realization of the set of conditions $\mathfrak{S}$). For example, if a pair of dice is thrown, then it is impossible for the sum thirteen to appear.

Clearly, all certain events are equivalent to one another. It is therefore permissible to denote all events that are certain by a single letter. We shall use the letter U for this purpose. All impossible events are likewise equivalent. Any impossible event will be denoted by the letter V.

7) Two events A and $\bar{A}$ are *complementary* if the two relations

$$A + \bar{A} = U, \, A\bar{A} = V$$

hold simultaneously. For example, if C denotes the event consisting of throwing an even number with one die, then

$$U - C = \bar{C}$$

is the event that an odd number is thrown.

8) Two events A and B are called *mutually exclusive* if their joint occurrence is impossible, i.e., if

$$AB = V.$$

§ 3. THE FIELD OF EVENTS

If
$$A = B_1 + B_2 + \ldots + B_n$$

and the events B_i are mutually exclusive in pairs, i.e.,
$$B_i B_j = V \text{ for } i \neq j,$$

then we say that the event A is *decomposable into the mutually exclusive events* $B_1, B_2, \ldots, B_n$. For example, in the throwing of a single die the event C consisting of the throw of an even number is decomposable into the mutually exclusive events E_2, E_4, and E_6, consisting respectively of the occurrence of a 2, a 4, and a 6.

The events $B_1, B_2, \ldots, B_n$ constitute a *complete group of events* if at least one of them is certain to happen (for every realization of the set $\mathfrak{S}$), i.e., if
$$B_1 + B_2 + \ldots + B_n = U.$$

Especially important for us in the following will be the *complete groups of pairwise mutually exclusive events*. Such a group, for example, is the family of events

$$E_1, E_2, E_3, E_4, E_5, E_6,$$

consisting, respectively, of the occurrence of a 1, 2, 3, 4, 5, and 6 in a single throw of a die.

9) In every problem in probability theory, one has to deal with some specific set of conditions $\mathfrak{S}$ and some specific family of events S which either do or do not occur after each realization of the set of conditions $\mathfrak{S}$. It is expedient to make the following assumptions concerning this family of events:

a) *If the events A and B belong to the family S, then so do the events AB, $A + B$, and $A - B$.*

b) *The family S contains the certain event and the impossible event.*

A family of events satisfying these assumptions is called a *field of events*.

In conclusion, we note once again that in all considerations in probability theory, equivalent events may be substituted for one another. Therefore in the sequel we shall agree simply to regard two equivalent events as being identical with one another.

§ 4. The Classical Definition of Probability

The classical definition of probability reduces the concept of probability to the concept of equiprobability (equal likelihood) of events, which is regarded as a primitive concept and hence not subject to formal definition. For example, in the throwing of a single perfectly cubical die made of completely homogeneous material, the equally likely events are the appearance of any of the specific number of points (from 1 through 6) marked on its faces, since by virtue of the symmetry no one face has objective preference over the other.

In the general case, we consider some group G consisting of n pairwise mutually exclusive, equally likely (equiprobable) events (these will be called *elementary events*)

$$E_1, E_2, \ldots, E_n.$$

Let us now form the family S consisting of the impossible event V, all of the events E_k of the group G, and all of the events A which can be decomposed into a sum of mutually exclusive events belonging to the group G.

For example, if the group G consists of the three events E_1, E_2, and E_3, then the events[5] in the family S are V, E_1, E_2, E_3, $E_1 + E_2$, $E_1 + E_3$, $E_2 + E_3$, and $U = E_1 + E_2 + E_3$.

It is easy to show that the family S is a field of events. In fact, it is clear that the sum, difference, and product of events in S are in S; the impossible event V is a member of S by definition, and the certain event U is in S because it is representable in the form

$$U = E_1 + E_2 + \ldots + E_n.$$

The classical definition of probability is given in terms of the events of the family S and may be formulated as follows:

If an event A is decomposable into the sum of m events belonging to a complete group of n pairwise mutually exclusive and equally likely events, then the probability $\mathsf{P}(A)$ of the event A is equal to

---

[5] These eight events exhaust the family S, provided we make no distinction between events which are equivalent to one another (as we agreed to do at the end of § 3). It is easily shown that in the general case, when the group G contains n events, the family S consists of 2^n events.

§ 4. The Classical Definition of Probability

$$P(A) = \frac{m}{n}.$$

For example, in the case of a single throw of a die, the complete group of pairwise mutually exclusive and equally likely events are the events

$$E_1, E_2, E_3, E_4, E_5, E_6,$$

consisting, respectively, in the throw of a 1, 2, 3, 4, 5, and 6. The event

$$C = E_2 + E_4 + E_6,$$

corresponding to the throw of an even number is divisible into the sum of three events belonging to the complete group of mutually exclusive and equally likely events. Therefore, the probability of the event C is equal to

$$P(C) = \frac{3}{6} = \frac{1}{2}.$$

By virtue of the definition, it is also evident that

$$P(E_i) = \frac{1}{6}, \quad 1 \leq i \leq 6,$$

$$P(E_1 + E_2) = \frac{2}{6} = \frac{1}{3},$$

and so forth.

In probability theory, the following terminology is widely used and we shall employ it frequently in the sequel. Let us imagine that to clear up the question as to whether an event A will or will not occur (for example, A may be the throwing of a number which is a multiple of three) it is necessary to make a certain *trial* (i.e., to realize the set of conditions $\mathfrak{S}$) which would yield the answer to our question. (In our example, a die has to be thrown.) The complete group of pairwise mutually exclusive and equally likely events which may occur when this experiment is performed is called the complete group of *possible outcomes* of the trial. The possible outcomes of the trial into which the event A can be subdivided are said to be the outcomes (or cases) favorable to A. Using this terminology, we can say that *the probability $P(A)$ of the event A is equal to the number of possible outcomes favorable to A divided by the total number of possible outcomes of the trial.*

Of course, this definition assumes that each possible outcome of the trial is equally probable.

I. THE CONCEPT OF PROBABILITY

Let us now consider the throwing of a pair of dice. If the dice are true, each of the 36 possible combinations of numbers on the two dice may be regarded as equally probable. Thus, the probability of throwing, say, a 12 is equal to 1/36. A sum of 11 may turn up in two ways: a 5 on the first die and a 6 on the second, or vice versa. Therefore, the probability of throwing a total of eleven is equal to $2/36 = 1/18$. The reader can easily verify that the probability of throwing any specific total is given by the following table:

TABLE 1

No. of pips	2	3	4	5	6	7	8	9	10	11	12
Probability ...	$\frac{1}{36}$	$\frac{2}{36}$	$\frac{3}{36}$	$\frac{4}{36}$	$\frac{5}{36}$	$\frac{6}{36}$	$\frac{5}{36}$	$\frac{4}{36}$	$\frac{3}{36}$	$\frac{2}{36}$	$\frac{1}{36}$

According to the definition given, every event belonging to the field of events S constructed above has a well-defined probability

$$P(A) = \frac{m}{n}$$

assigned to it, where m is the number of mutually exclusive events E_i of the original group G into the sum of which the event A is decomposable. Thus, the probability $P(A)$ may be regarded as *a function of the event A defined over the field of events S*.

This function possesses the following properties:

1. *For every event A of the field S,*

$$P(A) \geqq 0.$$

2. *For the certain event U,*

$$P(U) = 1.$$

3. *If the event A is decomposable into the mutually exclusive events B and C and all three of the events A, B, and C belong to the field S, then*

$$P(A) = P(B) + P(C).$$

This property is called the *theorem on the addition of probabilities*.

Property 1 is obvious, since the ratio m/n cannot be negative. The second property is equally obvious, since the n possible outcomes of the trial are all favorable to the certain event U and hence

§ 4. The Classical Definition of Probability

$$P(U) = \frac{n}{n} = 1.$$

Let us prove Property 3. Suppose that m' is the number of outcomes E_i of the group G favorable to the event B and m'' the number favorable to the event C. Since the events B and C are, by assumption, mutually exclusive, the outcomes E_i that are favorable to one of them are distinct from those that are favorable to the other. Thus, there are altogether $m' + m''$ events E_i that are favorable to the event $B + C = A$. Therefore,

$$P(A) = \frac{m' + m''}{n} = \frac{m'}{n} + \frac{m''}{n} = P(B) + P(C),$$

Q.E.D.

We confine ourselves at this point to indicating a few further properties of probability.

4. *The probability of the event $\bar{A}$ complementary to the event A is given by*

$$P(\bar{A}) = 1 - P(A).$$

In fact, since $A + \bar{A} = U$, it follows from Property 2 that

$$P(A + \bar{A}) = 1,$$

and since the events A and $\bar{A}$ are mutually exclusive, Property 3 implies that

$$P(A + \bar{A}) = P(A) + P(\bar{A}).$$

The last two equations prove our proposition.

5. *The probability of the impossible event is zero.*

In fact, the events U and V are mutually exclusive, so that

$$P(U) + P(V) = P(U),$$

from which it follows that

$$P(V) = 0.$$

6. *If the event A implies the event B, then*

$$P(A) \leq P(B).$$

Indeed, the event B can be represented as the sum of two events A and $\bar{A}B$. Using Properties 3 and 1, we obtain from this that

I. The Concept of Probability

$$P(B) = P(A + \bar{A}B) = P(A) + P(\bar{A}B) \geqq P(A).$$

7. *The probability of any event lies between zero and one.*

The relations

$$V \subset A + V = A = A \ U \subset U$$

hold for any event A, and from this and the preceding property it follows that the inequalities

$$0 = P(V) \leqq P(A) \leqq P(U) = 1$$

hold.

§ 5. Examples

At this point, we shall consider several examples of the calculation of the probabilities of events using the classical definition of probability. The examples cited are purely illustrative in character and do not pretend to impart to the reader all of the basic methods for computing probabilities.

Example 1. From an ordinary deck of 52 cards, three cards are drawn at random. Find the probability that there will be exactly one ace among them.

Solution. The complete group of equally likely and mutually exclusive events in this problem consists of all the possible combinations of three cards, and there are $\binom{52}{3}$ such combinations. The number of favorable cases (outcomes) can be computed as follows. One ace can be chosen in $\binom{4}{1}$ different ways and the two remaining cards (non-aces) in $\binom{48}{2}$ different ways. Since to each given ace there correspond $\binom{48}{2}$ ways in which the two remaining cards can be selected, the total number of favorable cases will be $\binom{4}{1}\binom{48}{2}$. The required probability is thus equal to

$$p = \frac{\binom{4}{1}\binom{48}{2}}{\binom{52}{3}} = \frac{\frac{4}{1} \cdot \frac{48 \cdot 47}{1 \cdot 2}}{\frac{52 \cdot 51 \cdot 50}{1 \cdot 2 \cdot 3}} = \frac{24 \cdot 47}{13 \cdot 17 \cdot 25} = \frac{1128}{5525} \approx 0.2041,$$

i.e., a little more than 0.20.

§ 5. EXAMPLES

Example 2. Three cards are drawn at random from a deck of 52 cards. Determine the probability that there will be at least one ace among them.

Solution. Let A denote the event in question. A is expressible as the sum of the following three mutually exclusive events: A_1, the occurrence of one ace; A_2, the occurrence of two aces; and A_3, the occurrence of three aces.

By arguments analogous to those carried out in the preceding example, we easily establish that the number of cases favorable to the event A_1 is $\binom{4}{1}\binom{48}{2}$; the number favorable to A_2, $\binom{4}{2}\binom{48}{1}$; and the number favorable to A_3, $\binom{4}{3}\binom{48}{0}$.

Since the total number of possible cases is equal to $\binom{52}{3}$, we have

$$P(A_1) = \binom{4}{1}\cdot\binom{48}{2} \Big/ \binom{52}{3} = \frac{24\cdot 47}{13\cdot 17\cdot 25} \approx 0.2042,$$

$$P(A_2) = \binom{4}{2}\cdot\binom{48}{1} \Big/ \binom{52}{3} = \frac{3\cdot 24}{13\cdot 17\cdot 25} \approx 0.0130,$$

$$P(A_3) = \binom{4}{3}\cdot\binom{48}{0} \Big/ \binom{52}{3} = \frac{1}{13\cdot 17\cdot 25} \approx 0.0002.$$

By virtue of the addition theorem,

$$P(A) = P(A_1) + P(A_2) + P(A_3) = \frac{1201}{13\cdot 17\cdot 25} \approx 0.2174.$$

This problem may be solved by another method. The complementary event $\bar{A}$ is the event that no aces occur among the cards drawn. It is obvious that three non-aces may be drawn from a deck of cards in $\binom{48}{3}$ different ways, and therefore

$$P(\bar{A}) = \frac{\binom{48}{3}}{\binom{52}{3}} = \frac{48\cdot 47\cdot 46}{52\cdot 51\cdot 50} \approx 0.7826.$$

The required probability is

$$P(A) = 1 - P(\bar{A}) \approx 0.2174.$$

Note: In both examples, the expression "at random" means that all possible combinations of three cards are equally probable.

Example 3. A deck of 52 cards is divided at random into two equal parts. What is the probability that an equal number of red and black cards turn up in both parts?

I. The Concept of Probability

The expression "at random" means that all possible divisions of the deck are equally likely.

Solution: We have to determine the probability that among 26 cards drawn at random from the deck 13 will be red and 13 black.

The total number of different ways in which 26 cards can be chosen out of 52 cards is $\binom{52}{26}$. The favorable cases will be all the different ways in which some group of 13 cards may be drawn from the 26 red cards and some group of 13 from the 26 black cards. The thirteen black cards may be drawn in $\binom{26}{13}$ different ways and the thirteen red cards may also be drawn in $\binom{26}{13}$ different ways. Since for each drawing of 13 given red cards, 13 black ones may be drawn in $\binom{26}{13}$ different ways, the total number of favorable cases is equal to $\binom{26}{13}\binom{26}{13}$. The desired probability is therefore equal to

$$p = \frac{\binom{26}{13} \cdot \binom{26}{13}}{\binom{52}{26}} = \frac{(26!)^4}{52!(13!)^4}.$$

In order to gain some idea of the magnitude of this probability without having to carry out a tedious computation, we use Stirling's formula, according to which the following asymptotic relation holds:

$$n! \sim \sqrt{2\pi n}\, n^n\, e^{-n}.$$

We thus find

$$26! \approx 26^{26} e^{-26} \sqrt{2\pi \cdot 26},$$
$$13! \approx 13^{13} e^{-13} \sqrt{2\pi \cdot 13},$$
$$52! \approx 52^{52} e^{-52} \sqrt{2\pi \cdot 52},$$

and this implies that

$$p \approx \frac{(26^{26}\, e^{-26}\, \sqrt{2\pi \cdot 26})^4}{52^{52} e^{-52}\, \sqrt{2\pi \cdot 52}\, (13^{13} e^{-13}\, \sqrt{2\pi \cdot 13})^4}.$$

On simplifying this expression, we obtain

$$p \approx \frac{2}{\sqrt{26\pi}} \approx 0.22.$$

Example 4. Consider n particles each of which may be found with the same probability $1/N$ in any one of N cells ($N > n$). Determine the

§ 5. EXAMPLES

probability that (1) each of n specified cells contain one particle, and (2) n arbitrary cells each contain one particle.

Solution: This problem plays an important role in modern statistical physics and—depending on how the complete group of equally probable events is formed—leads to the statistics of Boltzmann, of Bose-Einstein, or of Fermi-Dirac. (The word *statistics* is used here in a sense peculiar to physics.)

In the Boltzmann statistics, the equally likely events are all the conceivable distributions, taking all the particles to be distinguishable and allowing any number of them from 0 up to n to be found in each of the cells.

We compute the total number of possible distributions in the following way: Each particle may be found in any one of the N cells, so that the n particles may be distributed in N^n different ways.

In the first of the above questions, the number of favorable cases will clearly be $n!$ and this implies that the probability of a single particle falling in each of n specified cells is

$$p_1 = \frac{n!}{N^n}.$$

In the second question, the number of favorable cases will be $\binom{N}{n}$ times greater, and this means that the probability of a single particle falling in each of n arbitrary cells is

$$p_2 = \frac{\binom{N}{n} \cdot n!}{N^n} = \frac{N!}{N^n (N-n)!}.$$

In the Bose-Einstein statistics, those cases in which the various particles change places with each other are considered identical (all that counts is, how many particles are in a given cell and not, which ones), and the complete group of equally likely events consists of all possible distributions of the n particles in the N cells. Each such distribution corresponds to some entire class of Boltzmann distributions differing among themselves not as to the number of particles in each of the given cells but only as regards the identity of the particles themselves. In order to get a clear idea of the distinction between the Boltzmann and Bose-Einstein statistics, we shall examine the particular case $N = 4$, $n = 2$. The possible distributions in this example may be written in the form shown in the following table, where a and b are designations for the particles. In the Boltzmann statistics, all sixteen possibilities represent different equiprobable events; in the Bose-Einstein statistics, however, cases 5 and 11, 6 and 12, 7 and 13, 8 and 14, 9 and 13, and 10 and 16 are identified in pairs, and we obtain a group of ten equally likely events.

TABLE 2

Case	1	2	3	4	5	6	7	8	9	10	11	12	13	14	15	16
Cells	ab				a	a	a				b	b	b			
		ab			b			a	a		a			b	b	
			ab			b		b		a		a		a		b
				ab			b		b	b			a		a	a

Let us now compute the total number of equally likely events in the Bose-Einstein statistics. With this aim, observe that all of the possible distributions of the particles into cells can be obtained in the following way: We arrange the cells in sequence in a straight line and then we distribute our particles into the cells side by side along the same line. We then consider all possible permutations of the particles and dividing walls of the cells. As a little reflection shows, this will account for all the possible ways in which the cells can be filled, taking into account both the order of the particles and the order of the dividing walls.

The number of such permutations is equal to $(N + n - 1)!$. Of these permutations some are identical: every distribution among the cells is counted $(N - 1)!$ times, because we have distinguished the dividing walls of the cells and, over and above this, every distribution has been counted $n!$ times, because we have taken into account not only the number of particles in a cell but also the order in which all the particles have been distributed amongst the cells. Thus, each distribution has been counted $n!(N - 1)!$ times, and from this it follows that the number of distinguishable distributions of the particles in the Bose-Einstein statistics is equal to

$$\frac{(n + N - 1)!}{n!\,(N - 1)!}.$$

We have thus found the number of equally likely events in the complete

§ 5. EXAMPLES

group of events. We can now easily answer the question in our problem. In the Bose-Einstein statistics, the probabilities p_1 and p_2 are given by

$$p_1 = \frac{1}{\frac{(n+N-1)!}{n!\,(N-1)!}} = \frac{n!\,(N-1)!}{(n+N-1)!},$$

$$p_2 = \frac{\binom{N}{n}}{\frac{(n+N-1)!}{n!\,(N-1)!}} = \frac{N!\,(N-1)!}{(N-n)!\,(N+n-1)!}.$$

Finally, let us consider the Fermi-Dirac statistics. In this statistics, the particles are indistinguishable, and not more than one particle may be found in a cell.

The total number of distinct distributions of the particles among the cells is easily computed: the first particle may occupy any one of N different places; the second, only one of $N-1$, the third, $N-2$; and finally, the n-th, $N-n+1$ different places. But in doing this, we are taking as distinct the distributions which differ only by a permutation of the particles among the cells. Therefore, in order to take into account that the particles are indistinguishable, we still have to divide the number thus obtained by $n!$.

Thus, the number of different equally likely ways in which n particles can be distributed into N cells is

$$\frac{1}{n!} \cdot N\,(N-1) \ldots (N-n+1) = \frac{N!}{(N-n)!\,n!}.$$

It is easy to show that in the Fermi-Dirac statistics the desired probabilities are given by

$$p_1 = \frac{(N-n)!\,n!}{N!},$$

$$p_2 = 1.$$

The example just considered shows how important it is to define precisely which events in a given problem are to be regarded as equally likely.

I. THE CONCEPT OF PROBABILITY

Example 5. $2n$ people are queued up at a theater box office; n of them have only five-dollar bills and the remaining n only ten-dollar bills. There is no cash in the box office when it opens, and each patron in turn is going to purchase a single five-dollar ticket. What is the probability that no customer will be required to wait for change?

All the possible arrangements of the customers on line are equally probable. We make use of the following geometrical approach. Consider the xy-plane and suppose that the purchasers are arranged along the x-axis at the points with abscissas $1, 2, \ldots, 2n$ in the same order they occupy on line. The box office is located at the origin. To each person having a ten-dollar bill we assign an ordinate of $+1$ and to each having a five-dollar bill, an ordinate of -1. The ordinates defined in this way at the points with integral values of the abscissa are then summed from left to right, and at each point the partial sum is plotted (Fig. 2). It is easy

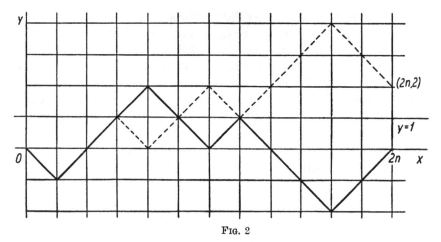

Fig. 2

to see that this sum is zero at the point with abscissa $2n$ (we have n terms equal to $+1$ and n equal to -1). We next connect adjacent points by means of line-segments and also connect the leftmost of these points to the origin in the same way. We shall call the broken line thus obtained, a trajectory.

The total number of distinct possible trajectories, as is readily seen, is equal to $\binom{2n}{n}$ (the number of possible distributions of n ascents among $2n$ ascents and descents). The trajectories favorable to the event in question will be those, and only those trajectories that do not go above the

x-axis (otherwise, there will be at least one occasion when a customer with a ten-dollar bill approaches the box office when no change is available).

Let us compute the total number of trajectories that touch or intersect the line $y=1$ at least once. For this purpose, we construct a new fictitious trajectory, as follows: The new trajectory coincides with the old one up to the first point of contact with the line $y=1$, and from that point on it is the mirror image of the old trajectory with respect to the line $y=1$ (see the broken polygonal line in Fig. 2). It is readily seen that a new trajectory is defined only for those trajectories that meet the line $y=1$ at least once. For the remaining trajectories (i.e., the trajectories favorable to the event we are interested in), the new trajectory coincides with the old. Moreover, a new trajectory, starting at the point $(0,0)$, ends at the point $(2n, 2)$. Thus, it has two more (single) ascents than descents. Consequently, the total number of new trajectories is equal to $\binom{2n}{n+1}$ (the number of possible distributions of $n+1$ ascents among $2n$ ascents and descents). Therefore, the number of favorable cases is $\binom{2n}{n} - \binom{2n}{n+1}$ and the desired probability is

$$p = \frac{\binom{2n}{n} - \binom{2n}{n+1}}{\binom{2n}{n}} = 1 - \frac{n}{n+1} = \frac{1}{n+1}.$$

§ 6. Geometrical Probability

Even at the very beginning of the development of the theory of probability, the "classical" definition of probability, based on the consideration of a finite group of equally likely events, was observed to be inadequate. Even at that time, particular examples led to some modification of this definition and to a formulation of a concept of probability applicable to situations in which the set of conceivable outcomes was infinite. Here, as usual, the notion of "equal likelihood" of certain events played a fundamental role.

The general problem which was posed, and which led to an extension of the concept of probability, can be formulated in the following way.

Suppose, for example, that there exists a region R in a plane and that it contains another region r with a rectifiable boundary. A point is tossed at random onto the region R (or, more prosaically: chosen at random in R), and we inquire as to the probability that the point fall in the region r. The expression "tossed at random onto the region R" is to be understood here to mean: the point on which it falls may be any point of the region R; the probability of its falling in some part of the region R is proportional to the measure of this part (its length, area, etc.) and is independent of its location and its shape.

Therefore, by definition,

$$p = \frac{\text{meas } r}{\text{meas } R}$$

is the probability that a point chosen at random in the region R fall within the region r.

Let us discuss several examples.

Example 1. *The encounter problem.* Two persons A and B have agreed to meet at a specified place between twelve and one o'clock. The one

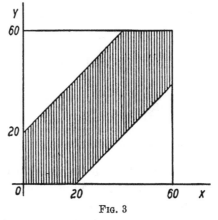

Fig. 3

who arrives first waits 20 minutes for the other, after which he leaves. What is the probability of a meeting between A and B if their arrivals during the indicated hour occur at random, and if their arrival times are independent?[6]

Solution: Let us denote the time of arrival of A and of B by x and y, respectively. In order for a meeting to occur, it is necessary and sufficient that

$$|x - y| \leq 20.$$

---

[6] That is, the time of arrival of one person has no effect on the time of arrival of the other. The concept of independence of events will be considered in detail in § 9.

§ 6. GEOMETRICAL PROBABILITY

We think of x and y as the cartesian coordinates of a point on a plane and take one minute as our unit of distance. All the possible outcomes will be represented by the points inside a square of side 60, and the outcomes favorable to a meeting by the points of the shaded region (Fig. 3).

The desired probability[7] is equal to the ratio of the area of the shaded figure to that of the entire square:

$$p = \frac{60^2 - 40^2}{60^2} = \frac{5}{9}.$$

This encounter problem has been applied to the solution of the following problem. A worker takes care of a number of machines of the same type, each of which may require his attention at a random instant of time. It may happen that at some instant when the worker is occupied with one machine his attention is needed at one of the other machines. It is required to determine the probability of this event, i.e., the average length of time that the machine waits for attention (or, the idle time of a machine). Let us observe, however, that the scheme of the encounter problem is of little use in solving this industrial problem, because an agreed-upon time in the course of which the machines are certain to require the worker's attention does not exist, and the amount of time he spends servicing any of the machines is not the same. Besides this fundamental reason, we should point out the complexity of the computations entailed in the encounter problem for the case of a large number of persons (machines). And it is not infrequently necessary to solve this problem for a large number of machines (in the textile industry, for example, some weavers operate up to 280 looms).

The theory of geometrical probability has repeatedly been subjected to criticism because of the arbitrary way it defines the probability of events. In this connection, many authors have arrived at the conviction that in the case of an infinite number of outcomes, no definition of probability can be given that is objective and independent of the method of calculation. As a particularly outstanding exponent of this skepticism, one may mention Joseph Bertrand, the French mathematician of the last century. In his book on the theory of probability, he cited a number of problems in geometrical probability in which the result depends on the method of

---

[7] In § 9 we shall see that since the arrival times of A and B are independent, the probability that A will arrive in the interval from x to $x + h$ and B in the interval from y to $y + s$ is $(h/60) \cdot (s/60)$, i.e., proportional to the area of the rectangle of sides h and s.

solution. As an example of this, let us discuss one of the problems considered by Bertrand.

Example 2. Bertrand's paradox. A chord of a circle is chosen at random. What is the probability that its length exceeds the length of a side of the inscribed equilateral triangle?

First Solution: By considerations of symmetry, we may fix the direction of the chord in advance. Let us draw the diameter perpendicular to this direction. Obviously, only the chords that intersect the diameter in the interval from one quarter to three quarters of its length will exceed the side of the equilateral triangle. Thus the required probability is 1/2.

Second Solution: By considerations of symmetry, one end-point of the chord may be fixed in advance. The tangent to the circle at this point and the two sides of the inscribed equilateral triangle with vertex at this point form three 60° angles. Only those chords falling within the middle angle are favorable cases. Thus, by this method of computation, the required probability turns out to be 1/3.

Third Solution: In order to fix the position of the chord, it suffices to give its midpoint. For the chord to satisfy the condition of the problem, it is necessary that its midpoint lie within the concentric circle with radius half that of the given circle. The area of this circle is one-fourth the area of the given circle; thus, the required probability is 1/4.

We ought now to find out the reason why the solution of our problem is not unique. Does it lie in the fact that it is fundamentally impossible to determine the probability when there are an infinite number of possible outcomes or does it lie in our having made use of some inadmissible premises in the process of solving?

As we can easily see, the trouble is this: The concept of drawing a chord at random was left undefined in the conditions of our problem, and we have taken advantage of this ambiguity so as to obtain solutions to three distinct problems rather than one.

As a matter of fact, in the first solution, we can let a circular cylindrical rod roll along one of the diameters (Fig. 4a). The set of all possible stopping points of the rod is the set of all points of the interval AB whose length is equal to the diameter. The equally likely events are taken to be the stopping of the rod in an interval of length h, regardless of where on the diameter the interval is located.

§ 6. GEOMETRICAL PROBABILITY

In the second solution, a rod with one end hinged at a point on the circle is made to perform oscillations of 180° (Fig. 4b). It is assumed here that the stopping of the rod on an arc of the circumference of length h depends only on the length of the arc and not on its location. Thus, the equally likely outcomes are taken to be the stopping of the rod upon any arcs of the circle of equal length. The inconsistency in the definition of

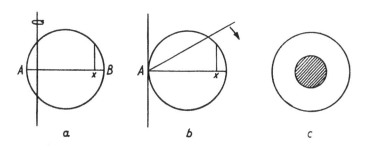

FIG. 4

the probability in the first and second solutions becomes quite clear after the following simple computation. According to the first setup, the probability of the rod stopping in the interval from A to x is x/D, where D is the length of the diameter. By elementary geometrical considerations, the probability that the projection of the point of intersection of the rod and the circle falls in the same interval is equal to

$$\frac{1}{\pi} \arccos \frac{D-2x}{D} \quad \text{for } x \leqq D/2$$

and

$$1 - \frac{1}{\pi} \arccos \frac{2x-D}{D} \quad \text{for } x \geqq D/2.$$

Finally, in the third solution, we choose a point at random in the given circle and ask what the probability is that the point fall within some smaller concentric circle.

The difference in the formulation of the problem in all three cases is now quite clear.

Example 3. Buffon's needle problem. A plane is ruled with a series of parallel lines a distance of $2a$ apart. A needle of length $2l$ ($l < a$) is tossed at random[8] onto the plane. Determine the probability that the needle will intersect one of the lines.

Solution: We denote by x the distance from the center of the needle to the closest line and by φ the angle which the needle forms with this line. The quantities x and φ completely determine the position of the needle. All possible positions of the needle are then given by the coordinates of the points inside a rectangle with sides of length a and π.

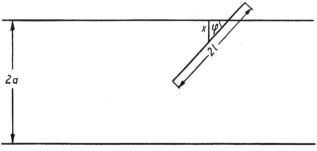

Fig. 5

From Fig. 5, we see that a necessary and sufficient condition for the needle to intersect the line is that

$$x \leq l \sin \varphi.$$

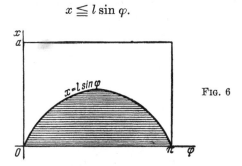

Fig. 6

By virtue of our assumptions, the required probability is the quotient of the area of the shaded region in Fig. 6 and the area of the rectangle:

$$p = \frac{1}{a\pi} \int_0^\pi l \sin \varphi \, d\varphi = \frac{2l}{a\pi}.$$

---

[8] The phrase "at random" in this example implies the following: First, the center of the needle falls at random in a segment of length $2a$ perpendicular to the lines, second, the angle φ which the needle makes with the lines will lie between φ_1 and $\varphi_1 + \Delta\varphi$, with a probability proportional to $\Delta\varphi$, and third, the quantities x and φ are independent (see § 9).

§ 6. Geometrical Probability

We note that the problem of Buffon is the starting point for the solution of some theoretical artillery problems which take into account the dimensions of the projectiles.

Example 4. On a horizontal plane ruled with a series of parallel lines a distance of $2a$ apart, a closed convex curve with a diameter less than $2a$ is thrown at random.[9] Determine the probability that this curve intersects one of the parallel lines.

Solution: Suppose first that the closed convex curve is an n-sided polygon. Let its sides be numbered from 1 to n. If the polygon intersects one of the lines, then the intersection is with two of the sides. Let us denote by $p_{ij} = p_{ji}$ the probability that the intersection is with the i-th and j-th sides. Obviously, the event A, that our polygon intersects one of the lines, can be expressed as the following sum of pairwise mutually exclusive events:

$$A = (A_{12} + A_{13} + \cdots + A_{1n}) + (A_{23} + A_{24} + \cdots + A_{2n}) + \cdots$$
$$\cdots + (A_{n-2,\, n-1} + A_{n-2,\, n}) + A_{n-1,\, n}\,,$$

in which A_{ij} ($i < j$, $i = 1, 2, \ldots$; $j = 1, 2, \ldots$) denotes the event consisting in the intersection of the i-th and j-th sides with the line. By the addition theorem for probabilities,

$$p = \mathsf{P}(A) = \mathsf{P}[(A_{12}) + \mathsf{P}(A_{13}) + \cdots + \mathsf{P}(A_{1n})] +$$
$$+ [\mathsf{P}(A_{23}) + \cdots + \mathsf{P}(A_{2n})] + \cdots + \mathsf{P}(A_{n-1,n}) =$$
$$= (p_{12} + p_{13} + \cdots + p_{1n}) + (p_{23} + p_{24} + \cdots + p_{2n}) + \cdots + p_{n-1,n}\,.$$

Using the equality $p_{ij} = p_{ji}$, we can write down the probability p in another way:

$$p = \frac{1}{2}\,[(p_{12} + p_{13} + \cdots + p_{1n}) + (p_{21} + p_{23} + \cdots + p_{2n}) + \cdots$$
$$\cdots + (p_{n1} + p_{n2} + \cdots + p_{n,\,n-1})]\,.$$

But the sum $\sum\limits_{j=1}^{n}{}' p_{ij}$, in which the term p_{ii} is set equal to zero, is nothing other than the probability of intersection of the i-th side of the polygon with one of the parallel lines. If we denote the length of the i-th side by $2l_i$, we find from the problem of Buffon that

---

[9] In this example, the phrase "at random" means that we take some cross-piece rigidly connected to the curve and throw it "at random" in the sense of the preceding example.

It is not difficult to show that the concept "at random" defined in this way is independent of the choice of cross-piece.

$$\sum_{j=1}^{n}{}' p_{ij} = \frac{2l_i}{\pi a},$$

and consequently,

$$p = \frac{\sum_{i=1}^{n} 2l_i}{2\pi a}.$$

If $2s$ is the perimeter of the polygon, then we finally obtain

$$p = s/\pi a.$$

We thus see that the probability p depends neither on the number of sides of the polygon nor on the length of the individual sides. From this we conclude that the formula also holds true for any convex closed curve, since we may always regard the latter as the limit of convex polygons the number of whose sides tends to infinity.

§ 7. The Statistical Definition of Probability

The classical definition of probability encounters insurmountable difficulties of a fundamental nature in passing from the simplest examples to a consideration of complex problems, in particular, those that occur in the natural sciences and in industry. First of all, the question arises, in a majority of cases, as to a reasonable way of selecting the "equally likely cases." Thus, for example, it is difficult, at least at present, to deduce the probability that an atom of some radioactive substance will disintegrate in a given interval of time from the considerations of symmetry on which we base our judgments concerning the equal likelihood of events, or else to determine the probability that a baby about to be born is a boy.

Lengthy observations as to the occurrence or non-occurrence of an event A in a large number of repeated trials under the same set of conditions $\mathfrak{S}$ show that for a wide class of phenomena the number of occurrences or non-occurrences of the event A is subject to a stable law. Namely, if we denote by μ the number of times the event A occurs in n independent trials, then it turns out that for sufficiently large n the ratio μ/n, in most of such series of observations, assumes an almost constant value, with large deviations being less frequent the greater the number of trials that are made.

§ 7. THE STATISTICAL DEFINITION OF PROBABILITY

This kind of stability of the relative frequency (i.e., the ratio μ/n) was first noticed in phenomena of a demographic nature. Thus, in ancient times, it was already observed that the ratio of male births to the total number of births in entire countries and in large cities remained almost unchanged from year to year. On the basis of censuses taken in ancient China 2238 years prior to our era, this number was found to be 1/2. Later on, particularly in the 17th and 18th centuries, a number of fundamental works devoted to the study of population statistics appeared. Apart from the stability of the ratio of male to female births, it was ascertained that stable laws of still another character existed: the percentage of deaths at a particular age in certain groups of the population (of a given economic and social background), the distribution of people (of a definite sex, age, and nationality) according to height, breadth of chest, length of footstep, etc.

In his famous book *Essai philosophique sur les probabilités*, Laplace tells of a very noteworthy episode which happened to him while studying the data on the birth of boys and girls. The voluminous statistical data which he analyzed for London, St. Petersburg, Berlin, and all of France yielded ratios of male births to all births that were almost identical. During one decade, all of these ratios fluctuated about one and the same number, which was approximately 22/43. At the same time, an investigation of the analogous statistical data for Paris during the forty years from 1745 to 1784 led to a different ratio, of 25/49. Laplace became interested in so significant a difference and he began to seek a rational explanation for it. In a detailed study of the archival data, it was discovered that all foundlings had also been included in the overall number of births in Paris. It turned out further that the local populace preferred to abandon infants mainly of one sex. This social practice was rather widespread at that time and materially distorted the true picture of births in Paris. When Laplace excluded the number of foundlings from the overall number of births, the birth ratio of boys in Paris proved to be fairly constant and close to 22/43, the same as for other peoples and for France as a whole.

Since the time of Laplace, an abundance of statistical data has accumulated which allows the quantitative characteristics of demographic statistics of importance to society to be computed in advance with great accuracy. As a final illustration of a statistical nature, we shall cite more modern data for which the relative frequency in large numbers of trials is practically constant: we consider the distribution of newborn children according to sex and month of the year. The data, which is adopted from the book *Mathematical Methods of Statistics* by H. Cramér, constitutes part of the official Swedish vital statistics for the year 1935.

TABLE 3

Month	1	2	3	4	5	6	7	8	9	10	11	12	Total
Total Births	7280	6957	7883	7884	7892	7609	7585	7393	7203	6903	6552	7132	88 273
Boys	3743	3550	4017	4173	4117	3944	3964	3797	3712	3512	3392	3761	45 682
Girls	3537	3407	3866	3711	3775	3665	3621	3596	3491	3391	3160	3371	42 591
Rel.freq. of Girls	0.486	0.489	0.490	0.471	0.478	0.482	0.462	0.484	0.485	0.491	0.482	0.473	0.4825

Fig. 7 shows how the monthly relative frequency of birth of girls

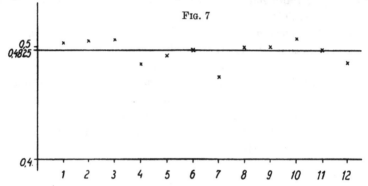

FIG. 7

deviates from the corresponding relative frequency for the year. We see that the relative frequency fluctuates about the number 0.482.

It turns out that in all those cases in which the classical definition of probability is applicable, the fluctuation of the relative frequency takes place in the neighborhood of the probability p of the event.

There exists a vast quantity of experimental material in verification of this fact. For instance, experiments have been performed in coin tossing, dice throwing, dropping a needle to determine empirically the value of π (see Example 3, § 6), and others as well. We quote here some of the results obtained, confining ourselves to experiments in coin tossing. [See the table on the following page.]

There exist today other examples of the verification of the above-mentioned empirical fact which are of scientific and practical importance. For example, in modern statistical practice, a considerable role is played by tables of random numbers, in which each entry is selected haphazardly

§ 7. THE STATISTICAL DEFINITION OF PROBABILITY

	Number of Tosses	Number of Heads	Relative Frequency
Buffon	4040	2048	0.5080
Karl Pearson . . .	12000	6019	0.5016
Karl Pearson . . .	24000	12012	0.5005

from the set of digits 0, 1, 2, 3, ..., 9. In one such table of random numbers, the digit 7 appears 968 times among the first 10,000 entries, i.e., its relative frequency is 0.0968 (the actual probability of occurrence of the digit 7 is 0.1). However, if we count up the number of times the digit 7 occurs in each succeeding thousand random numbers, then we obtain the following values:

Thousands	1	2	3	4	5	6	7	8	9	10
Number of 7's	95	88	95	112	95	99	82	89	111	102
Relative Frequency	0.095	0.088	0.095	0.112	0.095	0.099	0.082	0.089	0.111	0.102

The relative frequency of sevens in the various groups of a thousand fluctuates rather considerably but is nevertheless comparatively close to the probability.

The fact that in a number of instances the relative frequency of random events in a large number of trials is almost constant compels us to presume the existence of certain laws, independent of the experimenter, that govern the course of these phenomena and that manifest themselves in this near constancy of the relative frequency. Further, the fact that the relative frequency of an event to which the classical definition is applicable is, as a rule, close to its probability whenever the number of experiments is large compels us to presume that in the general case as well, there exists some constant about which the relative frequency fluctuates. Since this constant is an objective numerical characteristic of the phenomena, it is natural to call it the *probability* of the random event A under investigation.

Hence we shall say that an event A has a probability if this event has the following characteristics:

(a) It is possible, at least in principle, to make an unlimited number of mutually independent trials under the same set of conditions $\mathfrak{S}$, in each of which the event A may or may not occur;

(b) As the result of a sufficiently large number of trials, the relative frequency of the event A in almost every one of a large group of trials is observed to deviate only negligibly from a certain (generally speaking, unknown) constant.

As an approximation to the value of this constant, one may take the relative frequency of the event in a large number of trials, or else some value close to the relative frequency. Hence the probability of a random event as thus defined is called *statistical probability*. It is natural that statistical probability should be required to have the following properties:

1) The probability of the certain event is one;
2) The probability of the impossible event is zero;
3) If a random event C is the sum of a finite number of mutually exclusive events $A_1, A_2, \ldots, A_n$ each of which has a probability, then the probability of C exists and is equal to the sum of the probabilities of the events A_i:

$$\mathsf{P}(C) = \mathsf{P}(A_1) + \mathsf{P}(A_2) + \ldots + \mathsf{P}(A_n).$$

The statistical definition of probability which we have given is rather of a descriptive than of a formal mathematical character. One should note an inadequacy of still another type: namely, it does not bring to light the real nature of those phenomena for which the relative frequency is stable. By this we mean to stress the need for carrying out further research in the direction indicated. However, what is of particular importance is that in our definition probability retains its objective meaning, one that is independent of the observer. The fact that we can infer the existence of a probability for an event only after some preliminary observations have been made in no way lessens the value of our deductions, because a knowledge of laws of nature never has its origin in nothingness, but always follows upon experiment, or observation. The laws themselves existed, of course, before the intervention of the experimenting and thinking being and were merely unknown to science.

We have already stated above that we have not given here a formal mathematical definition of probability but have only postulated its existence under certain conditions and indicated a method of computing it approximately. Any objective property of the phenomena investigated, including the probability of the event A, should be determined solely

§ 7. THE STATISTICAL DEFINITION OF PROBABILITY

from the structure of the phenomenon itself irrespective of whether or not an experiment is performed and whether or not an experimenting intellect is present. Nevertheless, any experiment we perform does play an essential role: first, such an experiment permits us to observe the theoretical stochastic laws existing in nature; second, it permits us to approximate to the unknown probabilities of the events we study; and finally, it enables us to check up on the correctness of the theoretical premises that we have made in our investigations. This last requires elucidation.

Let us imagine that certain arguments have provided us with a basis for regarding the probability of an event A to be p. Suppose, further, that in several series of independent experiments it has turned out that the relative frequencies in an overwhelming number of these series deviate appreciably from the quantity p. This fact causes us to express doubt as to the correctness of our a priori judgments and justifies our undertaking a more detailed investigation of the premises underlying our a priori conclusions. Thus, for example, concerning a die, we make the assumption of its geometric regularity and the homogeneity of the substance of which it is made. From these preliminary premises we are justified in drawing the conclusion that when the die is thrown the probability that any one of its faces—for example, the one with the number 5—will come uppermost must be 1/6. If a repeated series of sufficiently numerous trials (throws) in our example systematically show that the relative frequency of occurrence of the number 5 deviates significantly from 1/6, then we should have our doubts not about the existence of a definite probability of throwing a five but rather about our assumptions concerning the regularity of the die or about whether our trials (throws) have taken place under suitably controlled conditions.

In conclusion, we must pause to discuss the interpretation of probability given by R. von Mises, which is very widely used, especially among the natural scientists. Since the relative frequency deviates less and less from the probability p as the number of experiments is continually increased, then, according to von Mises, one should have the limiting relation

$$p = \lim_{n \to \infty} \frac{\mu}{n}.$$

Von Mises proposes to regard this relation as defining the concept of probability. In his opinion, any apriori definition is doomed to failure and only his empirical definition is capable of serving the interests of natural science, mathematics, and philosophy. Since the statistical definition is applicable in all situations of scientific interest whereas the classical definition in terms of equal likelihood of events (based on symmetry) has only very limited applicability, von Mises proposes abandoning the latter entirely. Moreover, von Mises considers it altogether unnecessary to clarify the structure of the phenomena for which probability is an objective numerical property, since for him it suffices that the relative frequency is empirically stable.

According to von Mises, the theory of probability has to do with infinite sequences of observations, which he calls collectives. For example, all the throws with a pair of dice made in the course of a game form a collective. Each collective must satisfy the following two conditions.

1) *The Existence of limits*: The limits of the relative frequencies of events (e.g., the throws of the dice) with particular attributes (e.g., the number of points thrown) within the collective exist;

2) *The Principle of Randomness*: These limits are invariant with respect to the choice of any subsequence of the collective by some rule which is arbitrary except that it must not be based on distinguishing the elements of the collective in their relation to the attribute under consideration.

The construction of a mathematical theory based on the fulfillment of both these requirements encounters insurmountable difficulties. The fact is that the Principle of Randomness is inconsistent with the requirement of the existence of a limit.

We shall not stop to give the details of von Mises' theory. For these details, we refer the reader instead to his book *Probability, Statistics, and Truth* and, for a more extensive critique, to the article by Khintchine.[10] We limit ourselves here to merely a few remarks which should show the reader the inadmissibility of von Mises' concept.

The position advanced by von Mises has found warm admirers, especially among the representatives of the natural sciences, thanks, in great measure to its persuasive argument concerning the narrowness and limitation of the classical concepts and to its appeal to experiment as a means of determining probabilities.

---

[10] *The studies of Mises on probability and the principle of physical statistics*, Uspekhi Fiz. Nauk, Vol. IX, No. 2, 1929 [in Russian].

§ 8. AXIOMATIC CONSTRUCTION OF THE THEORY OF PROBABILITY 51

In the von Mises interpretation, probability loses its property of being an objective numerical characteristic of real phenomena. One cannot, in fact, even speak of the probability of an event until an infinite number of experiments have been carried out, and inasmuch as it is impossible to realize this, one is on the whole deprived of the possibility of making use of the theory of probability under any conditions. It should be noted here that in requiring of the relative frequency that it converge to the probability,[11] von Mises has set up a requirement such as is not satisfied in even a single field of natural science. None of us, after all, would renounce the concept of temperature merely because we cannot perform an infinite number of measurements and because we cannot verify whether the results of these measurements tend to a limit, supposing that we could actually perform them. Nor would we say of some object that it had no length, or other dimension, merely because our sequence of measurements did not tend to a limit. Furthermore, if we followed von Mises, we could not speak at all of the temperature of a body or the existence of the dimensions of an object until some thinking subject had made his appearance, had begun to do measuring, and had convinced himself that the results of the measurements tended to a limit.

§ 8. Axiomatic Construction of the Theory of Probability

Until recently, the theory of probability was an as yet unformulated mathematical science in which the fundamental concepts were not defined very clearly. This lack of clarity frequently led to paradoxical conclusions (recall Bertrand's paradox). As might be expected, the applications of the theory of probability to the study of natural phenomena rested on a shaky foundation, and at times encountered sharp and well-merited criticism. It must be admitted that this situation did not bother the natural scientists very much, and their naive probabilistic approach

---

[11] We should note, furthermore, that the concept of limit of a sequence ceases to have mathematical meaning in the theory of von Mises.

to various scientific fields indeed led to important successes. The development of natural science at the beginning of the current century put greater demands on the theory of probability. There grew a need to study the fundamental concepts of probability theory in a systematic way and to clarify the conditions under which it is possible to make use of the results of the theory. That is particularly why a formal logical foundation for the theory of probability—an axiomatic construction—acquired such very great importance. The theory of probability as a mathematical science must be based on certain premises that represent a generalization of centuries-old human experience. The further development of the theory should be accomplished by logical deduction from these basic assumptions, without recourse to intuitive notions or to "common sense" inferences. In other words, the theory of probability should be built up on the basis of axioms like any other well-developed mathematical discipline—geometry, theoretical mechanics, abstract group theory, etc.

Such a point of view was first expressed and developed by the Soviet mathematician S. N. Bernstein in 1917. In doing this, Bernstein proceeded from a qualitative comparison of random events according to their larger or smaller probability.

There exists another approach proposed by A. N. Kolmogorov. This closely relates probability theory to the theory of sets and the modern measure-theoretical aspects of the theory of functions of a real variable. The present book follows the path taken by Kolmogorov.

We shall see that the axiomatic treatment of the foundations of probability theory proceeds from the fundamental properties of probability observed in the examples illustrating the classical and statistical definitions. Thus, the axiomatic definition of probability includes as a special case both the classical and the statistical definitions and overcomes the shortcomings of each. It has been found possible to erect on this base a logically sound structure for modern probability theory and one that has at the same time met the increased demands made upon the theory by modern science.

In Kolmogorov's axiomatic treatment of probability theory, the concept of random event is not a primitive notion but is built up from more elementary concepts. We have already met such an approach in our discussion of some of the examples. Thus, in the problems involving the geometrical definition of probability, a region R in space (a line, a plane, etc.) is singled out in which a point is picked "at random." Here, the random events correspond to the point falling in some subregion of R.

§ 8. Axiomatic Construction of the Theory of Probability

Every random event is in this way a certain subset of the set of points of R. This idea is taken as a basis for the general concept of random event in the axiomatic treatment of Kolmogorov.

Kolmogorov starts with a set U consisting of *elementary events*. What the elements of this set are is immaterial for the logical development of probability theory. He then considers a certain family F of subsets of the set U; the elements of the family F are called *random events*. The following three conditions are imposed on the structure of the family F.

1) F contains the set U as one of its elements.

2) If the subsets A and B of the set U are elements of F, then the sets $A + B, AB, \bar{A}$, and $\bar{B}$ are also elements of F.

Here, $A + B$ is understood to be the set consisting of the elements of U that belong to either A or B or to both A and B; AB is the set consisting of the elements of U that belong to both A and B; and finally, $\bar{A}$ and $\bar{B}$ are, respectively, the set of elements of U not belonging to A and the set of elements of U not belonging to B.

Inasmuch as the set U itself belongs to F, the second requirement implies that F also contains $\bar{U}$, i.e., *the empty set is one of the elements of F*.

It is easy to see that the second requirement implies that sums, products, and complements of a finite number of random events of F belong to F. Thus, the elementary operations on random events do not take us outside the set of random events, i.e., F is *closed* under these three elementary operations on events. As in § 3, we shall call this family of events F a *field of events*.[12]

In many important problems, it is necessary for us to require more of a field of events, namely:

3) If the subsets $A_1, A_2, \ldots, A_n, \ldots$ of the set U are elements of the set F, then the sum $A_1 + A_2 + \ldots + A_n + \ldots$ of these subsets and the product $A_1 A_2 \ldots A_n \ldots$ of these subsets are also elements of F.

The set F formed in this way is called a *Borel field of events*.

The above method of defining a random event is in complete conformity with the idea we arrived at by considering concrete examples. For the sake of clarity, we consider two of the examples in greater detail from this point of view.

---

[12] Instead of the term *field of events* the term *σ-algebra of events* is now frequently used.

I. THE CONCEPT OF PROBABILITY

Example 1. A die is thrown. The set $U=(E_1, E_2, E_3, E_4, E_5, E_6)$ of *elementary events* consists of the six elements E_1, E_2, E_3, E_4, E_5, and E_6, where E_i denotes the throwing of the number i. The set F of *random events* consists of the following $2^6 = 64$ elements:

$(V), (E_1), (E_2), (E_3), (E_4), (E_5), (E_6), (E_1, E_2), (E_1, E_3), \ldots,$
$(E_5, E_6), (E_1, E_2, E_3), \ldots, (E_4, E_5, E_6), (E_1, E_2, E_3, E_4), \ldots,$
$(E_3, E_4, E_5, E_6), (E_1, E_2, E_3, E_4, E_5), \ldots,$
$(E_2, E_3, E_4, E_5, E_6), (E_1, E_2, E_3, E_4, E_5, E_6).$

Here, each pair of parentheses exhibits the elements of the set U that enter into the composition of a particular subset belonging to F; the symbol (V) denotes the empty set.

Example 2. *The encounter problem.* The set U consists of the points of the square: $0 \leq x \leq 60$, $0 \leq y \leq 60$.

The set F consists of all of the Borel sets that can be formed from the points of this square. In particular, the set of points in the closed region $|x-y| \leq 20$ belongs to F and is a random event.

It is natural to introduce the following definitions:

If two random events A and B are such that the elements of U that constitute A are distinct from the elements of U that constitute B, i.e., if there is no element of U that enters into the composition of both A and B, then we shall call them *mutually exclusive*.

The random events U and $\overline{U}$ (the empty set) are called, respectively, the *certain*, or *sure*, event and the *impossible event*. The events A and $\overline{A}$ are called *complementary*.

We can now proceed to formulate the axioms that define probability.

AXIOM 1: *With each random event A in a field of events F, there is associated a non-negative number $\mathsf{P}(A)$, called its probability.*

AXIOM 2: $\mathsf{P}(U) = 1$.

AXIOM 3 (Addition Axiom): *If the events $A_1, A_2, \ldots, A_n$ are pairwise mutually exclusive, then*

$$\mathsf{P}(A_1 + A_2 + \cdots + A_n) = \mathsf{P}(A_1) + \mathsf{P}(A_2) + \cdots + \mathsf{P}(A_n).$$

In the case of the classical definition of probability, it was not necessary to postulate the properties expressed by the second and third axioms, because we were able to prove them. The statement of Axiom 1, however, is contained in the classical definition of probability.

§ 8. Axiomatic Construction of the Theory of Probability

We now deduce from these axioms several important elementary consequences.[13]

First of all, from the obvious equality

$$U = V + U$$

and Axiom 3, we deduce that

$$\mathsf{P}(U) = \mathsf{P}(V) + \mathsf{P}(U).$$

Thus:

1. The probability of the impossible event is equal to zero.

In an analogous way, it is easy to show the following:

2. For any event A,

$$\mathsf{P}(\bar{A}) = 1 - \mathsf{P}(A).$$

3. No matter what random event A may be,

$$0 \leq \mathsf{P}(A) \leq 1.$$

4. If the event A implies the event B, then

$$\mathsf{P}(A) \leq \mathsf{P}(B).$$

5. Let A and B be two arbitrary events. Inasmuch as the summands in the sums $A + B = A + (B - AB)$ and $B = AB + (B - AB)$ are mutually exclusive events, it follows from Axiom 3 that

$$\mathsf{P}(A + B) = \mathsf{P}(A) + \mathsf{P}(B - AB), \qquad \mathsf{P}(B) = \mathsf{P}(AB) + \mathsf{P}(B - AB).$$

From this follows the addition theorem for two arbitrary (not necessarily mutually exclusive) events A and B:

$$\mathsf{P}(A + B) = \mathsf{P}(A) + \mathsf{P}(B) - \mathsf{P}(AB).$$

By virtue of the fact that $\mathsf{P}(AB)$ is non-negative, it follows that

$$\mathsf{P}(A + B) \leq \mathsf{P}(A) + \mathsf{P}(B).$$

---

[13] As we shall see in § 29, the study of probability is reduced by means of these axioms to the study of the measure defined on Borel fields of sets. Probability itself is a non-negative additive set function.

I. The Concept of Probability

We can now derive the inequality

$$P\{A_1 + A_2 + \cdots + A_n\} \leq P(A_1) + P(A_2) + \cdots P(A_n)$$

for arbitrary events $A_1, A_2, \ldots, A_n$, by means of mathematical induction.

The system of axioms of Kolmogorov is *consistent*, since there exist real objects that satisfy these axioms. For example, let U be an arbitrary set containing a finite number of elements, i.e., $U = \{a_1, a_2, \ldots, a_n\}$, and let F be the set of all the subsets $\{a_{i_1}, a_{i_2}, \ldots, a_{i_s}\}$, $0 \leq i_1 < i_2 < \ldots < i_s \leq n$, $0 \leq s \leq n$, of U. We can then satisfy all of the axioms of Kolmogorov by setting

$$P(a_i) = p_i \quad (i = 1, 2, \ldots, n),$$

where $p_1, p_2, \ldots, p_n$ are arbitrary non-negative numbers for which

$$p_1 + p_2 + \ldots + p_n = 1,$$

and

$$P(a_{i_1}, a_{i_2}, \ldots, a_{i_s}) = p_{i_1} + p_{i_2} + \ldots + p_{i_s}.$$

However, this system of axioms is *incomplete*, since for a given set U we may select the probabilities in the set F in different ways.

Thus, in the example of throwing a die which we considered above, we can set

$$P(E_1) = P(E_2) = \cdots = P(E_6) = \frac{1}{6} \quad (1)$$

or

$$P(E_1) = P(E_2) = P(E_3) = \frac{1}{4}, \quad P(E_4) = P(E_5) = P(E_6) = \frac{1}{12}, \quad (2)$$

etc.

The incompleteness of the system of axioms in the theory of probability does not mean that the choice of axioms is an unfortunate one or that not enough thought has been put into developing them, but derives from the very nature of things: In various problems situations arise where it is necessary to study the same sets of random events, but with different probabilities. For example, we may have a pair of dice one of which is a true die (a perfect cube made of homogeneous material) and the other of which is not true. In the first case, the probability is given by the system of equations (1) and in the second case by, say, the system of equations (2).

§ 8. Axiomatic Construction of the Theory of Probability

For the further development of the theory, an additional assumption is required which is referred to as the *extended axiom of addition*. The necessity for introducing a new axiom is motivated by the fact that in probability theory we constantly have to consider events that decompose into an infinite number of sub-events.

EXTENDED AXIOM OF ADDITION: *If the event A is equivalent to the occurrence of at least one of the pairwise mutually exclusive events $A_1, A_2, \ldots, A_n, \ldots,$ then*

$$P(A) = P(A_1) + P(A_2) + \cdots + P(A_n) + \cdots .$$

We note that the extended addition axiom can be replaced by the *axiom of continuity*, which is equivalent to it.

AXIOM OF CONTINUITY: *If the sequence of events $B_1, B_2, \ldots, B_n, \ldots$ is such that each succeeding event implies the preceding one and the product of all the events B_n is the impossible event, then*

$$P(B_n) \to 0 \quad \text{for} \quad n \to \infty .$$

Let us prove the equivalence of the last two propositions.

1. The extended addition axiom implies the axiom of continuity.

In fact, suppose that $B_1, B_2, \ldots, B_n, \ldots$ are events such that

$$B_1 \supset B_2 \supset \cdots B_n \supset \cdots$$

and for any $n \geq 1$,

$$\prod_{k \geq n} B_k = V. \tag{3}$$

Obviously,

$$B_n = \sum_{k=n}^{\infty} B_k \bar{B}_{k+1} + \prod_{k \geq n}^{\infty} B_k .$$

Since the events occurring in this sum are pairwise mutually exclusive, the extended axiom of addition gives

$$P(B_n) = \sum_{k=n}^{\infty} P(B_k \bar{B}_{k+1}) + P(\prod_{k \geq n} B_k) .$$

But by virtue of the condition (3),

$$P\left(\prod_{k \geq n} B_k\right) = 0 ,$$

and therefore

$$P(B_n) = \sum_{k=n}^{\infty} P(B_k \bar{B}_{k+1}) ,$$

i.e., $P(B_n)$ is the remainder of the convergent series

$$\sum_{k=1}^{\infty} P(B_k \bar{B}_{k+1}) = P(B_1).$$

Hence $P(B_n) \to 0$ as $n \to \infty$.

2. The axiom of continuity implies the extended axiom of addition.

Let the events $A_1, A_2, \ldots, A_n, \ldots$ be pairwise mutually exclusive and let

$$A = A_1 + A_2 + \cdots + A_n + \cdots .$$

Let us set

$$B_n = \sum_{k=n}^{\infty} A_k.$$

It is evident that $B_{n+1} \subset B_n$. If the event B_n has occurred, then some event, say A_i ($i \geq n$), has also occurred, and this implies, by virtue of the pairwise mutual exclusiveness of the events A_k, that the events $A_{i+1}, A_{i+2}, \ldots$ have not occurred. Thus, the events $B_{i+1}, B_{i+2}, \ldots$ are impossible and therefore the event $\prod_{k=n}^{\infty} B_k$ is impossible. By the axiom of continuity, $P(B_n) \to 0$ as $n \to \infty$. Since

$$A = A_1 + A_2 + \cdots + A_n + B_{n+1},$$

we have by the ordinary axiom of addition

$$P(A) = P(A_1) + P(A_2) + \cdots + P(A_n) + P(B_{n+1}) =$$
$$= \lim_{n \to \infty} \sum_{k=1}^{n} P(A_k) = \sum_{k=1}^{\infty} P(A_k).$$

In conclusion, we may say that from the standpoint of set theory, our axiomatic definition of probability is nothing other than the introduction into the set U of a normed, completely additive, non-negative measure defined on all the elements of the set F.

§ 9. Conditional Probability and the Simplest Basic Formulas

We have already said that a certain set of conditions $\mathfrak{S}$ underlies the definition of the probability of an event. If no restrictions other than the conditions $\mathfrak{S}$ are imposed when calculating the probability $P(A)$, then this probability is called *unconditional*.

§ 9. Conditional Probability and the Simplest Basic Formulas

However, in many cases, one has to determine the probability of an event under the condition that a certain event B whose probability is greater than zero has already occurred. We call such a probability *conditional*, and we shall denote it by the symbol $P(A/B)$; this stands for the probability of the event A given that the event B has occurred. Strictly speaking, unconditional probabilities are also conditional, since our theory assumed at the very outset that a certain fixed set of conditions $\mathfrak{S}$ existed.

Example 1. A pair of dice is thrown. What is the probability of throwing an eight with the two dice (event A) if it is known that the number thrown is even (event B)?

All the possible outcomes of throwing a pair of dice are indicated in Table 4, the entry in each box corresponding to one of the possible events: the first of the numbers in the parentheses is the one appearing on the first die, and the second is the one that appears on the second die.

TABLE 4

(1,1)	(2,1)	(3,1)	(4,1)	(5,1)	(6,1)
(1,2)	(2,2)	(3,2)	(4,2)	(5,2)	(6,2)
(1,3)	(2,3)	(3,3)	(4,3)	(5,3)	(6,3)
(1,4)	(2,4)	(3,4)	(4,4)	(5,4)	(6,4)
(1,5)	(2,5)	(3,5)	(4,5)	(5,5)	(6,5)
(1,6)	(2,6)	(3,6)	(4,6)	(5,6)	(6,6)

The total number of possible outcomes is 36, of which 5 are favorable to the occurrence of the event A. Therefore, we have the unconditional probability

$$P(A) = \frac{5}{36}.$$

If the event B has occurred, then one out of 18 (and not 36) possibilities has been realized, and therefore the conditional probability is

$$P(A/B) = \frac{5}{18}.$$

Example 2. Two cards are drawn in succession from a deck of playing cards. Determine (a) the unconditional probability that the second card is an ace (the first card drawn being unknown), and (b) the conditional probability that the second card is an ace if the first draw was an ace.

Let A denote the event that an ace appears in the second draw, and B the event that an ace appears in the first draw. It is clear that the equation
$$A = AB + A\overline{B}$$
holds.

Since the events AB and $A\overline{B}$ are mutually exclusive, we have
$$P(A) = P(AB) + P(A\overline{B}).$$

The drawing of two cards from a deck of 52 cards can be accomplished in $52 \cdot 51$ different ways (taking into account the order in which they are drawn). Of these, $4 \cdot 3$ cases are favorable to the event AB and $4 \cdot 48$ cases are favorable to the event $A\overline{B}$. Therefore,
$$P(A) = \frac{4 \cdot 3}{52 \cdot 51} + \frac{48 \cdot 4}{52 \cdot 51} = \frac{1}{13}.$$

If the first card is known to be an ace, then there are 51 cards remaining, of which three are aces. Hence,
$$P(A/B) = \frac{3}{51} = \frac{1}{17}.$$

The general solution to the problem of finding a conditional probability can be given without difficulty in the case of the classical definition of probability. In fact, suppose that of n exhaustive, mutually exclusive, and equally likely occurrences $A_1, A_2, \ldots, A_n$, m are favorable to the event A, k are favorable to the event B, and r are favorable to the event AB (clearly, $r \leq k$, $r \leq m$). If the event B has occurred, this implies that one of the events A_j favorable to B has occurred. Under this condition, r and only r of the events A_j are favorable to the occurrence of A. Thus,
$$P(A/B) = \frac{r}{k} = \frac{\frac{r}{n}}{\frac{k}{n}} = \frac{P(AB)}{P(B)}. \tag{1}$$

In exactly the same way, we can deduce that
$$P(B/A) = \frac{P(AB)}{P(A)}. \tag{1'}$$

§ 9. Conditional Probability and the Simplest Basic Formulas

Of course, if A or B is the impossible event, then the equations (1) and (1′) respectively cease to have meaning.

Each of formulas (1) and (1′) is equivalent to the so-called *Multiplication Theorem* (Theorem on Compound Probabilities), according to which

$$P(AB) = P(A)\, P(B/A) = P(B)\, P(A/B), \qquad (2)$$

i.e., *the probability of the product of two events is equal to the product of the probability of one of the events by the conditional probability of the other event, given that the first event has occurred.*

The Multiplication Theorem is also applicable if one of the events A or B is impossible since, in this case, one of the equations $P(A/B)=0$ and $P(AB)=0$ holds along with $P(A)=0$.

We say that an event A is *stochastically independent*—or (simply) *independent*—of an event B if the relation

$$P(A/B) = P(A) \qquad (3)$$

holds, i.e., if the occurrence of the event B does not affect the probability of the event A.

If the event A is independent of the event B, then it follows from (2) that

$$P(A)\, P(B/A) = P(B)\, P(A).$$

From this, we find

$$P(B/A) = P(B), \qquad (4)$$

i.e., the event B is also independent of A. Thus, *independence is a symmetrical relation.*

If the events A and B are independent, then the events A and $\bar{B}$ are also independent. Indeed, since

$$P(B/A) + P(\bar{B}/A) = 1$$

and since by assumption $P(B/A) = P(B)$, then

$$P(\bar{B}/A) = 1 - P(B) = P(\bar{B}).$$

Hence, we obtain the important conclusion: *If the events A and B are independent, then each pair of events $(\bar{A}, B)$, $(A, \bar{B})$, and $(\bar{A}, \bar{B})$ is independent.*

The concept of independence of events plays an important role in the theory of probability and its applications. In particular, the greater part of the results presented in this book is obtained on the assumption that the various events considered are independent.

In practical problems, we rarely resort to verifying that relations (3) and (4) are satisfied in order to determine whether or not the given events are independent. To ascertain independence, we usually make use of intuitive arguments based on experience.

Thus, for example, it is clear that the fact that one coin turns up heads does not affect the probability that other coins will turn up heads (or tails) provided only that these coins are not connected to one another (for example, not rigidly fastened) at the time they are tossed. In exactly the same way, the birth of a boy to one mother does not influence the probability of the birth of a boy (or a girl) to another mother. These are independent events.

The Multiplication Theorem takes on a particularly simple form for independent events, to wit: *If the events A and B are independent, then*

$$\mathsf{P}(AB) = \mathsf{P}(A) \cdot \mathsf{P}(B).$$

We next generalize the notion of the independence of two events to that of a collection of events.

The events $B_1, B_2, \ldots, B_n$ are called *collectively independent*, or *mutually independent*, if for any event B_p and any events $B_{i_1}, B_{i_2}, \ldots, B_{i_r}$ ($i_j \neq p$; $j = 1, 2, \ldots, r$) of this collection, the event B_p and the event $B_{i_1} B_{i_2} \ldots B_{i_r}$ are independent.

By virtue of what we have just said, this definition is equivalent to the following: The events $B_1, B_2, \ldots, B_n$ are mutually independent if for any i_j for which $1 \leq i_1 < i_2 < \ldots < i_n \leq n$ and any r ($1 \leq r \leq n$),

$$\mathsf{P}\{B_{i_1} B_{i_2} \ldots B_{i_r}\} = \mathsf{P}\{B_{i_1}\} \mathsf{P}\{B_{i_2}\} \ldots \mathsf{P}\{B_{i_r}\}.$$

We note that for several events to be mutually independent it is not sufficient that they be pairwise independent. The following simple example should convince us of this. Imagine the four faces of a tetrahedron to be colored red, green, blue, and a combination of all three colors, respectively. Suppose the tetrahedron to be thrown once and let A be the event that the face on which it lands contains red, B the event that it contains green, and C the event that it conains blue. It is easy to

§ 9. Conditional Probability and the Simplest Basic Formulas

see that the probability that the tetrahedron lands on a face containing red is 1/2: there are four faces and two of them contain red. Thus,

$$P(A) = \frac{1}{2}.$$

The probabilities

$$P(B) = P(C) = P(A/B) = P(B/C) = P(C/A) =$$
$$= P(B/A) = P(C/B) = P(A/C) = \frac{1}{2}$$

may be computed in exactly the same way; the events A, B, and C are therefore pairwise independent.

However, if we know that the events B and C have been realized, then the event A must also have been realized, i.e.,

$$P(A/BC) = 1.$$

The events A, B, and C are thus mutually dependent.

Formula (1'), which in the case of the classical definition was derived from the definition of conditional probability, will be taken as a definition in the case of the axiomatic definition of probability. Hence *in the general case—where $P(A) > 0$—we have by definition*

$$P(B/A) = \frac{P(AB)}{P(A)}.$$

(In the case $P(A) = 0$, the conditional probability $P(B/A)$ remains undefined.) This enables us to carry over automatically to the general concept of probability all the definitions and results of the present section.

Let us now suppose that the event B can occur together with one and only one of the n mutually exclusive events $A_1, A_2, \ldots, A_n$. In other words, let us assume that

$$B = \sum_{i=1}^{n} BA_i, \tag{5}$$

where events BA_i and BA_j with distinct subscripts i and j are mutually exclusive. By the Addition Theorem, we have

$$P(B) = \sum_{i=1}^{n} P(BA_i).$$

Using the Multiplication Theorem, we find:

64   I. The Concept of Probability

$$P(B) = \sum_{i=1}^{n} P(A_i) P(B/A_i).$$

This relation is known as the *formula of total probability*, and plays a basic role in the theory to follow.

For illustrative purposes, we discuss two examples.

Example 3. There are five urns:

Two urns with the contents A_1: two black balls and one white ball;
One urn with the contents A_2: ten black balls; and
Two urns with the contents A_3: three white balls and one black ball.

An urn is selected at random, and a ball is then drawn from it at random. What is the probability that the ball withdrawn is white (event B)?

Since the ball can only be taken from an urn of composition A_1, A_2, or A_3, we have

$$B = A_1 B + A_2 B + A_3 B.$$

By the formula on total probability,

$$P(B) = P(A_1) P(B/A_1) + P(A_2) P(B/A_2) + P(A_3) P(B/A_3).$$

But

$$P(A_1) = \frac{2}{5}, \qquad P(A_2) = \frac{1}{5}, \qquad P(A_3) = \frac{2}{5},$$

$$P(B/A_1) = \frac{2}{3}, \qquad P(B/A_2) = 0, \qquad P(B/A_3) = \frac{3}{4}.$$

Therefore,

$$P(B) = \frac{2}{5} \cdot \frac{2}{3} + \frac{1}{5} \cdot 0 + \frac{2}{5} \cdot \frac{3}{4} = \frac{17}{30}.$$

Example 4. It is known that the probability of receiving k calls at a telephone exchange in an interval of time t is $P_t(k)$ ($k = 0, 1, 2, \ldots$).

Assuming that the number of calls received during each of two consecutive intervals are independent, find the probability that s calls are received during an interval of time of length $2t$.

Solution. Let us denote by A_τ^k the event consisting in k calls being received during an interval of time τ. Obviously, we have the following equality:

$$A_{2t}^s = A_t^0 A_t^s + A_t^1 A_t^{s-1} + \cdots + A_t^s A_t^0;$$

this means that the event A_{2t}^s may be regarded as the sum of $s + 1$ mutually exclusive events consisting in i calls during the first interval of time

§ 9. Conditional Probability and the Simplest Basic Formulas

of length t and $s-i$ calls in the immediately following interval of time of the same length t ($i=0, 1, 2, \ldots, s$). By the theorem on the addition of probabilities,

$$P(A_{2t}^s) = \sum_{i=0}^{s} P(A_t^i A_t^{s-i}).$$

By the Multiplication Theorem for independent events,

$$P(A_t^i A_t^{s-i}) = P(A_t^i) P(A_t^{s-i}) = P_t(i) \cdot P_t(s-i).$$

Therefore, if we set

$$P_{2t}(s) = P(A_{2t}^s),$$

then

$$P_{2t}(s) = \sum_{i=0}^{s} P_t(i) \cdot P_t(s-i). \tag{6}$$

We shall later see that under certain very general conditions,

$$P_t(k) = \frac{(at)^k}{k!} e^{-at} \tag{7}$$

($k=0, 1, 2, \ldots$), where a is some constant.

From formula (6), we find:

$$P_{2t}(s) = \sum_{i=0}^{s} \frac{(at)^s e^{-2at}}{i!(s-i)!} = (at)^s e^{-2at} \sum_{i=0}^{s} \frac{1}{i!(s-i)!}.$$

But

$$\sum_{i=0}^{s} \frac{1}{i!(s-i)!} = \frac{1}{s!} \sum_{i=0}^{s} \frac{s!}{i!(s-i)!} = \frac{1}{s!}(1+1)^s = \frac{2^s}{s!}.$$

Hence,

$$P_{2t}(s) = \frac{(2at)^s e^{-2at}}{s!} \qquad (s=0, 1, 2, \ldots).$$

Thus, if formula (7) holds for an interval of time of length t, then it also holds for an interval twice as long and, as one can easily convince himself, it continues to hold for an interval which is any arbitrary multiple of t.

We are now in a position to derive the important *formula of Bayes*, or as it is sometimes called, the *formula for probabilities of hypotheses*.

Suppose, as before, that (5) holds. It is required to find the probability of the event A_i if it is known that B has occurred. According to the Multiplication Theorem, we have

$$P(A_i\,B) = P(B)\,P(A_i/B) = P(A_i)\,P(B/A_i).$$

Hence,

$$P(A_i/B) = \frac{P(A_i)\,P(B/A_i)}{P(B)};$$

using the formula of total probability, we then find:

$$P(A_i/B) = \frac{P(A_i)\,P(B/A_i)}{\sum_{j=1}^{n} P(A_j)\,P(B/A_j)}.$$

This formula, or the preceding one, is referred to as *Bayes' formula*. The general scheme for applying Bayes' formula in the solution of practical problems is as follows. Suppose that an event B can occur under a number of different conditions concerning the nature of which n hypotheses $A_1, A_2, \ldots, A_n$ can be made. For one reason or another, the probabilities of these hypotheses $P(A_i)$ are known beforehand. We also know that the hypothesis A_i assigns a conditional probability $P(B/A_i)$ to the event B. An experiment is performed, in which the event B occurs. This should result in a re-appraisal of the probabilities of the hypotheses A_i; Bayes' formula gives a quantitative solution of this problem.

In artillery practice, we perform so-called ranging fire for the purpose of making our knowledge of firing conditions (for example, the accuracy of the gunsight) more precise. Bayes' formula finds wide use in the theory of ranging fire. However, we shall merely content ourselves with a purely schematic example for the sake of illustrating the nature of the problems that are solvable by this formula.

Example 5. There are five urns, as follows:

Two urns, each containing 2 white balls and 3 black balls (composition A_1);

Two urns, each containing 1 white ball and 4 black balls (composition A_2); and

One urn, containing 4 white balls and 1 black ball (composition A_3).

From one of the urns, chosen at random, a ball is withdrawn. It turns out to be white (event B). What is the probability, after the experiment has been performed (the aposteriori probability), that the ball was taken from an urn of the third type of composition?

By hypothesis,

$$P(A_1) = \frac{2}{5}, \qquad P(A_2) = \frac{2}{5}, \qquad P(A_3) = \frac{1}{5};$$

$$P(B/A_1) = \frac{2}{5}, \qquad P(B/A_2) = \frac{1}{5}, \qquad P(B/A_3) = \frac{4}{5}.$$

Bayes' formula then gives:

$$P(A_3/B) = \frac{P(A_3)\,P(B/A_3)}{P(A_1)\,P(B/A_1) + P(A_2)\,P(B/A_2) + P(A_3)\,P(B/A_3)} =$$

$$= \frac{\frac{1}{5} \cdot \frac{4}{5}}{\frac{2}{5} \cdot \frac{2}{5} + \frac{2}{5} \cdot \frac{1}{5} + \frac{1}{5} \cdot \frac{4}{5}} = \frac{4}{10} = \frac{2}{5}.$$

In exactly the same way, we find:

$$P(A_1/B) = \frac{2}{5}, \qquad P(A_2/B) = \frac{1}{5}.$$

§ 10. Examples

We now give several somewhat more complicated examples of the use of the above theory.

Example 1. Two players A and B continue a certain game until one of them is completely ruined. The capital of the first player is a dollars and that of the second, b dollars. The probability of winning at each play is p for player A and q for player B; $p + q = 1$ (there are no draws). In each play one of the players wins (and so the other loses) the sum of one dollar. Find the probability of ruin of each player (the results of the individual plays are assumed to be independent).[14]

Solution: Before proceeding to the analytic solution of the problem, let us ascertain what meaning attaches to the concept of elementary event and how the probability of the event we are investigating is to be defined.

---

[14] We are retaining the classical formulation of this problem of the ''ruin of a gambler,'' but other formulations are also possible, for example: A material particle is located at the point O on a line; it is subjected to a random impulse every second, resulting in its displacement 1 cm. to the right with probability p or 1 cm. to the left with probability $q = 1 - p$. What is the probability that the particle will get to the right of the point with coordinate b ($b > 0$) before it reaches a position to the left of the point with coordinate a ($a < 0$, a and b integers) ?

By an elementary event we shall mean any infinite sequence of outcomes of the individual plays. For example, $(A, \bar{A}, A, \ldots)$ is the elementary event in which all the odd plays are won by the player A and all the even ones by the player B. The random event corresponding to the ruin of player A consists of all the elementary events in which player A loses his capital before player B does. We note that each elementary event can be thought of as a denumerable sequence of the letters A and $\bar{A}$, and therefore, in every elementary event which forms part of the random event we are investigating—the ruin of the player A—the letters A and $\bar{A}$ still appear a denumerable number of times after the play which has ended in the ruin of the player A.

Let us first restrict ourselves to N plays, with a corresponding modification in our definition of elementary event.

Let $p_n(N)$ denote the probability of the ruin of player A in N plays if he started the game with a capital of n dollars. This probability can easily be determined, since the set of elementary events consists of a finite number of elements. It is natural to take the probability of each elementary event to be $p^m q^{N-m}$, where m and $N-m$ are the respective number of occurrences of A and $\bar{A}$ out of the total of N occurrences of both these letters. In exactly the same way, let $q_n(N)$ and $r_n(N)$ be, respectively, the probability of the ruin of player B and the probability of no result being reached within N plays.

It is clear that as N increases the numbers $p_n(N)$ and $q_n(N)$ do not decrease and $r_n(N)$ does not increase. Hence, the following limits exist:

$$p_n = \lim_{N \to \infty} p_n(N), \qquad q_n = \lim_{N \to \infty} q_n(N), \qquad r_n = \lim_{N \to \infty} r_n(N).$$

We shall call these respective limits the probabilities that A loses, that B loses, and that nobody loses, under the condition that at the start of the game A had n dollars and B had $a+b-n$ dollars. Since for any $N > 0$

$$p_n(N) + q_n(N) + r_n(N) = 1,$$

we also have in the limit that

$$p_n + q_n + r_n = 1.$$

Furthermore, it is obvious that

1) If A already has all the money at the start of the game and B has nothing, then

§ 10. Examples

$$p_{a+b} = 0, \qquad q_{a+b} = 1, \qquad r_{a+b} = 0; \tag{1}$$

2) If A had nothing and B had all the money at the start of the game, then

$$p_0 = 1, \qquad q_0 = 0, \qquad r_0 = 0. \tag{1'}$$

If A had n dollars just prior to a certain play, then his ruin can come about in two distinct ways: either he first wins the next play and then goes on to lose the entire game, or he loses both the next play and the game. Therefore, by the formula on total probability,

$$p_n = p \cdot p_{n+1} + q \cdot p_{n-1}.$$

We have thus obtained a finite difference equation for p_n; it is easy to see that it may be expressed in the following form:

$$q(p_n - p_{n-1}) = p(p_{n+1} - p_n). \tag{2}$$

Let us first solve this equation for $p = q = 1/2$. Under this assumption,

$$p_{n+1} - p_n = p_n - p_{n-1} = \cdots = p_1 - p_0 = c,$$

where c is a constant. Hence, we find:

$$p_n = p_0 + nc.$$

Inasmuch as $p_0 = 1$ and $p_{a+b} = 0$, we have

$$p_n = 1 - \frac{n}{a+b}.$$

Thus the probability of the ruin of A is

$$p_a = 1 - \frac{a}{a+b} = \frac{b}{a+b}.$$

In an analogous way, we find that for $p = 1/2$ the probability of the ruin of B is

$$q_a = \frac{a}{a+b}.$$

Hence, for $p = q = 1/2$, it follows that

$$r_a = 0.$$

I. The Concept of Probability

In the general case, where $p \neq q$, we find from (2) that

$$q^n \prod_{k=1}^{n} (p_k - p_{k-1}) = p^n \prod_{k=1}^{n} (p_{k+1} - p_k).$$

After simplifying, and making use of (1'), we obtain:

$$p_{n+1} - p_n = \left(\frac{q}{p}\right)^n (p_1 - 1).$$

Let us consider the difference $p_{a+b} - p_n$; obviously,

$$p_{a+b} - p_n = \sum_{k=n}^{a+b-1} (p_{k+1} - p_k) = \sum_{k=1}^{a+b-1} \left(\frac{q}{p}\right)^k (p_1 - 1) =$$

$$= (p_1 - 1) \frac{\left(\frac{q}{p}\right)^n - \left(\frac{q}{p}\right)^{a+b}}{1 - \frac{q}{p}}.$$

Inasmuch as $p_{a+b} = 0$,

$$p_n = (1 - p_1) \frac{\left(\frac{q}{p}\right)^n - \left(\frac{q}{p}\right)^{a+b}}{1 - \frac{q}{p}};$$

and because $p_0 = 1$,

$$1 = (1 - p_1) \frac{\left(\frac{q}{p}\right)^0 - \left(\frac{q}{p}\right)^{a+b}}{1 - \frac{q}{p}}.$$

Eliminating the quantity p_1 from the last two equations, we obtain:

$$p_n = \frac{\left(\frac{q}{p}\right)^{a+b} - \left(\frac{q}{p}\right)^n}{\left(\frac{q}{p}\right)^{a+b} - 1}.$$

Hence, the probability of the ruin of player A is

$$p_a = \frac{q^{a+b} - q^a p^b}{q^{a+b} - p^{a+b}} = \frac{1 - \left(\frac{p}{q}\right)^b}{1 - \left(\frac{p}{q}\right)^{a+b}}.$$

§ 10. EXAMPLES

In an analogous way, we find that the probability of the ruin of player B for $p \neq q$ is

$$q_a = \frac{1 - \left(\frac{q}{p}\right)^a}{1 - \left(\frac{q}{p}\right)^{a+b}}.$$

The last two formulas indicate that in the general case the probability of a draw is zero:

$$r_a = 0.$$

From the above formulas we can draw the following conclusions: If the capital of one of the players, say B, exceeds by far the capital of the player A, so that b may be regarded as infinite in comparison with a, and if the players are equally skillful, then the ruin of B is practically impossible. This conclusion will be quite different if A plays better than B, and so $p > q$. Assuming $b \sim \infty$, we then find:

$$q_a \sim 1 - \left(\frac{q}{p}\right)^a$$

and

$$p_a \sim \left(\frac{q}{p}\right)^a.$$

From this we can draw the conclusion that a skillful gambler, even with small capital, stands less of a chance of being ruined than a gambler with a large amount of capital who is less skillful.

The solution of certain problems in physics and engineering can be reduced to the problem of the gambler's ruin.

Example 2. Find the probability that a machine that is operating at the time t_0 will not stop before the time $t_0 + t$, if it is known that: 1) This probability depends only on the length of the interval of time $(t_0, t_0 + t)$; 2) The probability that the machine will stop in the interval of time Δt is proportional to Δt except for infinitesimals of higher order[15] with respect to Δt; 3) The events that consist in the stopping of the machine in non-overlapping intervals of time are mutually independent.

---

[15] In the following, we shall use the notation $\alpha = o(\beta)$ to indicate the fact that a quantity α is an infinitesimal of higher order than the quantity β. However, if $|\alpha/\beta|$ is bounded, then we shall write $\alpha = O(\beta)$.

Solution: Let us denote the required probability by $p(t)$. The probability that the machine will stop in the interval of time Δt is

$$1 - p(\Delta t) = a\,\Delta t + o(\Delta t),$$

where a is some constant.

We now determine the probability that the machine, which was operating at the time t_0, is still running at $t_0 + t + \Delta t$. In order for this event to take place, it is necessary for the machine not to have stopped for intervals of length t and Δt; therefore, by the Multiplication Theorem

$$p(t + \Delta t) = p(t) \cdot p(\Delta t) = p(t)\left(1 - a\,\Delta t - o(\Delta t)\right).$$

Hence

$$\frac{p(t + \Delta t) - p(t)}{\Delta t} = -a\,p(t) - o(1). \tag{3}$$

Let us now pass to the limit, letting $\Delta t \to 0$; since the limit of the right-hand side of equation (3) exists, it follows that the limit of the left-hand side also exists. As a result, we have:

$$\frac{dp(t)}{dt} = -a\,p(t).$$

The solution of this differential equation is the function

$$p(t) = Ce^{-at},$$

where C is a constant. This constant is determined from the obvious condition that $p(0) = 1$. Therefore,

$$p(t) = e^{-at}.$$

The first condition of the problem imposes a strong restriction on the conditions under which the machine is operated; however, there are industries where it is realized with a great degree of accuracy. As an example, one could cite the operation of an automatic loom. We might note that many other questions are reducible to the problem under consideration, for example, the question of the probability distribution of the mean free path of a molecule, in the kinetic theory of gases.

§ 10. Examples

Example 3. The construction of mortality tables is often based on the following assumptions:

1) The probability that a person will die during the interval of time from t to $t + \Delta t$ is equal to

$$p(t, t + \Delta t) = a(t)\, \Delta t + o(\Delta t),$$

where $a(t)$ is a non-negative continuous function;

2) The death of a particular person (or his survival) in a given interval of time (t_1, t_2) does not depend on what happened prior to the time t_1;

3) The probability of death at the moment of birth is zero.

Starting from the above assumptions, determine the probability that person A will die before attaining age t.

Solution: Letting $\pi(t)$ denote the probability that A will live to age t, we compute $\pi(t + \Delta t)$. It is obvious from the assumptions made in the problem that we have the equality

$$\pi(t + \Delta t) = \pi(t)\, \pi(t + \Delta t;\, t),$$

where $\pi(t + \Delta t;\, t)$ denotes the probability that A will be alive at age $t + \Delta t$ if he already has lived to age t. In accordance with the first and second assumptions,

$$\pi(t + \Delta t;\, t) = 1 - p(t, t + \Delta t) = 1 - a(t)\, \Delta t - o(\Delta t);$$

therefore,

$$\pi(t + \Delta t) = \pi(t)\, [1 - a(t)\, \Delta t - o(\Delta t)].$$

Hence, we find that $\pi(t)$ satisfies the following differential equation:

$$\frac{d\pi(t)}{dt} = -\, a(t)\, \pi(t).$$

The solution of this equation, taking the third assumption in the problem into account, is the function

$$\pi(t) = e^{-\int_0^t a(z)\, dz}.$$

The probability of death before attaining age t is thus

$$1 - \pi(t) = 1 - e^{-\int_0^t a(z)\, dz}.$$

In constructing mortality tables for an adult population, one frequently makes use of Makeham's formula, which states that

$$a(t) = \alpha + \beta e^{\gamma t},$$

where α, β, and γ are positive constants.[16] This formula was derived on the assumption that an adult may die from causes that do not depend upon age and from causes that do depend on age, where the probability of death from the second type of cause increases geometrically with increasing age. Under such an additional assumption, we obtain

$$\pi(t) = e^{-\alpha t - \frac{\beta}{\gamma}(e^{\gamma t} - 1)}.$$

Example 4. In modern nuclear physics, a Geiger-Müller counter is used to measure the intensity of a source of radiation. A particle that hits the counter causes an electrical discharge of duration τ during which the counter does not register any further particle that may hit it. Find the probability that the counter will register all the particles that hit it during the time t if the following conditions are fulfilled:

1) The probability that k particles hit the counter in an interval of time t does not depend on how many hit it before the commencement of the interval;

2) The probability that k particles hit the counter in the interval of time t_0 to $t_0 + t$ is given by the formula:[17]

$$p_k(t_0, t_0 + t) = \frac{(at)^k e^{-at}}{k!},$$

where a is a positive constant; and

3) τ is a constant.

Solution: Let $A(t)$ denote the event that all particles hitting the counter during the time t have been registered, and $B_k(t)$ the event that k particles have hit the counter in the time t.

---

[16] Their values are determined by the conditions under which the group of persons in question exist and, above all, by social conditions.

[17] Later on, we shall make clear why we have assumed that

$$p_k = \frac{(at)^k e^{-at}}{k!}$$

both in this example and in Example 4 of the preceding section.

§ 10. EXAMPLES

By virtue of the first condition of the problem, for $t \geq \tau$, we have
$$P\{A(t+\Delta t)\} = P\{A(t)\} P\{B_0(\Delta t)\} + P\{A(t-\tau)\} P\{B_0(\tau)\} P\{B_1(\Delta t)\} + o(\Delta t),$$
and for $0 \leq t \leq \tau$,
$$P\{A(t+\Delta t)\} = P\{A(t)\} P\{B_0(\Delta t)\} + P\{B_0(t)\} P\{B_1(\Delta t)\} + o(\Delta t).$$

For brevity, let us write $\pi(t) = P\{A(t)\}$; then, by the second and third conditions, for $0 \leq t \leq \tau$,
$$\pi(t+\Delta t) = \pi(t) e^{-a\Delta t} + e^{-a\Delta t} a \Delta t e^{-at} + o(\Delta t)$$
and for $t \geq \tau$,
$$\pi(t+\Delta t) = \pi(t) e^{-a\Delta t} + \pi(t-\tau) e^{-a\Delta t} a \Delta t e^{-a\tau} + o(\Delta t).$$

Letting $\Delta t \to 0$, we obtain in the limit the differential equation
$$\frac{d\pi(t)}{dt} = -a\pi(t) + a e^{-at} \tag{4}$$
for $0 \leq t \leq \tau$, and the differential equation
$$\frac{d\pi(t)}{dt} = -a[\pi(t) - \pi(t-\tau) e^{-a\tau}] \tag{5}$$
for $t \geq \tau$.

From equation (4), we find that for $0 \leq t \leq \tau$,
$$\pi(t) = e^{-at}(c + at).$$

Using the condition $\pi(0) = 1$, we can determine the constant c. Finally, for $0 \leq t \leq \tau$,
$$\pi(t) = e^{-at}(1 + at). \tag{6}$$

When $\tau \leq t \leq 2\tau$, the probability $\pi(t)$ is determined from the equation
$$\frac{d\pi(t)}{dt} = -a[\pi(t) - \pi(t-\tau) e^{-a\tau}]$$
$$= -a[(\pi t) - e^{-a(t-\tau)}(1 + a(t-\tau)) e^{-a\tau}]$$
$$= -a[\pi(t) - e^{-at}(1 + a(t-\tau))].$$
Solving this equation, we get:
$$\pi(t) = e^{-at}\left(c_1 + at + \frac{a^2(t-\tau)^2}{2!}\right).$$

According to (6),
$$\pi(\tau) = e^{-a\tau}(1 + a\tau),$$
and this can be used to evaluate the constant c_1. Thus, $c_1 = 1$, and for $\tau \leq t \leq 2\tau$,
$$\pi(t) = e^{-at}\left[1 + at + \frac{a^2(t-\tau)^2}{2!}\right].$$

It can be shown by mathematical induction that for $(n-1)\tau \leq t \leq n\tau$ the equality
$$\pi(t) = e^{-at} \sum_{k=0}^{n} \frac{a^k [t - (k-1)\tau]^k}{k!}$$
holds.

EXERCISES

A, B, and C are random events.

1. Explain the meaning of the relations:
   a) $ABC = A$;
   b) $A + B + C = A$.

2. Simplify the expressions:
   a) $(A + B)(B + C)$;
   b) $(A + B)(A + \bar{B})$;
   c) $(A + B)(\bar{A} + B)(A + \bar{B})$.

3. Prove the relations:
   a) $\overline{\bar{A}\bar{B}} = A + B$; b) $\overline{\bar{A} + \bar{B}} = AB$;
   c) $\overline{A_1 + A_2 + \cdots + A_n} = \bar{A}_1 \bar{A}_2 \cdots \bar{A}_n$;
   d) $\overline{A_1 A_2 \cdots A_n} = \bar{A}_1 + \bar{A}_2 + \cdots + \bar{A}_n$.

4. A four-volume work is placed on a shelf in random order. What is the probability that the books are in proper order from right to left or left to right?

5. The numbers 1, 2, 3, 4, 5 are written down on five cards. Three cards are drawn at random in succession, and the digits thus obtained are written from left to right in the order in which they were drawn. What is the probability that the resulting three-digit number turns out to be even?

6. In a lot consisting of N items, M are defective. n items are selected at random from the lot $(n < N)$. What is the probability that m $(m \leq M)$ of them will prove to be defective?

7. A quality control inspector examines the articles in a lot consisting of m items of first grade and n items of second grade. A check of the first b articles chosen at

random from the lot has shown that all of them are of second grade $(b < n-1)$. What is the probability that of the next two items selected at random from those remaining at least one proves to be of second grade?

8. Using probability-theoretic arguments, verify the identity $(a < A)$:

$$1 + \frac{A-a}{A-1} + \frac{(A-a)(A-a-1)}{(A-1)(A-2)} + \cdots + \frac{(A-a)\ldots 2\cdot 1}{(A-1)\ldots(a+1)a} = \frac{A}{a}.$$

Hint: Balls are drawn at random without replacement from an urn containing A balls a of which are white. Find the probability that sooner or later a white ball is drawn.

9. From a box containing m white balls and n black balls $(m > n)$, one ball after another is drawn at random. What is the probability that at some point the number of white balls and black balls drawn will be the same?

10. A man writes letters to n addressees and prepares addressed envelopes; someone inserts one letter per envelope at random and mails them. What is the probability that at least one letter is in the correct envelope?

11. In an urn are n tickets numbered from 1 to n. The tickets are withdrawn one by one at random, without replacement. What is the probability that in at least one of the drawings the number on the ticket coincides with the number of the drawing?

12. An even number of balls is drawn at random from an urn containing n white and n black balls (all of the different ways in which an even number of balls may be drawn are regarded as equally likely irrespective of their number). Find the probability that the same number of black balls and white balls are drawn.

13. *The problem of Chevalier de Méré.* Which is more probable: Throwing at least one ace with four dice or at least one double ace in 24 throws of a pair of dice?

14. Three points are selected at random on the interval $(0, a)$. Find the probability that three line-segments of length equal to the distances from the point 0 to each of the three points chosen can form a triangle.

15. A rod of length l is broken at two points chosen at random. What is the probability that a triangle can be formed from the segments obtained?

16. A point is chosen at random on a line-segment AB of length a, and a point is also chosen at random on a line-segment BC of length b. What is the probability that a triangle can be formed from the line-segments 1) from A to the first point, 2) from the first point to the second, and 3) from the second point to the point C?

17. N points are scattered at random and independently of one another inside a sphere of radius R.

a) What is the probability that the distance from the center of the sphere to the nearest point will not be less than r?

b) What is the limit of the probability found in part a) if

$$R \to \infty \text{ and } N/R^3 \to 4\pi\lambda/3?$$

Note: This problem is taken from astronomy: in the neighborhood of the sun, $\lambda \approx 0.0063$ if R is measured in parsecs.

18. The events $A_1, A_2, \ldots, A_n$ are independent, and $P(A_k) = p_k$. Find the probability of

a) the occurrence of at least one of these events;
b) the non-occurrence of all of them; and
c) the occurrence of exactly one—no matter which—of them.

19. Let $A_1, A_2, \ldots, A_n$ be any random events. Derive the formula

$$P\left\{\sum_{k=1}^{n} A_k\right\} = \sum_{i=1}^{n} P(A_i) - \sum_{1 \leq i < j \leq n} P(A_i A_j) + \\ + \sum_{1 \leq i < j < k \leq n} P(A_i A_j A_k) \mp \ldots \pm P(A_1 A_2 \ldots A_n).$$

Using this formula, solve Exercises 10 and 11.

20. The probability that a molecule which has collided with another at time $t = 0$ and undergone no further collisions with other molecules up to the time t will have a collision with another molecule in the interval of time t to $t + \Delta t$ is $\lambda \Delta t + o(\Delta t)$. Determine the probability that the time of free motion (i.e., the time between two successive collisions) is greater than t.

21. A bacterium reproduces by dividing into two bacteria. Assuming that a bacterium divides in an interval of time Δt with a probability $a \Delta t + o(\Delta t)$ which does not depend on the number of previous divisions nor on the number of existing bacteria, determine the probability that if at time $t = 0$ there was one bacterium there will be i bacteria at time t.

22. Prove that if the events A and B are mutually exclusive, and if both $P(A)$ and $P(B)$ are positive, then A and B are dependent.

CHAPTER II

SEQUENCES OF INDEPENDENT TRIALS

In the present chapter we investigate the fundamental laws governing one of the most important schemes in probability theory—that of a sequence of independent trials. By this concept we mean the following.

We consider a sequence of complete groups of events $A_1^{(s)}, A_2^{(s)}, \ldots, A_k^{(s)}$, $s=1, 2, \ldots, n$. The probability of occurrence of the event $A_i^{(s)}$ ($i=1, 2, 3, \ldots, k$) in the group with superscript s does not depend on either this superscript or on what events have occurred or will occur in the other groups.[1]

It is often convenient to use a different terminology and to say that n independent trials are performed each having k mutually exclusive outcomes $A_1^{(s)}, A_2^{(s)}, \ldots, A_k^{(s)}$, where the probability p_i of the outcome $A_i^{(s)}$ ($i=1, 2, \ldots, k$) does not depend on the number s of the trial; since the outcomes $A_i^{(s)}$ are mutually exclusive and exhaustive, we obviously have $\Sigma p_i = 1$. In this setting, one also speaks of *repeated independent trials*. This scheme was first considered by James Bernoulli in the most important special case $k=2$; for this reason, the case $k=2$ is referred to as the *Bernoulli scheme of trials*, or for short, *Bernoulli trials*. In Bernoulli trials, it is customary to set $p_1 = p$ and $p_2 = 1 - p = q$.

A detailed investigation of such sequences of trials merits special attention not only because of the immediate value of such sequences in probability theory and its applications, but also because the course of development of probability theory has shown that the laws which were first discovered in studying sequences of independent trials and, in particular, Bernoulli trials are capable of generalization. Many of the facts noticed in this particular scheme have subsequently served as clues in the study

---

[1] The scheme of a sequence of independent trials could be given a still broader interpretation by assuming that the number of outcomes and their probabilities depend on the number of the trial. We shall set these more general formulations aside for the time being.

of more complex schemes. This remark concerns both the past and the present development of the theory of probability. We shall convince ourselves of this later on in the case of the law of large numbers and in that of the DeMoivre-Laplace Theorem.

§ 11. The Probability $P_n(m_1, m_2, \ldots, m_k)$

The simplest problem concerning a scheme of independent trials consists in the determination of the probability $P_n(m_1, m_2, \ldots, m_k)$ that in n trials the event $A_1^{(s)}$ will take place m_1 times, the event $A_2^{(s)}$ m_2 times, ..., the event $A_k^{(s)}$ m_k times. Here, of course, the relation

$$m_1 + m_2 + \ldots + m_k = n$$

has to be satisfied. We shall confine ourselves to a detailed consideration of this problem for the case of Bernoulli trials, leaving it to the reader to derive and investigate the formula in the general case. In discussing Bernoulli trials, we denote the event $A_1^{(s)}$ by $A^{(s)}$, the event $A_2^{(s)}$ by $\bar{A}^{(s)}$, the number of occurrences of the event $A^{(s)}$ by μ, and the probability $P_n(m_1, m_2)$ by $P_n(m)$, where $m = m_1$.

We first determine the probability that the event $A^{(s)}$ occurs in m given trials (for example, in the trials numbered $s_1, s_2, \ldots, s_m$) and does not occur in the remaining $n - m$ trials. By the Multiplication Theorem for independent events, this probability is equal to

$$p^m q^{n-m}.$$

By the addition theorem for probabilities, the desired probability $P_n(m)$ is equal to the probability just calculated summed for all the different ways in which m occurrences and $n - m$ non-occurrences of the event can happen in n trials. As is well-known from combinatorial analysis, the number of such combinations is equal to $\binom{n}{m} = \dfrac{n!}{m!\,(n-m)!}$; consequently, the required probability is given by

$$P_n(m) = \binom{n}{m} p^m q^{n-m}. \tag{1}$$

Since all the possible mutually exclusive outcomes consist in the occurrence of the event $A^{(s)}$ zero times, 1 time, 2 times, ..., and n times, it is obvious that

§ 11. THE PROBABILITY $P_n(m_1, m_2, \ldots, m_k)$ 81

$$\sum_{m=0}^{n} P_n(m) = 1.$$

This relation can also be derived, without recourse to probability-theoretical reasoning, from the equation

$$\sum_{m=0}^{n} P_n(m) = (p+q)^n = 1^n = 1.$$

It is easy to see that the probability $P_n(m)$ is the coefficient of x^m in the binomial expansion of $(q + px)^n$ in powers of x; because of this property, the set of probabilities $P_n(m)$ is called the *binomial probability distribution law*.

By a slight modification of the above reasoning, the reader can easily satisfy himself that

$$P_n(m_1, m_2, \ldots, m_k) = \frac{n!}{m_1! \, m_2! \ldots m_k!} p_1^{m_1} p_2^{m_2} \ldots p_k^{m_k} \tag{1'}$$

and also that this probability is the coefficient of $x_1^{m_1} x_2^{m_2} \ldots x_k^{m_k}$ in the multinomial expansion of

$$(p_1 x_1 + p_2 x_2 + \ldots + p_k x_k)^n.$$

We shall now discuss several numerical examples, having in mind the formulation of general problems concerning the scheme of independent trials. In these examples, the computations of the probabilities will not be carried to immediate completion but will be postponed until we shall have developed convenient methods of computation.

Example 1. There are two containers A and B each having a volume of 1,000 cc. Each of them contains 2.7×10^{22} molecules of a gas. The containers are connected so that a free exchange of molecules can occur between them. Find the probability that after 24 hours there will be at least 0.00000001% more molecules in one container than in the other.

The probability that after 24 hours a given molecule is in one container or the other is the same and is equal to 1/2. Thus the situation is as if 5.4×10^{22} trials were being performed in each of which the probability of a molecule being in container A is 1/2. Let μ be the number of molecules in container A after 24 hours. This means that there will be

$5.4 \times 10^{22} - \mu$ molecules in container B. We must determine the probability that

$$|\mu - (5.4 \cdot 10^{22} - \mu)| \geq \frac{5.4 \cdot 10^{22}}{10^{10}} = 5.4 \cdot 10^{12};$$

in other words, we have to find the probability

$$p = \mathsf{P}\{|\mu - 2.7 \cdot 10^{22}| \geq 2.7 \cdot 10^{12}\}.$$

According to the addition theorem,

$$p = \Sigma \mathsf{P}\{\mu = m\},$$

where the sum is taken over those values of m for which

$$|m - 2.7 \cdot 10^{22}| \geq 2.7 \cdot 10^{12}.$$

Example 2. The probability of a certain manufactured item being defective is equal to 0.005. What is the probability that, of 10,000 items chosen at random, a) exactly 40, b) no more than 70 will prove to be defective?

In our example, $n = 10{,}000$, $p = 0.005$; therefore, by formula (1), we find:

a) $P_{10,000}(40) = \binom{10{,}000}{40} (0.995)^{9960} (0.005)^{40}$.

The probability $\mathsf{P}\{\mu \leq 70\}$ that there will be no more than seventy defective items is equal to the sum of the probabilities of the number of defective items being equal, respectively, to 0, 1, 2, ..., and 70. Thus,

b) $\mathsf{P}\{\mu \leq 70\} = \sum\limits_{m=0}^{70} P_n(m) = \sum\limits_{m=0}^{70} \binom{10{,}000}{m}(0.995)^{10{,}000-m}(0.005)^m$.

The examples considered show that a direct calculation of probabilities by formula (1) (and also formula (1′)) frequently involves great technical difficulties, and the problem thus arises of seeking simple approximative formulas for the probabilities $P_n(m)$ and also for sums of the form

$$\sum_{m=s}^{t} P_n(m)$$

for large values of n. This problem will be solved in §§ 12 and 13. But

§ 11. The Probability $P_n(m_1, m_2, \ldots, m_k)$

we shall now establish some elementary facts concerning the behavior of the probabilities $P_n(m)$ for fixed n. Let us begin by studying how $P_n(m)$ behaves as a function of m. For $0 \leq m < n$, a simple calculation shows that

$$\frac{P_n(m+1)}{P_n(m)} = \frac{n-m}{m+1} \cdot \frac{p}{q};$$

hence it follows that

$$P_n(m+1) > P_n(m)$$

if $(n-m)p > (m+1)q$, i.e., if $np - q > m$;

$$P_n(m+1) = P_n(m)$$

if $m = np - q$ and finally,

$$P_n(m+1) < P_n(m)$$

if $m > np - q$.

We see that the probability $P_n(m)$ first increases with increasing m, then attains a maximum, and with the further increase of m, decreases. If $np - q$ is an integer, then the probability $P_n(m)$ takes on its maximum value for two values of m, namely, $m_0 = np - q$ and $m_0' = np - q + 1 = np + p$. If, however, $np - q$ is not an integer, then $P_n(m)$ attains a maximum for the smallest integer $m = \overline{m}_0$ greater than m_0. The number $\overline{m}_0$ is called the *most probable value of* μ. We have just seen that if $np - q$ is an integer, then μ has two most probable values: m_0 and $m_0' = m_0 + 1$.

We note that if $np - q < 0$, then

$$P_n(0) > P_n(1) > \cdots > P_n(n),$$

and if $np - q = 0$, then

$$P_n(0) = P_n(1) > P_n(2) > \cdots > P_n(n).$$

In the sequel, we shall see that for large values of n all the probabilities $P_n(m)$ tend close to zero, and only for m near the most probable value of μ do they differ at all noticeably from zero. This fact will be proved later on, but at this point we shall just illustrate it with a numerical example.

Example 3. Let $n = 50$, $p = 1/3$.

There are two most probable values: $m_0 = np - q = 16$ and $m_0 + 1 = 17$.

The values of the probabilities $P_n(m)$ to four decimal places are given in Table 5.

TABLE 5

m	$P_n(m)$	m	$P_n(m)$	m	$P_n(m)$
<5	0.0000	12	0.0679	23	0.0202
5	0.0001	14	0.0898	24	0.0114
6	0.0004	15	0.1077	25	0.0059
7	0.0012	16	0.1178	26	0.0028
8	0.0033	17	0.1178	27	0.0013
9	0.0077	18	0.1080	28	0.0005
10	0.0157	19	0.0910	29	0.0002
11	0.0286	20	0.0705	30	0.0001
12	0.0465	21	0.0503	>30	0.0000
		22	0.0332		

§ 12. The Local Limit Theorem

In our discussion of numerical applications in the preceding section, we came to the conclusion that for large values of $n, m_1, m_2, \ldots, m_k$, the calculation of the probabilities $P_n(m_1, m_2, \ldots, m_k)$ by means of formula (1′) of § 11 involves considerable difficulties. Thus there is a need for asymptotic formulas that would enable one to determine these probabilities to a sufficient degree of accuracy. A formula of this kind was first discovered by DeMoivre in 1730 for the special case of Bernoulli trials in which $p = q = 1/2$ and was subsequently generalized by Laplace to the case of arbitrary p differing from 0 and 1.

This formula has been called the Local Laplace Theorem; we shall refer to it as the Local DeMoivre-Laplace Theorem for the sake of doing historical justice. We shall begin the exposition with a proof of the analogous theorem for the general scheme of independent trials and we shall then derive the DeMoivre-Laplace Theorem from it as a special case.

Before formulating the theorem, let us introduce some notation: Let us set

$$q_i = 1 - p_i \quad (i = 1, 2, \ldots, k),$$

$$x_i = \frac{m_i - np_i}{\sqrt{n p_i q_i}}. \qquad (1)$$

§ 12. The Local Limit Theorem

The quantities x_i depend not only on i (i.e., on p_i) but also on n and m_i. However, for the sake of brevity we refrain from introducing additional subscripts.

THE LOCAL THEOREM: *If the probabilities $p_1, p_2, \ldots, p_k$ that the respective events $A_1^{(s)}, A_2^{(s)}, \ldots, A_k^{(s)}$ occur in the s-th trial do not depend on the number of the trial and differ from 0 and 1 ($0 < p_i < 1$, $i = 1, 2, \ldots, k$), then the probability $P_n(m_1, m_2, \ldots, m_k)$ that in n independent trials the events A_i occur m_i times ($m_1 + m_2 + \ldots + m_k = n$) satisfies the asymptotic relation*[2]

$$\sqrt{n^{k-1}}\, P_n(m_1, m_2, \ldots, m_k) : \frac{e^{-\frac{1}{2} \sum_{i=1}^{k} q_i x_i^2}}{(2\pi)^{\frac{k-1}{2}} \sqrt{p_1 p_2 \cdots p_k}} \to 1 \quad (n \to \infty) \quad (2)$$

uniformly in all the m_i for which the corresponding x_i lie in the arbitrary finite intervals $a_i \leq x_i \leq b_i$.

Proof: The proof we shall give is based on the well-known formula of Stirling

$$s! = \sqrt{2\pi s} \cdot s^s\, e^{-s}\, e^{\theta_s},$$

in which the remainder exponent θ_s satisfies the inequality

$$|\theta_s| \leq \frac{1}{12s}. \qquad (3)$$

Formula (1) can be written in the form

$$m_i = n p_i + x_i \sqrt{n p_i q_i}\,; \qquad (1')$$

hence it follows that if the x_i remain bounded, then the quantities m_i corresponding to them tend to infinity with n.

The application of Stirling's formula yields

---

[2] This limiting relation is expressed in terms of homogeneous coordinates; the variables x_i are connected by the relation $\sum_{i=1}^{k} x_i \sqrt{n p_i q_i} = 0$, which is easily deduced from the relations $\Sigma m_i = n$ and $\Sigma p_i = 1$. If we want the x_i to be independent variables, then it is necessary to eliminate one of the arguments in expression (2).

$$P_n(m_1, m_2, \ldots, m_k) = \frac{n!}{m_1! m_2! \ldots m_k!} p_1^{m_1} p_2^{m_2} \ldots p_k^{m_k}$$

$$= \frac{\sqrt{2\pi n}\, n^n e^{-n}}{\prod_{i=1}^{k} \{\sqrt{2\pi m_i}\, m_i^{m_i} e^{-m_i}\}} \prod_{i=1}^{k} p_i^{m_i} e^{\theta_n - \sum_{i=1}^{k} \theta_{m_i}}$$

$$= \frac{1}{(2\pi)^{\frac{k-1}{2}}} \sqrt{\frac{n}{m_1 m_2 \ldots m_k}} \prod_{i=1}^{k} \left(\frac{n p_i}{m_i}\right)^{m_i} \cdot e^{\theta}, \qquad (4)$$

where

$$\theta = \theta_n - \theta_{m_1} - \theta_{m_2} - \ldots - \theta_{m_k}.$$

By virtue of the estimate (3), we have

$$|\theta| < \frac{1}{12}\left(\frac{1}{n} + \frac{1}{m_1} + \cdots + \frac{1}{m_k}\right).$$

If $a_i \leq x_i \leq b_i$, the value m_i corresponding to x_i satisfies the inequality

$$m_i \geq n p_i + a_i \sqrt{n p_i q_i} = n p_i \left(1 + a_i \sqrt{\frac{q_i}{n p_i}}\right);$$

this implies that the estimate

$$|\theta| < \frac{1}{12n}\left(1 + \sum_{i=1}^{k} \frac{1}{p_i} \frac{1}{1 + a_i \sqrt{\frac{q_i}{n p_i}}}\right) \qquad (5)$$

holds for the stipulated values of m_i. Hence, we see that for x_i contained in the intervals (a_i, b_i), the quantity θ tends to zero, as $n \to \infty$, uniformly in the x_i, and consequently, the factor e^{θ} approaches 1 uniformly in the x_i.

Let us now consider the expression

$$\log A_n = \log \prod_{i=1}^{k} \left(\frac{n p_i}{m_i}\right)^{m_i} = -\sum_{i=1}^{k} m_i \log \frac{m_i}{n p_i}$$

$$= -\sum_{i=1}^{k} (n p_i + x_i \sqrt{n p_i q_i}) \log\left(1 + x_i \sqrt{\frac{q_i}{n p_i}}\right).$$

By the conditions of the theorem, the quantities $x_i \sqrt{\frac{q_i}{n p_i}}$ ($1 \leq i \leq k$) can be made as small as desired by choosing n sufficiently large and therefore we can expand the function $\log\left(1 + x_i \sqrt{\frac{q_i}{n p_i}}\right)$ in a power series.

§ 12. THE LOCAL LIMIT THEOREM

Retaining the first two terms, we find:

$$\log\left(1 + x_i \sqrt{\frac{q_i}{n\,p_i}}\right) = x_i \sqrt{\frac{q_i}{n\,p_i}} - \frac{1}{2}\frac{q_i\,x_i^2}{n\,p_i} + O\left(\frac{1}{n^{3/2}}\right).$$

The estimate of the remainder term is uniform for x_i in any finite interval. Hence,

$$\log A_n = -\sum_{i=1}^{k}(n\,p_i + x_i\sqrt{n\,p_i\,q_i})\left[x_i\sqrt{\frac{q_i}{n\,p_i}} - \frac{1}{2}\frac{q_i\,x_i^2}{n\,p_i} + O\left(\frac{1}{n^{3/2}}\right)\right]$$

$$= -\sum_{i=1}^{k} x_i\sqrt{n\,p_i\,q_i} - \frac{1}{2}\sum_{i=1}^{k} q_i\,x_i^2 + O\left(\frac{1}{\sqrt{n}}\right).$$

Since by formula (1),

$$\sum_{i=1}^{k} x_i\sqrt{n\,p_i\,q_i} = \sum_{i=1}^{k}(m_i - n\,p_i) = \sum_{i=1}^{k} m_i - n\sum_{i=1}^{k} p_i = n - n = 0,$$

it follows that

$$\log A_n = -\sum_{i=1}^{k}\frac{1}{2} q_i\,x_i^2 + O\left(\frac{1}{\sqrt{n}}\right);$$

and therefore the relation

$$\prod_{i=1}^{n}\left(\frac{n\,p_i}{m_i}\right)^{m_i} : e^{-\frac{1}{2}\sum_{i=1}^{k} q_i\,x_i^2} \to 1. \tag{6}$$

holds, as $n \to \infty$, uniformly in the x_i for x_i contained in the arbitrary finite intervals (a_i, b_i). We have, further,

$$\sqrt{\frac{n}{m_1 m_2 \ldots m_k}} = \sqrt{\frac{n}{n^k\,p_1 p_2 \cdots p_k}}\sqrt{\prod_{i=1}^{k}\frac{n\,p_i}{m_i}}$$

$$= \sqrt{\frac{n}{n^k\,p_1 p_2 \cdots p_k}}\frac{1}{\sqrt{\prod_{i=1}^{k}\left(1 + x_i\sqrt{\frac{q_i}{n\,p_i}}\right)}}. \tag{7}$$

By the conditions of the theorem, the second factor on the right-hand side of this equation approaches 1 as $n \to \infty$, and moreover, does so uniformly for all x_i ($i = 1, 2, \ldots, k$) contained in any finite interval.

It is readily apparent that relations (5), (6), and (7) prove the theorem.

II. Sequences of Independent Trials

The Local DeMoivre-Laplace Theorem: *If the probability of occurrence of some event A in n independent trials is constant and equal to p ($0 < p < 1$), then the probability $P_n(m)$ that the event A occurs exactly m times in these trials satisfies the relation*

$$\sqrt{npq}\, P_n(m) : \frac{1}{\sqrt{2\pi}} e^{-\frac{1}{2}x^2} \to 1 \tag{8}$$

as $n \to \infty$, uniformly in all m for which the corresponding x are contained in some finite interval.

Proof: As we said at the beginning of the present section, the DeMoivre-Laplace Theorem is a special case of the theorem just proved. Here, $k = 2$, $q_1 = q$, $q_2 = 1 - q$, $p_1 = p$, $p_2 = 1 - p = q$,

$$m_1 = m, \quad m_2 = n - m, \quad x = x_1 = \frac{m - np}{\sqrt{npq}},$$

$$x_2 = \frac{n - m - nq}{\sqrt{npq}} = -\frac{m - np}{\sqrt{npq}} = -x.$$

Substituting these values in relation (2), we find:

$$\sqrt{npq}\, P_n(m) : \frac{1}{\sqrt{2\pi}} e^{-\frac{1}{2}px^2 - \frac{1}{2}qx^2} \to 1.$$

It is obvious that this relation is equivalent to (8).

We are now in a position to complete the computations in the **Example** 2a) of the preceding section.

Example 1. In Example 2 of the preceding section, it was necessary to determine $P_n(m)$ for $n = 10{,}000$, $m = 40$, and $p = 0.005$. By the DeMoivre-Laplace Theorem just proved, we have:

$$P_n(m) \sim \frac{1}{\sqrt{2\pi npq}} e^{-\frac{1}{2}\left(\frac{m-np}{\sqrt{npq}}\right)^2}.$$

In our example,

$$\sqrt{npq} = \sqrt{10000 \cdot 0.005 \cdot 0.995} = \sqrt{49.75} \approx 7.05,$$

$$\frac{m - np}{\sqrt{npq}} \approx -1.42.$$

Consequently,

§ 12. The Local Limit Theorem

$$P_n(m) \approx \frac{1}{7.05\sqrt{2\pi}} e^{-\frac{1.42^2}{2}}.$$

The values of the function

$$\varphi(x) = \frac{1}{\sqrt{2\pi}} e^{-\frac{x^2}{2}}$$

have been tabulated, and a short table is given at the end of the book (see page 426). From this table, we find:

$$P_n(m) \approx \frac{0.1456}{7.05} \approx 0.0206.$$

The exact computations, not using the DeMoivre-Laplace Theorem, give:

TABLE 6

$n = 4$

m	0	1	2	3	4
$P_n(m)$	0.4096	0.4096	0.1536	0.0256	0.0016
x	—1.00	0.25	1.50	2.75	4.00
$\sqrt{npq}\, P_n(m)$	0.3277	0.3277	0.1229	0.0205	0.0013
$\varphi(x)$	0.2420	0.3867	0.1295	0.0091	0.0001

TABLE 7

$n = 25$

m	x	$P_n(m)$	$P_n(m)\cdot\sqrt{npq}$	$\varphi(x)$	m	x	$P_n(m)$	$P_n(m)\cdot\sqrt{npq}$	$\varphi(x)$
0	—2.5	0.0037	0.0075	0.0175	8	1.5	0.0623	0.1247	0.1295
1	—2.0	0.0236	0.0472	0.0540	9	2.0	0.0294	0.0589	0.0540
2	—1.5	0.0708	0.1417	0.1295	10	2.5	0.0118	0.0236	0.0175
3	—1.0	0.1358	0.2715	0.2420	11	3.0	0.0040	0.0080	0.0044
4	—0.5	0.1867	0.3734	0.3521	12	3.5	0.0012	0.0023	0.0009
5	0.0	0.1960	0.3920	0.3989	13	4.0	0.0003	0.0006	0.0001
6	0.5	0.1633	0.3267	0.3521	14	4.5	0.0000	0.0000	0.0000
7	1.0	0.1108	0.2217	0.2420	>14	>4.5	0.0000	0.0000	0.0000

II. Sequences of Independent Trials

Table 8 $n = 100$

m	x	$P_n(m)$	$\sqrt{npq} \cdot P_n(m)$	$\varphi(x)$	m	x	$P_n(m)$	$\sqrt{npq} \cdot P_n(m)$	$\varphi(x)$
8	—3.00	0.0006	0.0023	0.0044	21	0.25	0.0946	0.3783	0.3867
9	—2.75	0.0015	0.0059	0.0091	22	0.50	0.0849	0.3396	0.3521
10	—2.50	0.0034	0.0134	0.0175	23	0.75	0.0720	0.2879	0.3011
11	—2.25	0.0069	0.0275	0.0317	24	1.00	0.0577	0.2309	0.2420
12	—2.00	0.0127	0.0510	0.0540	25	1.25	0.0439	0.1755	0.1826
13	—1.75	0.0216	0.0863	0.0862	26	1.50	0.0316	0.1266	0.1295
14	—1.50	0.0335	0.1341	0.1295	27	1.75	0.0217	0.0867	0.0862
15	—1.25	0.0481	0.1923	0.1826	28	2.00	0.0141	0.0565	0.0540
16	—1.00	0.0638	0.2553	0.2420	29	2.25	0.0088	0.0351	0.0317
17	—0.75	0.0788	0.3154	0.3011	30	2.50	0.0052	0.0208	0.0175
18	—0.50	0.0909	0.3636	0.3521	31	2.75	0.0029	0.0117	0.0091
19	—0.25	0.0981	0.3923	0.3867	32	3.00	0.0016	0.0063	0.0044
20	—0.00	0.0993	0.3972	0.3989					

Table 9 $n = 400$

m	x	$P_n(m)$	$\sqrt{npq} \cdot P_n(m)$	$\varphi(x)$	m	x	$P_n(m)$	$\sqrt{npq} \cdot P_n(m)$	$\varphi(x)$
56	—3.000	0.0004	0.0034	0.0044	81	0.125	0.0492	0.3936	0.3957
57	—2.875	0.0006	0.0051	0.0064	82	0.250	0.0478	0.3828	0.3867
58	—2.750	0.0009	0.0076	0.0091	83	0.375	0.0458	0.3667	0.3719
59	—2.625	0.0014	0.0100	0.0127	84	0.500	0.0432	0.3460	0.3521
60	—2.500	0.0019	0.0156	0.0175	85	0.625	0.0402	0.3215	0.3282
61	—2.375	0.0027	0.0218	0.0238	86	0.750	0.0368	0.2944	0.3011
62	—2.250	0.0037	0.0298	0.0317	87	0.875	0.0332	0.2656	0.2721
63	—2.125	0.0050	0.0399	0.0417	88	1.000	0.0295	0.2362	0.2420
64	—2.000	0.0066	0.0526	0.0540	89	1.125	0.0259	0.2070	0.2119
65	—1.875	0.0085	0.0679	0.0684	90	1.250	0.0223	0.1788	0.1826
66	—1.750	0.0108	0.0862	0.0862	91	1.375	0.0190	0.1523	0.1550
67	—1.625	0.0134	0.1075	0.1065	92	1.500	0.0160	0.1279	0.1295
68	—1.500	0.0164	0.1316	0.1295	93	1.625	0.0132	0.1059	0.1065
69	—1.375	0.0198	0.1583	0.1550	94	1.750	0.0108	0.0865	0.0862
70	—1.250	0.0234	0.1871	0.1827	95	1.875	0.0087	0.0696	0.0684
71	—1.125	0.0271	0.2175	0.2119	96	2.000	0.0069	0.0553	0.0540
72	—1.000	0.0310	0.2484	0.2420	97	2.125	0.0054	0.0433	0.0417
73	—0.875	0.0349	0.2790	0.2721	98	2.250	0.0042	0.0355	0.0317
74	—0.750	0.0385	0.3081	0.3011	99	2.375	0.0032	0.0255	0.0238
75	—0.625	0.0419	0.3349	0.3282	100	2.500	0.0024	0.0192	0.0175
76	—0.500	0.0447	0.3580	0.3521	101	2.625	0.0018	0.0143	0.0127
77	—0.375	0.0471	0.3766	0.3719	102	2.750	0.0013	0.0105	0.0091
78	—0.250	0.0487	0.3899	0.3867	103	2.875	0.0009	0.0075	0.0064
79	—0.125	0.0497	0.3973	0.3957	104	3.000	0.0007	0.0054	0.0044
80	—0.000	0.0498	0.3986	0.3989					

§ 12. The Local Limit Theorem

$$P_n(m) \approx 0.0197.$$

In order to illustrate the kind of approximation given by the DeMoivre-Laplace Theorem, and also to give a geometrical interpretation of the analytical transformations used in the proof, we discuss a numerical example.

Let the probability p be equal to 0.2. In Tables 6-9 are given the values of m, $x = \dfrac{m-np}{\sqrt{npq}}$, the probabilities $P_n(m)$, the quantities $\sqrt{npq}\, P_n(m)$, and also the function $\varphi(x) = \dfrac{1}{\sqrt{2\pi}} e^{-\frac{x^2}{2}}$ to four decimal places, for $n = 4$, 25, 100, and 400 trials, respectively. In Fig. 8, the ordinates depict the probabilities $P_n(m)$ for various integral values of the abscissa m. It is

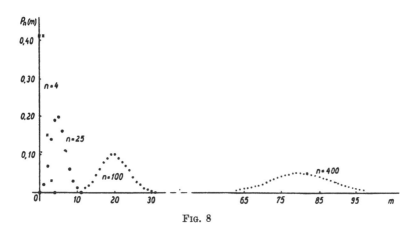

Fig. 8

apparent from the diagram that with increasing n the quantities $P_n(m)$ decrease uniformly. Even for the values of n under consideration, in order to keep the points $[m, P_n(m)]$ in the figure from being so close to the m-axis as to be practically indistinguishable from it, we have chosen radically different scales for the coordinate axes.

If $x_n = (m - nq)/\sqrt{npq}$ and $y_n(m) = \sqrt{npq}\, P_n(m)$ are introduced in place of m and $P_n(m)$, respectively, as abscissa and ordinate, we obtain

1) a translation of the origin to the point $(np, 0)$, which lies near the abscissa corresponding to the maximum of the ordinates $P_n(m)$;

2) an increase in the unit of length along the abscissa axis by a factor $\sqrt{npq}$ (in other words, a contraction of the diagram by a factor of $\sqrt{npq}$ in the direction of the abscissa); and

3) a diminution in the unit of length along the ordinate axis by a

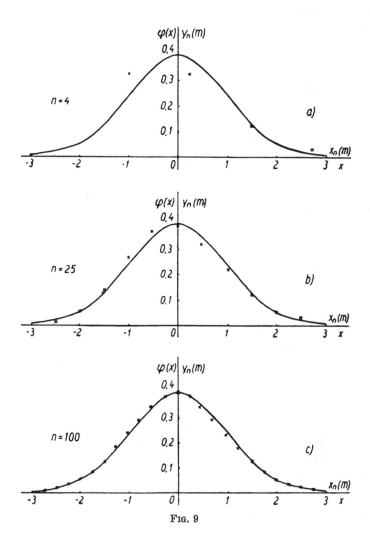

Fig. 9

factor of $\sqrt{npq}$ (in other words, an expansion of the diagram in the direction of the ordinate by a factor of $\sqrt{npq}$).

Figs. 9 a, b, c depict the curve $y = \varphi(x)$ and the points $[m, P_n(m)]$ after the transformation just described has been made, i.e., the points $[x_n, y_n(m)]$. We see that for $n = 25$, the points $[x_n, y_n(m)]$ already begin to coincide with corresponding points on the graph of the function $y = \varphi(x)$. This agreement becomes even better for values of n larger than 25.

In order to get a clear idea of the extent to which it is possible to make use of the asymptotic formula of DeMoivre-Laplace for finite values of n,[3] i.e., to replace the binomial law for computing the probabilities $P_n(m)$ by the function $y = \varphi(x)$, let us consider an example. For simplicity, let us examine the case $p = q = 1/2$ and let us take only those values of n for which the value $x_{nm} = 1$ is possible; for example, such values could be $n = 25, 100, 400$, and 1156. For these n, $x_{nm} = 1$ when $m = 15, 55, 210$, and 595.

For brevity, let us set

$$P_n(m) = P_n$$

and

$$\frac{1}{\sqrt{2\pi npq}} e^{-\frac{x_{nm}^2}{2}} = Q_n$$

in the case $p = q = 1/2$ and $x_{nm} = 1$.

By the local DeMoivre-Laplace Theorem, the ratio P_n/Q_n must tend to the value 1 when $n \to \infty$. The computation for the above-mentioned values of n gives

TABLE 10

n	P_n	Q_n	$P_n - Q_n$	P_n/Q_n
25	0.09742	0.09679	0.00063	1.0065
100	0.04847	0.04839	0.00008	1.0030
400	0.024207	0.024194	0.000013	1.0004
1156	0.014236	0.014234	0.000002	1.0001

§ 13. The Integral Limit Theorem

The Local Limit Theorem proved in the preceding section will now be used to establish another limit relation of probability theory — the In-

---

[3] Very precise estimates of the remainder term are given in the paper by S. N. Bernstein, "Re-examination of the question of the accuracy of the Laplace limit formula," Izv. Akad. Nauk SSSR, vol. 7, 1943 [in Russian].

tegral Limit Theorem. We begin the exposition with the simplest special case of this theorem—the Integral Theorem of DeMoivre-Laplace.

INTEGRAL THEOREM OF DEMOIVRE-LAPLACE: *If μ is the number of occurrences of an event in n independent trials in each of which the probability of this event is p, where $0 < p < 1$, then the relation*

$$\mathsf{P}\left\{a \leq \frac{\mu - np}{\sqrt{npq}} < b\right\} \to \frac{1}{\sqrt{2\pi}} \int_a^b e^{-\frac{z^2}{2}} dz$$

holds uniformly in a and b ($-\infty \leq a \leq b \leq \infty$) *as* $n \to \infty$.

Proof: For brevity, we introduce the notation

$$P_n(a, b) = \mathsf{P}\left\{a \leq \frac{\mu - np}{\sqrt{npq}} < b\right\}.$$

This probability is evidently equal to $\Sigma P_n(m)$ taken over all those values of m for which $a \leq x_m < b$ where, as before, $x_m = (m - np)/\sqrt{npq}$.

Let us now define a function $y = \Pi_n(x)$, as follows:

$$y = \Pi_n(x) = \begin{cases} 0 & \text{for } x < x_0 = -\frac{np}{\sqrt{npq}}, \\ 0 & \text{for } x \geq x_n + \frac{1}{\sqrt{npq}} = \frac{1+nq}{\sqrt{npq}}, \\ \sqrt{npq}\, P_n(m) & \text{for } x_m \leq x < x_{m+1} \ (m = 0, 1, \ldots, n). \end{cases}$$

The probability $P_n(m)$ is obviously equal to the area bounded by the curve $y = \Pi_n(x)$, the x-axis, and the ordinate lines $x = x_m$ and $x = x_{m+1}$, i.e.,

$$P_n(m) = \sqrt{npq}\, P_n(m)\, (x_{m+1} - x_m) = \int_{x_m}^{x_{m+1}} \Pi_n(x)\, dx.$$

Hence, it follows that the desired probability $P_n(a, b)$ is equal to the area bounded by the curve $y = \Pi_n(x)$, the x-axis, and the ordinate lines $x = x_{\underline{m}}$ and $x = x_{\overline{m}}$, where $\overline{m}$ and $\underline{m}$ are defined by the inequalities

$$a \leq x_{\underline{m}} < a + \frac{1}{\sqrt{npq}}, \qquad b \leq x_{\overline{m}} < b + \frac{1}{\sqrt{npq}}.$$

Thus,

§ 13. The Integral Limit Theorem

$$P_n(a, b) = \int_{x_{\underline{m}}}^{x_{\overline{m}}} \Pi_n(x)\, dx = \int_a^b \Pi_n(x)\, dx + \int_b^{x_{\overline{m}}} \Pi_n(x)\, dx - \int_a^{x_{\underline{m}}} \Pi_n(x)\, dx.$$

Since the maximum value of the probability $P_n(m)$ occurs at $m_0 = [(n+1)p]$,[4] the maximum value of $\Pi_n(x)$ lies in the interval

$$0 \leqq \frac{m_0 - np}{\sqrt{npq}} \leqq x < \frac{m_0 + 1 - np}{\sqrt{npq}} \leqq \frac{2}{\sqrt{npq}}.$$

The Local DeMoivre-Laplace Theorem is applicable in this interval and we can therefore conclude that for all sufficiently large values of n,

$$\max \Pi_n(x) < 2 \frac{1}{\sqrt{2\pi}} \max e^{-\frac{x^2}{2}} = \sqrt{\frac{2}{\pi}}.$$

From this, we deduce first of all that

$$|\varrho_n| = \left| \int_b^{x_{\overline{m}}} \Pi_n(x)\, dx - \int_a^{x_{\underline{m}}} \Pi_n(x)\, dx \right| \leqq \int_b^{x_{\overline{m}}} \max \Pi_n(x)\, dx +$$

$$+ \int_a^{x_{\underline{m}}} \max \Pi_n(x)\, dx < \sqrt{\frac{2}{\pi}} (-b + x_{\overline{m}} + x_{\underline{m}} - a) \leqq 2 \sqrt{\frac{2}{\pi n p q}}$$

and therefore that

$$\lim_{n \to \infty} \varrho_n = 0.$$

Thus, $P_n(a, b)$ only differs from $\int_a^b \Pi_n(x)\, dx$ by an infinitesimal.

We shall first assume that a and b are finite. Under this assumption, we have by the local limit theorem,

$$\Pi_n(x_m) = \frac{1}{\sqrt{2\pi}} e^{-\frac{x_m^2}{2}} [1 + \alpha_n(x_m)]$$

for $a \leqq x_m < b$, where $\alpha_n(x_m) \to 0$ uniformly in x_m as $n \to \infty$. Obviously, for intermediate values of the argument as well,

$$\Pi_n(x) = \frac{1}{\sqrt{2\pi}} e^{-\frac{x^2}{2}} [1 + \alpha_n(x)],$$

---

[4] The greatest integer contained in $(n+1)p$.

where $\lim\limits_{n\to\infty} \max\limits_{a \leq x < b} \alpha_n(x) = 0$. In fact, we have, for any x in the interval $x_m \leq x < x_{m+1}$,

$$\Pi_n(x) = \Pi_n(x_m) = \frac{1}{\sqrt{2\pi}} e^{-\frac{x^2}{2}} [1 + \alpha_n(x)],$$

where

$$\alpha_n(x) = e^{\frac{x^2 - x_m^2}{2}} [\alpha_n(x_m) + 1] - 1.$$

Since

$$\frac{x^2 - x_m^2}{2} \leq |x| \cdot |x - x_m| < \frac{\max(|a|, |b|)}{\sqrt{npq}},$$

it follows that

$$\lim_{n\to\infty} \max_{a < x < b} \alpha_n(x) = 0.$$

Collecting our various estimates, we obtain

$$P_n(a, b) = \frac{1}{\sqrt{2\pi}} \int_a^b e^{-\frac{x^2}{2}} dx + R_n,$$

where

$$R_n = \frac{1}{\sqrt{2\pi}} \int_a^b e^{-\frac{x^2}{2}} \alpha_n(x) \, dx + \varrho_n.$$

Since

$$|R_n| \leq \max_{a \leq x < b} |\alpha_n(x)| \cdot \frac{1}{\sqrt{2\pi}} \int_a^b e^{-\frac{x^2}{2}} dx + \varrho_n,$$

it is clear from the above that

$$\lim_{n\to\infty} R_n = 0.$$

The theorem is now proved, under the assumption, made in the course of the proof, that a and b are finite. It still remains for us to free ourselves of this restriction.

To this end, we first of all observe that[5]

---

[5] Here and in the sequel, integrals in which the limits of integration are not indicated are to be taken from $-\infty$ to ∞.

§ 13. The Integral Limit Theorem

$$\frac{1}{\sqrt{2\pi}} \int e^{-\frac{z^2}{2}} dz = 1.$$

Therefore, for any $\varepsilon > 0$, it is possible to choose A sufficiently large that

$$\frac{1}{\sqrt{2\pi}} \int_{-A}^{A} e^{-\frac{z^2}{2}} dz > 1 - \frac{\varepsilon}{4},$$

and

$$\frac{1}{\sqrt{2\pi}} \int_{-\infty}^{-A} e^{-\frac{z^2}{2}} dz = \frac{1}{\sqrt{2\pi}} \int_{A}^{\infty} e^{-\frac{z^2}{2}} dz < \frac{\varepsilon}{8}.$$

We then choose n, in accordance with what has been shown above, so large that for $-A \leq a \leq b < A$

$$\left| P_n(a, b) - \frac{1}{\sqrt{2\pi}} \int_a^b e^{-\frac{z^2}{2}} dz \right| < \frac{\varepsilon}{4}.$$

It is then clear that $\quad P_n(-A, A) > 1 - \frac{\varepsilon}{2},$

$$P(-\infty, -A) + P(A, +\infty) = 1 - P(-A, A) < \frac{\varepsilon}{2}.$$

We shall now show that, for any a and b $(-\infty \leq a \leq b \leq \infty)$,

$$\left| P_n(a, b) - \frac{1}{\sqrt{2\pi}} \int_a^b e^{-\frac{z^2}{2}} dz \right| < \varepsilon,$$

whereupon the proof of the Laplace Theorem will clearly be complete.

To do this, it is necessary to consider separately each of the different ways in which the points a and b can be situated with respect to the interval $(-A, A)$. For example, let us select the case $a \leq -A$, $b \geq A$ (the remaining cases being left to the reader).

In this case,

$$\frac{1}{\sqrt{2\pi}} \int_a^b e^{-\frac{z^2}{2}} dz = \frac{1}{\sqrt{2\pi}} \left(\int_a^{-A} + \int_{-A}^{A} + \int_{A}^{b} e^{-\frac{z^2}{2}} dz \right),$$

$$P_n(a, b) = P_n(a, -A) + P_n(-A, A) + P_n(A, b).$$

Therefore,

$$\left|P_n(a,b) - \frac{1}{\sqrt{2\pi}} \int_a^b e^{-\frac{z^2}{2}} dz\right| \leq \left|P_n(a,-A) - \frac{1}{\sqrt{2\pi}} \int_a^{-A} e^{-\frac{z^2}{2}} dz\right| +$$

$$+ \left|P_n(-A,A) - \frac{1}{\sqrt{2\pi}} \int_{-A}^A e^{-\frac{z^2}{2}} dz\right| + \left|P_n(A,b) - \frac{1}{\sqrt{2\pi}} \int_A^b e^{-\frac{z^2}{2}} dz\right|$$

$$\leq P_n(-\infty,-A) + \frac{1}{\sqrt{2\pi}} \int_{-\infty}^{-A} e^{-\frac{z^2}{2}} dz +$$

$$+ \left|P_n(-A,A) - \frac{1}{\sqrt{2\pi}} \int_{-A}^A e^{-\frac{z^2}{2}} dz\right| + P_n(A,+\infty) + \frac{1}{\sqrt{2\pi}} \int_A^\infty e^{-\frac{z^2}{2}} dz$$

$$< \frac{\varepsilon}{2} + \frac{\varepsilon}{4} + \frac{\varepsilon}{8} + \frac{\varepsilon}{8} = \varepsilon.$$

We now proceed to a derivation of the Integral Limit Theorem for the general case of a scheme of repeated independent trials. As before, let μ_i ($i = 1, 2, \ldots, k$) denote the number of occurrences of the event $A_i^{(s)}$ ($s = 1, 2, \ldots, n$) in n repeated trials. Depending on chance, the numbers

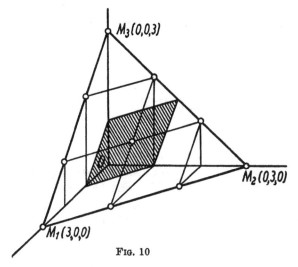

Fig. 10

μ_i may take on only the values $0, 1, 2, \ldots, n$. Moreover, since there are k outcomes possible in each trial and these outcomes are mutually exclusive, the equation

$$\mu_1 + \mu_2 + \ldots + \mu_k = n \tag{1}$$

§ 13. THE INTEGRAL LIMIT THEOREM

must hold. We shall now regard the quantities $\mu_1, \mu_2, \ldots, \mu_k$ as the rectangular coordinates of a point in k-dimensional euclidean space.

In this way, the outcomes of the n trials are represented by a point whose coordinates are integers not less than 0 and not greater than n. From now on we shall call these points *lattice points*. Equation (1) shows that the outcomes of the trials are representable not by any lattice points whatever within the hypercube $0 \leq \mu_i \leq n$ ($i = 1, 2, \ldots, k$) but merely by those points that lie on the hyperplane (1). Fig. 10 depicts the positions of the possible outcomes of the trials on the hyperplane (1) for the case $n = 3$, $k = 3$.

We now make a transformation of coordinates by means of the formulas

$$x_i = \frac{\mu_i - n p_i}{\sqrt{n p_i q_i}} \quad (i = 1, 2, \ldots, k; \; q_i = 1 - p_i).$$

In the new coordinates, the equation of the hyperplane (1) takes on the following form:

$$\sum_{i=1}^{k} x_i \sqrt{n p_i q_i} = 0. \tag{2}$$

Let us agree that the points of the hyperplane (2) which are the transforms of the lattice points of the hyperplane (1) are also to be called "lattice points."

Let $P_n(R)$ denote the probability that as the result of n trials the number of occurrences μ_i ($i = 1, 2, \ldots, k$) of each of the possible outcomes will be such that the point with the coordinates

$$x_i = \frac{\mu_i - n p_i}{\sqrt{n p_i q_i}}$$

falls inside some region R.

We then have the following theorem.

THEOREM: *If in a scheme of repeated independent trials each having k possible outcomes, the probability of each outcome does not depend on the number of the trial and differs from 0 and 1, then for any region R of the hyperplane (2) for which the $(k-1)$-dimensional volume of its boundary is equal to zero, the relation*

$$P_n\{R\} \to \sqrt{\frac{q_1 q_2 \cdots q_k}{(2\pi)^{k-1} \sum_{i=1}^{k} p_i q_i}} \int_R e^{-\frac{1}{2} \sum q_i x_i^2} dv$$

holds uniformly in R as $n \to \infty$, where dv denotes the element of volume of the region R and the integration is taken over the region R.

Proof: Since the proof proceeds almost exactly along the lines of the proof of the Integral Theorem of DeMoivre-Laplace, copying both the idea of that proof and the way it is carried out, we can limit ourselves to a brief sketch, leaving to the reader the task of filling in all of the details. We begin the exposition by carrying out two simple computations. First of all, we shall find the number $S_k(n)$ of possible outcomes of n trials in each of which k mutually exclusive events may occur.

The quantity $S_k(n)$ satisfies the following relation:[6]

$$S_k(n) = \sum_{r=0}^{n} S_{k-1}(r). \tag{3}$$

In fact, this equation means that the event A_k can occur $n-r$ times, $0 \leq n-r \leq n$; and in each of these cases, there can be $S_{k-1}(r)$ outcomes for the $k-1$ remaining events. Since these outcomes are distinct for distinct values of r, equation (3) follows from this. For Bernoulli trials ($k=2$),

$$S_2(n) = n+1,$$

and therefore for $k=3$, we have

$$S_3(n) = \sum_{r=0}^{n} (r+1) = \frac{(n+1)(n+2)}{1 \cdot 2}.$$

This expression suggests that $S_k(n)$ is given by the formula

$$S_k(n) = \frac{(n+1)(n+2)\cdots(n+k-1)}{(k-1)!} = \binom{n+k-1}{n}. \tag{4}$$

Let us assume that this relation is true for all $k \leq N$; we show that it is then also true for $k = N+1$, i.e., that the equality

$$S_{N+1}(n) = \sum_{r=0}^{n} S_N(r) = \sum_{r=0}^{n} \binom{r+N-1}{r} = \binom{n+N}{n}$$

holds. This last statement is a consequence of the following easily verified chain of equations:

---

[6] Another proof of this theorem follows from the discussion on pp. 34-35.

§ 13. The Integral Limit Theorem

$$\binom{n+N}{n} = \frac{(n+N)!}{n!N!} = \frac{(n+1)(n+2)\cdots(n+N)}{N!}$$

$$= \frac{n(n+1)\cdots(n+N-1)}{N!} + \frac{n(n+1)\cdots(n+N-1)}{(N-1)!}$$

$$= \binom{n+N-1}{n} + \binom{(n-1)+N}{n-1}$$

$$= \binom{n+N-1}{n} + \binom{n-1+N-1}{n-1} + \binom{(n-2)+N}{(n-2)}$$

$$\cdots = \sum_{r=0}^{n} \binom{r+N-1}{r}, \quad \text{Q.E.D.}$$

It is also necessary for us to know the $(k-1)$-dimensional volume of the portion of the hyperplane (2) defined by the inequalities

$$-\sqrt{\frac{n\,p_i}{q_i}} \leq x_i \leq \sqrt{\frac{n\,q_i}{p_i}} \quad (i = 1, 2, \ldots, k).$$

This is the region of the hyperplane (2) which contains all of the "lattice" points representing outcomes of the trials. Its volume, as the reader can easily verify by integration, is

$$v = \frac{\sqrt{n^{k-1}}}{(k-1)!} \sqrt{\frac{\sum_{i=1}^{k} p_i q_i}{\prod_{i=1}^{k} p_i q_i}}.$$

Now, with the points $(x_1, x_2, \ldots, x_k)$ of the hyperplane (2), where $x_i = (m_i - np_i)/\sqrt{np_iq_i}$ and m_i is a non-negative integer $(i = 1, 2, \ldots, n)$, we associate certain disjoint parallelepipedal regions containing these points. These are the parallelepipeds whose vertices are at the centers of the "lattice parallelepipeds" surrounding the lattice points in question. (In Fig. 10, the shaded area depicts one of these regions around the point $(1, 1, 1)$.)

All of these regions have the same volume dv and their sum will give a volume v' slightly larger[7] than v. But it is not difficult to show that the ratio of the volumes v and v' approaches 1 as $n \to \infty$. Therefore,

$$dv = \frac{v}{S_k(n)} [1 + o(1)].$$

It is easy to see that when $n \to \infty$, the volume dv tends to zero as $1/\sqrt{n^{k-1}}$.

---

[7] Since the regions surrounding "lattice" points lying on the boundary fall outside the volume v.

We now define a function $\pi(x_1, x_2, \ldots, x_k)$ on the points of the hyperplane (2) in the following way: At every "lattice" point $(x_1, x_2, \ldots, x_k)$ and in the region surrounding it, we set

$$\pi(x_1, x_2, \ldots, x_k) = \frac{1}{dv} P_n(m_1, m_2, \ldots, m_k)$$

$$= \frac{\prod_{i=1}^{k-1}(n+i)}{\sqrt{n^{k-1}}} \sqrt{\frac{\prod_{i=1}^{k} p_i q_i}{\sum p_i q_i}} P_n(m, \ldots, m_k) [1 + o(1)].$$

Outside of these regions we set

$$\pi(x_1, x_2, \ldots, x_k) = 0.$$

We now plot the values of the function $\pi(x_1, x_2, \ldots, x_k)$ along the perpendiculars to the hyperplane at the points $(x_1, x_2, \ldots, x_k)$, where the perpendiculars are erected in the direction of that part of the space for which $\Sigma x_i \sqrt{np_i q_i} > 0$. Geometrically, the function constructed is a "step hypersurface"; the k-dimensional volume bounded by the hypersurface and the hyperplane is given by

$$\sum \pi(x_1, x_2, \ldots, x_k) \, dv = \sum P_n(m_1, m_2, \ldots, m_k) = 1.$$

The probability $P(R)$ that the outcomes of the trials will fall in region R is clearly equal to

$$P(R) = \sum_{(x_i) \subset R} P_n(m_1, m_2, \ldots, m_k),$$

where the summation extends over all "lattice" points lying in region R. By writing $P(R)$ in the form

$$P(R) = \sum_{(x_i) \subset R} \pi(x_1, x_2, \ldots, x_k) \, dv,$$

we observe that the required probability is almost equal to the volume of a cylinder with generators passing through the boundary of the region R perpendicularly to the hyperplane (2) and bounded on one end by the hyperplane (2) and on the other end by the hypersurface $\pi(x_1, x_2, \ldots, x_k)$. The discrepancy is a quantity $|\alpha_n - \beta_n|$, where α_n is the total volume lying over the portions of the regions belonging to R surrounding lattice points not in R and β_n is the total volume lying over the portions of the regions not belonging to R but surrounding lattice points in R.

§ 13. The Integral Limit Theorem

Let us now assume that the region R is situated in a bounded portion of the hyperplane (2) and that its boundary is an integrable $(k-2)$-dimensional surface. Under these conditions we have, by the Local Limit Theorem,

$$\pi(x_1, x_2, \ldots, x_k) = \frac{e^{-\frac{1}{2}\Sigma q_i x_i^2}}{dv \sqrt{(2\pi n)^{k-1} p_1 p_2 \ldots p_k}} (1+\gamma_n) \tag{5}$$

$$= \sqrt{\frac{q_1 q_2 \ldots q_k}{(2\pi)^{k-1} \Sigma p_i q_i}} \prod_{i=1}^{k-1}\left(1+\frac{i}{n}\right) e^{-\frac{1}{2}\Sigma q_i x_i^2}(1+\gamma_n),$$

where $\gamma_n = \gamma(x_1, x_2, \ldots, x_k)$ tends to zero uniformly for all points of the region R as $n \to \infty$. Thus,

$$P(R) = \sum_R \frac{e^{-\frac{1}{2}\Sigma q_i x_i^2}}{\sqrt{\frac{\Sigma p_i q_i}{q_1 q_2 \ldots q_k}(2\pi)^{k-1}}} \prod_{i=1}^{k-1}\left(1+\frac{i}{n}\right)(1+\gamma_n)\, dv - \alpha_n + \beta_n. \tag{6}$$

By virtue of the choice of the regions which surround the lattice points, the quantity $|\alpha_n - \beta_n|$ can not exceed the volume lying above those regions which have points in common with the boundary of R. This volume does not exceed the product of the three quantities: the $(k-2)$-dimensional volume of the boundary of R (which we shall denote by u), twice the diameter of the region surrounding a "lattice" point of the plane (2), and the maximum value of the function $\pi(x_1, x_2, \ldots, x_k)$. By virtue of (5), the maximum value of $\pi(x_1, x_2, \ldots, x_k)$ for all sufficiently large values of n does not exceed

$$2\pi(0, 0, \ldots, 0) = 2\sqrt{\frac{q_1 q_2 \ldots q_k}{(2\pi)^{k-1} \Sigma p_i q_i}}.$$

The largest diameter of the regions surrounding the lattice points does not exceed the largest distance between neighboring lattice points and consequently is not greater than

$$\sqrt{\sum_{i=1}^{k} \frac{1}{n p_i q_i}}.$$

Thus,

$$|\alpha_n - \beta_n| \leq 4u \sqrt{\frac{q_1 q_2 \ldots q_k}{(2\pi)^{k-1} \Sigma p_i q_i}} \sum \frac{1}{p_i q_i} \cdot \frac{1}{\sqrt{n}},$$

and therefore, as $n \to \infty$,

$$|\alpha_n - \beta_n| \to 0. \tag{7}$$

Letting $n \to \infty$, we obtain, by the definition of a multiple integral,

$$\sum_R \frac{e^{-\frac{1}{2}\Sigma q_i x_i^2}}{\sqrt{\frac{\Sigma p_i q_i}{q_1 q_2 \cdots q_k}}(2\pi)^{k-1}} dv \to \sqrt{\frac{q_1 q_2 \cdots q_k}{(2\pi)^{k-1} \Sigma p_i q_i}} \int_R e^{-\frac{1}{2}\Sigma q_i x_i^2} dv. \tag{8}$$

The difference

$$P(R) - \sum_R \frac{e^{-\frac{1}{2}\Sigma q_i x_i^2}}{\sqrt{\frac{\Sigma p_i q_i}{q_1 q_2 \cdots q_k}}(2\pi)^{k-1}} dv = \alpha_n - \beta_n + \sum_R \left[\prod_{i=1}^k \left(1 + \frac{i}{n}\right) \gamma_n + \right. \tag{9}$$

$$\left. + \left(\prod_{i=1}^k \left(1 + \frac{i}{n}\right) - 1 \right) \right] \frac{e^{-\frac{1}{2}\Sigma q_i x_i^2}}{\sqrt{\frac{\Sigma p_i q_i}{q_1 q_2 \cdots q_k}}(2\pi)^{k-1}}$$

approaches zero as $n \to \infty$, since relation (7) holds as $n \to \infty$ and also because

$$\left|\prod_{i=1}^k \left(1 + \frac{i}{n}\right) \gamma_n \right| \leq 2^k \max |\gamma_n| \to 0$$

and

$$\prod_{i=1}^k \left(1 + \frac{i}{n}\right) - 1 \to 0.$$

Relations (6), (8), and (9) prove the theorem for the case of a bounded region with integrable boundary.

For unbounded regions, we reason as follows. Let ε be an arbitrary positive number. In the hyperplane (2) we describe a $(k-1)$-dimensional sphere S about the origin of radius r so large that[8]

$$\int_{(S)} \sqrt{\frac{q_1 q_2 \cdots q_k}{(2\pi)^{k-1} \Sigma p_i q_i}} e^{-\frac{1}{2}\Sigma q_i x_i^2} dv > 1 - \frac{\varepsilon}{4} \tag{10}$$

By the theorem just proved,

---

[8] It is possible to do this because the value of this integral when taken over the entire space is equal to one.

§ 13. The Integral Limit Theorem

$$P_n(S) > 1 - \frac{\varepsilon}{2} \tag{10'}$$

for sufficiently large n. Now, let R be an unbounded region such that in every finite portion of the hyperplane (2) its boundary is integrable. We divide the region R into two regions—a region (R_1) interior to the sphere S and a region (R_2) exterior to S; it is obvious that

$$P(R) = P(R_1) + P(R_2).$$

By (9) and (10), given any $\varepsilon > 0$, we can choose r so large that for all sufficiently large values of n, we have:

$$\left| P(R_2) - \int_{R_2} \sqrt{\frac{q_1 \cdots q_k}{(2\pi)^{k-1} \sum p_i q_i}} \, e^{-\frac{1}{2} \sum q_i z_i^2} \, dv \right| < \frac{\varepsilon}{2}. \tag{11}$$

Moreover, for sufficiently large n, as we have just proved,

$$\left| P(R_1) - \int_{R_1} \sqrt{\frac{q_1 \cdots q_k}{(2\pi)^{k-1} \sum p_i q_i}} \, e^{-\frac{1}{2} \sum q_i z_i^2} \, dv \right| < \frac{\varepsilon}{2}. \tag{12}$$

Since ε is arbitrary, the inequalities (11) and (12) complete the proof of our theorem.

Remark: The theorem just proved has been formulated in such a way that all the variables $x_1, x_2, \ldots, x_k$ play an identical role.

In the Integral Theorem of DeMoivre-Laplace, however, we preferred to conduct our arguments using only one variable $x = x_1$, disregarding the homogeneity of the variables x_1 and x_2. Geometrically, this means that we considered not the outcomes of the trials themselves (the lattice points on the line $x_1 + x_2 = 0$) but rather their projections onto the x-axis. In an analogous way, we can avoid homogeneous variables in the general case by integrating not over the region R but over its projection R' onto some coordinate hyperplane, say, the hyperplane $x_k = 0$. The element of volume dv' in the hyperplane $x_k = 0$ is related to the element of volume dv in the hyperplane (2) by the equation

$$dv' = dv \cos \varphi$$

where φ is the angle between the two hyperplanes in question. It is easy to show that

$$\cos \varphi = \frac{\sqrt{p_k q_k}}{\sqrt{\Sigma\, p_i q_i}}.$$

In the coordinate plane, the element of volume $dv' = dx_1\, dx_2 \ldots dx_{k-1}$ and therefore the equation

$$\sqrt{\frac{q_1 q_2 \cdots q_k}{(2\pi)^{k-1} \Sigma\, p_i q_i}} \int_R e^{-\frac{1}{2} \Sigma\, q_i x_i^2}\, dv = \sqrt{\frac{q_1 q_2 \cdots q_{k-1}}{(2\pi)^{k-1} p_k}} \int_{R'} e^{-\frac{1}{2} \sum_1^k q_i x_i^2}\, dx_1 \ldots dx_{k-1}$$

holds. We now must replace the variable x_k in the integrand by its value in terms of $x_1, x_2, \ldots, x_{k-1}$:

$$x_k = -\frac{1}{\sqrt{p_k q_k}} \sum_{i=1}^{k-1} \sqrt{p_i q_i}\, x_i.$$

As a result of this substitution, we obtain:

$$\sum_{i=1}^{k} q_i x_i^2 = \sum_{i=1}^{k-1} q_i \left(1 + \frac{p_i}{p_k}\right) x_i^2 + 2 \sum_{1 \leq i < j \leq k-1} x_i x_j \frac{\sqrt{p_i q_i p_j q_j}}{p_k} = Q(x_1, x_2, \ldots, x_{k-1}).$$

The Integral Limit Theorem can thus be formulated in another way, namely:

Under the conditions of the Integral Limit Theorem, as $n \to \infty$,

$$P(R) \to \sqrt{\frac{q_1 q_2 \cdots q_{k-1}}{(2\pi)^{k-1} p_k}} \int_{R'} e^{-\frac{1}{2} Q(x_1, x_2, \ldots, x_{k-1})}\, dx_1\, dx_2 \ldots dx_{k-1}. \quad (13)$$

The Integral Theorem of DeMoivre-Laplace is a special case of the theorem just proved; it can easily be obtained from formula (13).

To do this, it suffices to observe that in Bernoulli trials $k = 2$, $p = p_1$, and $q = p_2 = 1 - p$.

For $k = 3$, formula (13) assumes the following form:

$$P(R) \to \sqrt{\frac{q_1 q_2}{(2\pi)^2 p_3}} \int_{R'} e^{-\frac{1}{2} Q(x_1, x_2)}\, dx_1\, dx_2,$$

where

§ 14. Applications of Integral Theorem of DeMoivre-Laplace 107

$$p_3 = 1 - p_1 - p_2,$$

$$Q(x_1, x_2) = q_1\left(1 + \frac{p_1}{p_3}\right)x_1^2 + q_2\left(1 + \frac{p_2}{p_3}\right)x_2^2 + 2\frac{\sqrt{p_1 q_1 p_2 q_2}}{p_3}x_1 x_2$$

$$= \frac{q_1 q_2}{p_3}\left(x_1^2 + x_2^2 + 2\sqrt{\frac{p_1 p_2}{q_1 q_2}}\,x_1 x_2\right).$$

A simple calculation shows that $p_3 = 1 - p_1 - p_2 = q_1 q_2 - p_1 p_2$,[9] and thus

$$Q(x_1, x_2) = \frac{1}{1 - \frac{p_1 p_2}{q_1 q_2}}\left(x_1^2 + x_2^2 + 2\sqrt{\frac{p_1 p_2}{q_1 q_2}}\,x_1 x_2\right).$$

§ 14. Applications of the Integral Theorem of DeMoivre-Laplace

As a first application of the Integral Theorem of DeMoivre-Laplace, we estimate the probability of the inequality

$$\left|\frac{\mu}{n} - p\right| < \varepsilon,$$

where ε is a positive constant.

We have

$$\mathsf{P}\left\{\left|\frac{\mu}{n} - p\right| < \varepsilon\right\} = \mathsf{P}\left\{-\varepsilon\sqrt{\frac{n}{pq}} < \frac{\mu - np}{\sqrt{npq}} < \varepsilon\sqrt{\frac{n}{pq}}\right\}$$

and this implies, by virtue of the Integral Theorem of DeMoivre-Laplace, that

$$\lim_{n \to \infty} \mathsf{P}\left\{\left|\frac{\mu}{n} - p\right| < \varepsilon\right\} = \frac{1}{\sqrt{2\pi}}\int e^{-\frac{z^2}{2}}dz = 1.$$

Thus, for any positive constant ε, the probability of the inequality $|\mu/n - p| < \varepsilon$ approaches 1.

This fact was first discovered by James Bernoulli; it is called the *law of large numbers*, or *Bernoulli's Theorem*. Bernoulli's Theorem and its many generalizations are among the most important theorems of probability theory. It is through these theorems that the theory comes

---

[9] In fact, since $p_1 + q_1 = 1$ and $p_2 + q_2 = 1$, it follows that

$$1 - p_1 - p_2 = q_1(p_2 + q_2) - p_2(p_1 + q_1) = q_1 q_2 - p_1 p_2.$$

into contact with practice and it is they that are responsible for the success of the application of probability theory to diverse scientific and engineering problems. This will be discussed in greater detail in the chapter devoted to the law of large numbers; there we shall give a proof of Bernoulli's Theorem using a simpler method, distinct both from the one just presented and from the one given by Bernoulli.

We now examine some typical problems leading to the DeMoivre-Laplace Theorem.

Let n independent trials be performed in each of which the probability of occurrence of an event A is p.

I. A question that arises is: What is the probability that the relative frequency of occurrence of the event A will differ from the probability p by not more than α? This probability is given by

$$P\left\{\left|\frac{\mu}{n}-p\right|\leq\alpha\right\}=P\left\{-\alpha\sqrt{\frac{n}{pq}}\leq\frac{\mu-np}{\sqrt{npq}}\leq\alpha\sqrt{\frac{n}{pq}}\right\}\sim$$

$$\sim\frac{1}{\sqrt{2\pi}}\int_{-\alpha\sqrt{\frac{n}{pq}}}^{\alpha\sqrt{\frac{n}{pq}}}e^{-\frac{x^2}{2}}dx=\frac{2}{\sqrt{2\pi}}\int_{0}^{\alpha\sqrt{\frac{n}{pq}}}e^{-\frac{x^2}{2}}dx.$$

II. What is the smallest number of trials one must carry out in order that, with a probability not less than β, the relative frequency differ from p by no more than α? To answer this, we have to determine n from the inequality

$$P\left\{\left|\frac{\mu}{n}-p\right|\leq\alpha\right\}\geq\beta.$$

We replace the probability appearing on the left-hand side of this inequality by the approximate value given by the DeMoivre-Laplace Integral Theorem. As a result, we obtain the inequality

$$\frac{2}{\sqrt{2\pi}}\int_{0}^{\alpha\sqrt{\frac{n}{qp}}}e^{-\frac{x^2}{2}}dx\geq\beta$$

for the determination of n.

§ 14. Applications of Integral Theorem of DeMoivre-Laplace

III. For a given probability β and a given number of trials n, it is required to find a bound for the possible values of $|\mu/n - p|$. In

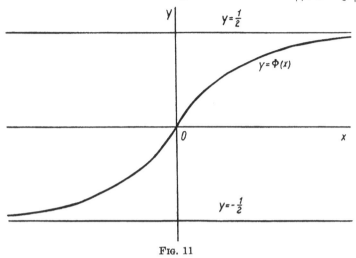

Fig. 11

other words, knowing β and n, we are to determine an α for which

$$P\left\{\left|\frac{\mu}{n} - p\right| < \alpha\right\} = \beta.$$

The application of the Laplace Theorem yields the equation

$$\frac{2}{\sqrt{2\pi}} \int_0^{\alpha\sqrt{\frac{n}{pq}}} e^{-\frac{x^2}{2}} dx = \beta$$

for the determination of α.

The numerical solution of each of these problems entails the computation of the values of the integral

$$\Phi(x) = \frac{1}{\sqrt{2\pi}} \int_0^x e^{-\frac{z^2}{2}} dz \qquad (1)$$

for arbitrary values of x, and the solution of the inverse problem: given the value of the integral $\Phi(x)$, to find the corresponding value of the argument x. These computations necessitate special tables, since for

II. Sequences of Independent Trials

$0 < x < \infty$ the integral (1) is not expressible in closed form in terms of elementary functions. Such tables have been compiled, and one is to be found at the end of the book.

Fig. 11 gives a graphical representation of the function $\Phi(x)$. By use of the table of values of the function $\Phi(x)$ in conjunction with the formula $J(a,b) = \Phi(b) - \Phi(a)$, we can also compute the value of the integral

$$J(a,b) = \frac{1}{\sqrt{2\pi}} \int_a^b e^{-\frac{z^2}{2}} dz.$$

The table of values of $\Phi(x)$ are compiled for positive x only; for negative x, the function $\Phi(x)$ can be determined from the identity

$$\Phi(-x) = -\Phi(x).$$

We are now in a position to complete the solution of Example 1 of § 11.

Example 1. In Example 1 of § 11, we had to find the probability

$$p = \Sigma \mathsf{P}\{\mu = m\},$$

the summation extending over those values of m for which

$$|m - 2.7 \cdot 10^{22}| \geq 2.7 \cdot 10^{12},$$

under the condition that the total number of trials $n = 5.4 \times 10^{22}$ and $p = 1/2$. Since

$$p = \mathsf{P}\left\{\frac{|\mu - np|}{\sqrt{npq}} \geq \frac{2.7 \cdot 10^{12}}{\sqrt{5.4 \cdot 10^{22} \cdot \frac{1}{4}}}\right\} \sim \mathsf{P}\left\{\frac{|\mu - np|}{\sqrt{npq}} \geq 2.33 \cdot 10\right\},$$

by virtue of the Laplace Theorem, we have

$$p \sim \frac{2}{\sqrt{2\pi}} \int_{2.33 \cdot 10}^{\infty} e^{-\frac{x^2}{2}} dx.$$

Because

$$\int_z^{\infty} e^{-\frac{x^2}{2}} dx < \frac{1}{z} \int_z^{\infty} x e^{-\frac{x^2}{2}} dx = \frac{1}{z} e^{-\frac{z^2}{2}},$$

§ 14. APPLICATIONS OF INTEGRAL THEOREM OF DEMOIVRE-LAPLACE

it follows that

$$p < \frac{1}{\sqrt{2\pi \cdot 10}} e^{-2.7 \cdot 100} < 10^{-100}.$$

We may judge how small this probability is by the following analogy. Suppose that a sphere of radius 6000 km. is filled with white sand in which there is a single grain of black sand. Each grain has a volume of 1 mm.^{3}. One grain is taken at random from the entire mass. What is the probability that it will be black?

It is easily shown that the volume of a sphere of radius 6,000 km. is a little less than 10^{30} mm.^{3}, and consequently the probability of drawing the black grain is a little more than 10^{-30}.

Example 2. In Example 2 of § 11, we had to determine the probability that the number of defective items would be no larger than seventy if the probability of each article being defective was $p = 0.005$, and if the number of articles was 10,000. By the DeMoivre-Laplace Integral Theorem, this probability is given by

$$P\{\mu \leq 70\} = P\left\{-\frac{50}{\sqrt{49.75}} \leq \frac{\mu - np}{\sqrt{npq}} \leq \frac{20}{\sqrt{49.75}}\right\}$$

$$= P\left\{-7.09 \leq \frac{\mu - np}{\sqrt{npq}} \leq 2.84\right\} \sim \frac{1}{\sqrt{2\pi}} \int_{-7.09}^{2.84} e^{-\frac{z^2}{2}} dz$$

$$= \Phi(2.84) - \Phi(-7.09) = \Phi(2.84) + \Phi(7.09) = 0.9975.$$

The value of $\Phi(x)$ is not tabulated for $x = 7.09$; we have taken it as $1/2$, committing, in so doing, an error less than 10^{-10}.

Of course, in the examples of this and the preceding sections, as well as in any other problem involving the determination of the probabilities $P_n(m)$ for arbitrary finite values of m and n by means of the asymptotic formula of DeMoivre-Laplace, it is necessary to estimate the error introduced by making such a substitution. For a long time, the theorems of DeMoivre-Laplace were applied to the solution of problems of this kind without any satisfactory estimate being given for the remainder term. A purely empirical feeling of certainty arose that, as long as n is of the order of several hundreds or larger and p not too close to 0 or 1, the use of the DeMoivre-Laplace Theorem leads to satisfactory results. At present, there exist sufficiently good estimates of the error committed in using the asymptotic formula of DeMoivre-Laplace.[10]

---

[10] See, for example, the paper of S. N. Bernstein cited on page 93.

We shall pause here to extend the Bernoulli Theorem to the case of a general scheme of repeated independent trials. Let there be k possible outcomes in each trial, with the respective probabilities $p_1, p_2, \ldots, p_k$, and let $\mu_1, \mu_2, \ldots, \mu_k$ be the respective number of occurrences of each outcome in n repeated independent trials. We determine the probability of the simultaneous realization of the inequalities

$$\left|\frac{\mu_1}{n} - p_1\right| < \varepsilon_1, \quad \left|\frac{\mu_2}{n} - p_2\right| < \varepsilon_2, \ldots, \quad \left|\frac{\mu_k}{n} - p_k\right| < \varepsilon_k, \qquad (2)$$

i.e., of the inequalities

$$|x_1| < \varepsilon_1 \sqrt{\frac{n}{p_1 q_1}}, \quad |x_2| < \varepsilon_2 \sqrt{\frac{n}{p_2 q_2}}, \ldots, \quad |x_k| < \varepsilon_k \sqrt{\frac{n}{p_k q_k}}.$$

Strictly speaking, the last of these inequalities is a consequence of the preceding ones, since by equation (2) of § 13, the first $k-1$ inequalities in (2) yield the estimate

$$|x_k| = \left|-\sum_{i=1}^{k-1} \sqrt{\frac{p_i q_i}{p_k q_k}}\, x_i\right| \leq \sum_{i=1}^{k-1} \sqrt{\frac{n}{p_k q_k}}\, \varepsilon_i. \qquad (3)$$

By (14) of § 13, the probability of the first $k-1$ inequalities of (2) (and consequently the inequality (3) as well) approaches the integral

$$\sqrt{\frac{q_1 q_2 \cdots q_{k-1}}{(2\pi)^{k-1} p_k}} \int \cdots \int e^{-\frac{1}{2} Q(x_1, \ldots, x_{k-1})}\, dx_1\, dx_2 \cdots dx_{k-1} = 1$$

in the limit, as $n \to \infty$.

§ 15. Poisson's Theorem

An examination of the proof of the Local DeMoivre-Laplace Theorem shows that the asymptotic approximation of the probability $P_n(m)$ by means of the function $\frac{1}{\sqrt{2\pi}} e^{-\frac{x^2}{2}}$ becomes increasingly poorer as the probability p differs more and more from one-half, i.e., the smaller the values of p or q under consideration, and this approximation ceases to be applicable both for $p=0$, $q=1$ and for $p=1$, $q=0$. However, a considerable range of problems calls for our being able to compute the

§ 15. Poisson's Theorem

probabilities $P_n(m)$ for just such small values of p.[11] In order for the DeMoivre-Laplace Theorem to give results with negligible error in these cases, it is necessary that the number of trials be very large. The problem thus arises of seeking an asymptotic formula which is especially suited to the case of small p. Such a formula was found by Poisson.

Let us consider the double sequence

$$E_{11},$$
$$E_{21}, E_{22},$$
$$E_{31}, E_{32}, E_{33},$$
$$\ldots\ldots\ldots\ldots\ldots,$$
$$E_{n1}, E_{n2}, E_{n3}, \ldots, E_{nn},$$
$$\ldots\ldots\ldots\ldots\ldots,$$

in which the events in any one of the rows are mutually independent and each of them has a probability p_n depending only on the number of the row. We denote by μ_n the number of times an event actually occurs in the n-th row.

POISSON'S THEOREM: *If $p_n \to 0$ as $n \to \infty$, then*

$$P\{\mu_n = m\} - \frac{a_n^m}{m!} e^{-a_n} \to 0, \tag{1}$$

where

$$a_n = n p_n.$$

Proof: It is obvious that

$$P_n(m) = P\{\mu_n = m\} = \binom{n}{m} p_n^m (1-p_n)^{n-m}$$
$$= \frac{n!}{m!(n-m)!} \left(\frac{a_n}{n}\right)^m \left(1 - \frac{a_n}{n}\right)^{n-m}$$
$$= \frac{a_n^m}{m!} \left(1 - \frac{a_n}{n}\right)^n \frac{\left(1-\frac{1}{n}\right)\left(1-\frac{2}{n}\right)\cdots\left(1-\frac{m-1}{n}\right)}{\left(1-\frac{a_n}{n}\right)^m}. \tag{2}$$

Let m be fixed. Let us choose an arbitrary $\varepsilon > 0$. Then, we can choose $A = A(\varepsilon)$ so large that for $a \geq A$ we have

---

[11] And for small values of q as well; but it is clear that the problems of seeking asymptotic formulas for $P_n(m)$ for small values of p or q are reducible to each other.

$$\frac{a^m}{m!} e^{-\frac{1}{2}a} \leqq \frac{\varepsilon}{2}.$$

We first consider those values of n for which $a_n \geqq A$. For these n, by the inequality $1 - x \leqq e^{-x}$, $0 \leqq x \leqq 1$, we get

$$P_n(m) \leqq \frac{a_n^m}{m!} e^{-\frac{n-m}{n} a_n} \leqq \frac{\varepsilon}{2} \quad \text{for } n \geqq 2m$$

and

$$\frac{a_n^m}{m!} e^{-a_n} < \frac{\varepsilon}{2}.$$

Therefore, for the values of n indicated,

$$\left| P_n(m) - \frac{a_n^m}{m!} e^{-a_n} \right| < \frac{\varepsilon}{2} + \frac{\varepsilon}{2} = \varepsilon.$$

Let us now consider those values of n for which $a_n \leqq A$. Since for $a_n \leqq A$, $\lim_{n \to \infty} \left\{ \left(1 - \frac{a_n}{n}\right)^n - e^{-a_n} \right\} = 0$, and for fixed m,

$$\lim_{n \to \infty} \frac{\left(1 - \frac{1}{n}\right)\left(1 - \frac{2}{n}\right)\cdots\left(1 - \frac{m-1}{n}\right)}{\left(1 - \frac{a_n}{n}\right)^m} = 1,$$

by formula (2), we have, for $n \geqq n_0(\varepsilon)$,

$$\left| P_n(m) - \frac{a_n^m}{m!} e^{-a_n} \right| < \varepsilon,$$

Q.E.D.

We note that the theorem of Poisson is also valid for the case where the probability of the event A_n is zero in every trial. In this case, $a_n = 0$. If $\lim np_n = a$, we define

$$P(m) = \lim_{n \to \infty} P_n(m) = \frac{a^m e^{-a}}{m!} \quad (m = 0, 1, 2, \ldots).$$

This probability distribution is called the *Poisson law*.

§ 15. POISSON'S THEOREM

It is easily shown that the quantities $P(m)$ satisfy the relation $\sum_m P(m) = 1$. Let us study the behavior of $P(m)$ as a function of m. For this purpose, we consider the ratio

$$\frac{P(m)}{P(m-1)} = \frac{a}{m}.$$

We see that if $m > a$, then $P(m) < P(m-1)$; however, if $m < a$, then $P(m) > P(m-1)$; and finally, if $m = a$, then $P(m) = P(m-1)$. From this we conclude that $P(m)$ increases from $m = 0$ up to $m_0 = [a]$, and with the further increase of m decreases monotonically. If a is an integer, then $P(m)$ has two maximum values, at $m_0 = a$ and at $m_0' = a - 1$. We now discuss some examples.

Example 1. At each firing, the probability of hitting a target is 0.001. Find the probability of hitting a target with two or more bullets if the total number of shots fired is 5,000.[12]

Regarding each firing as one trial and a target hit as our event, we can utilize Poisson's Theorem for calculating the probability $\mathsf{P}\{\mu_n \geq 2\}$. In the example under consideration,

$$a_n = np = 0.001 \cdot 5{,}000 = 5.$$

The required probability is equal to

$$\mathsf{P}\{\mu_n \geq 2\} = \sum_{m=2}^{\infty} P_n(m) = 1 - P_n(0) - P_n(1).$$

By Poisson's Theorem,

$$P_n(0) \sim e^{-5}, \quad P_n(1) \sim 5e^{-5}.$$

Therefore,

$$\mathsf{P}\{\mu_n \geq 2\} \sim 1 - 6\,e^{-5} \approx 0.9596.$$

The probability $P(m)$ takes on its maximum value for $m = 4$ and $m = 5$.

---

[12] In World War II, the conditions of our problem were actually realized in the use of small-arms fire against airplanes. An airplane could be put out of action by bullets only if hit in one of a few vulnerable spots: the motor, the fuel tank, the pilot himself, etc. The probability of a hit in these vulnerable spots was extremely small but, as a rule, an entire division directed its fire at a plane and the overall number of shots fired at an airplane was considerable. As a result, the probability of hitting the plane with, at any rate, one or two bullets was appreciably high. This fact came to attention as a purely practical observation.

This probability correct to four decimal places is given by

$$P(4) = P(5) \approx 0.1754.$$

Computations using the exact formula give the values $P_{5000}(0) = 0.0067$ and $P_{5000}(1) = 0.0336$ correct to four decimal places, and therefore

$$\mathsf{P}\{\mu_n \geq 2\} = 0.9597.$$

The error introduced by using the asymptotic formula is thus less than 0.02% of the exact value.

Example 2. In a spinning factory, a worker attends several hundred spindles each spinning its own skein. As a spindle turns, the yarn breaks at certain chance moments because of irregularity in tension, unevenness, and other reasons. For production purposes, it is important to know how frequently breaks can occur under a variety of operational conditions (quality of yarn, spindle speed, etc.).

Assuming that the worker attends 800 spindles and that the probability of a break in the yarn during a certain interval of time τ is 0.005 for each spindle, find the most probable number of breaks and the probability that no more than 10 breaks will occur during the interval of time τ.

Since

$$a_n = np = 0.005 \cdot 800 = 4,$$

$[p(n+1)] = [a_n + p] = 4$, and this will be the most probable number of breaks in the interval of time τ. Its probability is

$$P_{800}(4) = 0.1959.$$

By Poisson's formula, we have:

$$P_{800}(3) \sim \frac{4^3}{3!} e^{-4} = 0.1954 \sim P_{800}(4) \sim \frac{4^4}{4!} e^{-4} = 0.1954.$$

The exact value $P_{800}(4) = 0.1959$. The probability that there will be no more than 10 breaks in an interval of time τ is equal to

$$\mathsf{P}\{\mu_n \leq 10\} = \sum_{m=0}^{10} P_{800}(m) = 1 - \sum_{m=11}^{\infty} P_{800}(m).$$

By virtue of Poisson's Theorem,

§ 16. ILLUSTRATION OF SCHEME OF INDEPENDENT TRIALS   117

$$P_{800}(m) \approx \frac{4^m}{m!} e^{-4} \quad (m = 0, 1, 2, \ldots),$$

and therefore

$$\mathsf{P}\{\mu_n \leq 10\} = 1 - \sum_{m=11}^{\infty} \frac{4^m}{m!} e^{-4}.$$

But

$$\sum_{m=11}^{\infty} \frac{4^m}{m!} e^{-4} > \left(\frac{4^{11}}{11!} + \frac{4^{12}}{12!} + \frac{4^{13}}{13!}\right) e^{-4} = \frac{4^{12} \cdot 14}{11!\, 39} e^{-4} = 0.00276.$$

On the other hand,

$$\sum_{m=11}^{\infty} \frac{4^m}{m!} e^{-4} < \frac{4^{11}}{11!} e^{-4} + \frac{4^{12}}{12!} e^{-4} + \frac{4^{13}}{13!} e^{-4} \left[1 + \frac{4}{14} + \left(\frac{4}{14}\right)^2 + \cdots \right] =$$

$$= \frac{4^{12} \cdot 24}{11!\, 65} e^{-4} = 0.00284.$$

Thus,

$$0.99716 \leq \mathsf{P}\{\mu_n \leq 10\} \leq 0.99724.$$

Just as in the application of the DeMoivre-Laplace Theorem, the question arises of estimating the error made in the replacement of the exact formula for calculating $P_n(m)$ by the asymptotic formula of Poisson.

From the equality

$$P_n(0) = \left(1 - \frac{a_n}{n}\right)^n = e^{n \log\left(1 - \frac{a_n}{n}\right)}$$

$$= \exp\left\{-n \sum_{k=1}^{\infty} \frac{1}{k} \left(\frac{a_n}{n}\right)^k\right\} = e^{-a_n}(1 - R_n),$$

where

$$R_n = 1 - \exp\left\{-n \sum_{k=2}^{\infty} \frac{1}{k} \left(\frac{a_n}{n}\right)^k\right\},$$

we can easily find this estimate for the case $m = 0$. In fact, since for any positive x, $0 < 1 - e^{-x} < x$, we have:

$$0 < R_n < n \sum_{k=2}^{\infty} \frac{1}{k} \left(\frac{a_n}{n}\right)^k$$

whatever values a_n and n may have. Since

$$\sum_{k=2}^{\infty}\frac{1}{k}\left(\frac{a_n}{n}\right)^k \leq \frac{a_n^2}{2\,n^2}+\frac{1}{3}\sum_{k=3}^{\infty}\left(\frac{a_n}{n}\right)^k =$$

$$= \frac{a_n^2}{2\,n^2}+\frac{a_n^3}{3\,n^3\left(1-\frac{a_n}{n}\right)} = \frac{a_n^2}{6\,n^2}\cdot\frac{3\,n-a_n}{n-a_n} < \frac{a_n^2}{2\,n(n-a_n)},$$

we obtain:

$$0 < R_n < \frac{a_n^2}{2\,(n-a_n)}.$$

From the fact that R_n is non-negative, we conclude that the replacement of $P_n(0)$ by e^{-a_n} slightly increases the value of the probability $P_n(0)$.

§ 16. Illustration of the Scheme of Independent Trials

As an illustration of how the preceding results can be used for scientific purposes, we shall consider, in highly schematic form, the problem of the random walk of a particle on a straight line. This problem may be regarded as the prototype of actual physical problems in the theory of diffusion, Brownian motion, etc.

Imagine that a particle, initially located at the point $x = 0$, undergoes random collisions at certain moments of time in consequence of each of which it is displaced one unit to the right or to the left. Hence, at each of these times, the particle is displaced one unit to the right with probability 1/2 or one unit to the left with probability 1/2. As the result of n impacts, the particle will be displaced a distance μ. It is clear that in this problem we are dealing with Bernoulli trials in its purest form. It follows from this that for any n and m we can calculate the probability that $\mu = m$, to wit[13]

---

[13] If the particle is $\mu = m$ units from the origin, then it must have moved through, say, α steps of $+1$ and β steps of -1, so that $m = \alpha - \beta$ and $n = \alpha + \beta$. Thus, the possible paths are the number of different ways in which α positive unit steps can be chosen from n steps, i.e., $\binom{n}{\alpha} = \binom{n}{(m+n)/2}$.

§ 16. ILLUSTRATION OF SCHEME OF INDEPENDENT TRIALS

$$P\{\mu = m\} = \begin{cases} \binom{n}{(m+n)/2} \left(\frac{1}{2}\right)^n, & \text{for } -n \leq m \leq n, \\ 0, & \text{for } |m| > n. \end{cases}$$

For large values of n, the Local DeMoivre-Laplace Theorem implies

$$P\{\mu = m\} \sim \frac{\sqrt{2}}{\sqrt{\pi n}} e^{-\frac{m^2}{2n}}. \tag{1}$$

We may interpret the formula obtained as follows. Let there be a great number of particles initially located at $x = 0$. All particles, in consequence of the random impacts, begin to move along the line independently

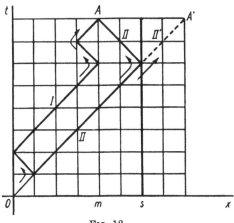

FIG. 12

of one another. Then the fraction of particles that has been displaced a distance m after n impacts is given by formula (1).

Of course, we are assuming idealized conditions for the motion of the particles; the conditions under which molecules actually move is far more complex; the result obtained does, however, give an accurate *qualitative* picture of the phenomenon.

In physics, one has to consider more complicated examples of random walks. We confine ourselves here to a schematic treatment of the effect on a particle of 1) a reflecting barrier and 2) an absorbing barrier.

Let us imagine a reflecting barrier located s units to the right of the point $x = 0$, so that if at any time a particle reaches the barrier, the next impact causes it to return in the same direction from which it came, with probability one.

For the sake of clarity, let us plot the position of the particle in the xt-plane (Fig. 12). In this way, the path of the particle will be represented in the form of a polygonal line. For $x < s$, the particle moves under each impact one unit "upward" and one unit to the right or left (with probability 1/2 in each case). If $x = s$, however, the particle moves one unit to the left.

To compute the probability $\mathsf{P}\{\mu = m\}$, we proceed as follows: We imagine the barrier removed, and we allow the particle to move freely, as if there were no barrier. Fig. 12 shows such idealized paths leading to the points A and A' which are symmetrically located with respect to the barrier. In order for a physical particle that has undergone reflections to reach the point A, it is both necessary and sufficient that it reach either the point A or the point A' while moving under the above idealized setup (without a reflecting barrier). But the probability of the particle reaching the point A in the idealized setup is obviously equal to

$$\mathsf{P}\{\mu = m\} = \frac{n!}{\left(\frac{m+n}{2}\right)!\left(\frac{n-m}{2}\right)!}\left(\frac{1}{2}\right)^n.$$

In exactly the same way, the probability of the particle reaching the point A' (which has the abscissa $2s - m$) is equal to

$$P\{\mu = 2s - m\} = \frac{n!}{\left(s + \frac{n-m}{2}\right)!\left(\frac{n+m}{2} - s\right)!}\left(\frac{1}{2}\right)^n.$$

The required probability is therefore given by

$$P_n(m, s) = \mathsf{P}\{\mu = m\} + \mathsf{P}\{\mu = 2s - m\}.$$

Taking advantage of the local DeMoivre-Laplace Limit Theorem, we find:

$$P_n(m, s) \sim \frac{2}{\sqrt{2\pi n}}\left\{e^{-\frac{m^2}{2n}} + e^{-\frac{(2s-m)^2}{2n}}\right\}.$$

§ 16. ILLUSTRATION OF SCHEME OF INDEPENDENT TRIALS

This is the well-known formula in the theory of Brownian motion. It can be put in more symmetric form by shifting the origin to the point $x = s$, which is accomplished by making the change of variables $z = x - s$. As a result of this substitution, we obtain:

$$P(z = k) = P_n\{k + s, s\} = \frac{2}{\sqrt{2\pi n}} \left\{ e^{-\frac{(k+s)^2}{2n}} + e^{-\frac{(k-s)^2}{2n}} \right\}.$$

We now proceed to discuss the third problem, in which an absorbing barrier exists in the path of the particle at the point $x = s$. A particle that collides with the wall makes no further contribution to the motion of the particles. It is obvious that in the present example the probability of the particle being at the point $x = m$ ($m < s$) after n impacts will be less than $P_n(m)$ (i.e., less than the probability of its reaching this point if there were no absorbing barrier present); let us denote the required probability by $\overline{P}_n(m, s)$.

To compute the probability $\overline{P}_n(m, s)$, we again imagine the absorbing barrier removed, and we allow the particle to move freely along the line.

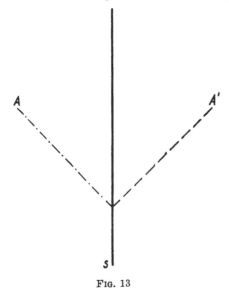

FIG. 13

A particle that reaches the position $x = s$ at some instant, is found in the following instant either to the right or to the left of the line $x = s$ (Fig. 13) with the same probability in each case. That is, after reaching the line $x = s$, the particle may be found at the point $A(m, n)$ or at the point

$A'(2s-m, n)$ with equal probability. But the particle can reach the point A' only by having first reached the position $x = s$, and therefore, for any path leading to the point A', there exists a path symmetric to it with respect to the line $x = s$ which leads to the point A; in exactly the same way, for any path, inadmissible in the actual motion, which leads via $x = s$ to the point A, there exists a path symmetric to it with respect to the line $x = s$ leading to the point A'. Note that we consider the symmetry of the paths only from the moment the particle reaches the line $x = s$. This reasoning shows that the probability of the particle being at A in the actual motion can be computed by subtracting from the total number of paths leading to the point A in the idealized motion the number of paths leading to the point A'. It is obvious from this that

$$\bar{P}_n(m, s) = \mathsf{P}\{\mu = m\} - \mathsf{P}\{\mu = 2s - m\}.$$

In view of the Local DeMoivre-Laplace Theorem, we have

$$\bar{P}_n(m, s) \sim \frac{2}{\sqrt{2\pi n}}\left\{e^{-\frac{m^2}{2n}} - e^{-\frac{(2s-m)^2}{2n}}\right\}.$$

Exercises

1. A workman attends 12 machines of the same type. The probability that a machine will require his attention during an interval of time of length τ is equal to 1/3. What is the probability that

a) 4 machines will require the workman's attention in the time τ;

b) The number of times his attention is required will be between 3 and 6 inclusive?

2. There are 10 children in a certain family. Assuming that the probability of the birth of a boy, or of a girl, is 1/2, determine the probability that

a) There are 5 boys and 5 girls in the family;

b) The number of boys in the family lies between 3 and 8, inclusive.

3. In a group of 4 people, the birthdays of three occur in the same month and that of the fourth, in one of the other eleven months. Assuming that the probability of anyone being born in any particular month is equal to 1/12, find the probability that

a) The three designated persons were born in January, and the fourth in October;

b) The three persons were born in some one month and the fourth in some other month.

4. In 14,400 tosses of a coin, heads has come up 7,428 times. What is the probability of the number of heads differing from the quantity np by an amount equal to or greater than the difference in this experiment if the coin is symmetric (i.e., the probability of tossing a head in each trial is 1/2)?

EXERCISES

5. Plugged into an electric circuit are n electrical appliances each requiring a kilowatts; the probability that any particular one of them is on at any given moment is p. Find the probability that at a given moment the power used

a) will be less than nap;

b) will exceed $rnap$ $(r>0)$ under the condition that np is large.

6. 730 students are enrolled at an educational institution. The probability that the birthday of a student chosen at random from the student register occurs on a specific day of the year is equal to 1/365 for each day of the year. Find

a) the most probable number of students born on January 1;

b) the probability that there are at least three students having the same birthday.

7. It is known that the probability of turning out a brittle (defective) drill bit is 0.02. Bits are packed in boxes of 100. What is the probability that

a) There will be no defective bits in a box;

b) The number of defective bits will not exceed 3?

c) How many bits is it necessary to place in a box so that the probability is not less than 0.9 that it will contain no fewer than 100 good ones?

Hint: Use the Poisson distribution.

8. 10,000 persons of the same age and social group have policies with an insurance company. The probability of death during the year is 0.006 for each one of them. Each insured pays a premium of $12 on January 1, and in the event he dies, his beneficiaries receive $1000 from the company. What is the probability that

a) The company will lose money;

b) The company will make a profit of not less than $40,000? $60,000? $80,000?

9. *Banach's matchbox problem*. A certain mathematician carries two matchboxes with him. Each time he wants to use a match, he selects one of the boxes at random. Find the probability that when the mathematician discovers that one box is empty the other box contains r matches $(r=0,1,2,\ldots,n$; n being the number of matches initially contained in each box).

10. There are n machines hooked up to an electrical circuit. The probability that a machine which is drawing power from the circuit at time t will cease doing so before time $t+\Delta t$ is equal to $\alpha \Delta t + o(\Delta t)$. If a machine is not drawing power at time t, then the probability that it will begin to do so before time $t+\Delta t$ is $\beta \Delta t + o(\Delta t)$, independent of the operation of the other machines. Derive the differential equations satisfied by $P_r(t)$, the probability that r machines are drawing power at time t.

One can easily find concrete situations in which the conditions of this problem are realized: trolley busses, electric welding, the consumption of power by machines with automatic cutoffs, etc.

11. A single workman attends n automatic machines of the same type. If a machine is operating at time t, then the probability that it will require his attention before time $t+\Delta t$ is $\alpha \Delta t + o(\Delta t)$. If the workman is attending one of the machines at time t, then the probability that he will finish attending to it before time $t+\Delta t$ is $\beta \Delta t + o(\Delta t)$. Derive the differential equations that are satisfied by $P_r(t)$, the probability that there are $n-r$ machines operating at time t; i.e., one is being attended

to and $r-1$ are awaiting attention. ($P_0(t)$ is the probability that all of the machines are in operation.)

Note: In an analogous way, it is not difficult to derive the differential equations for the more complicated problem in which N machines are attended by a team of k workmen. For practical reasons, it is important to compare the efficiency of the various systems of work organization. For this purpose, it is necessary to investigate the steady-state condition, i.e., to consider the probabilities $P_r(t)$ as $t \to \infty$. It turns out that the work of a team attending to kn machines is more profitable than the maintenance of n machines by a single workman, both in the sense of better utilization of the running time of the machine and the work time of the workmen.

12. Prove the following theorem: If P and P' are the respective probabilities of the most probable number of occurrences of an event A in n and $n+1$ independent trials (in each trial $\mathsf{P}(A) = p$), then $P' \leq P$. The equality can hold only if $(n+1)p$ is an integer.

13. In Bernoulli trials $p = 1/2$. Prove that

a) $$\frac{1}{2\sqrt{n}} \leq P_{2n}(n) \leq \frac{1}{\sqrt{2n+1}};$$

b) $$\lim_{n \to \infty} \frac{P_{2n}(n \pm h)}{P_{2n}(n)} = e^{-z^2} \quad \text{if} \quad \lim_{n \to \infty} \frac{h}{\sqrt{n}} = z \quad (0 \leq z \leq \infty).$$

14. Prove that for $npq \geq 25$

$$P_n(m) = \frac{1}{\sqrt{2\pi npq}} e^{-\frac{t^2}{2}} \left[1 + \frac{(q-p)(t^3 - 3t)}{6\sqrt{npq}} \right] + \Delta,$$

where

$$t = \frac{m - np}{\sqrt{npq}}, \qquad |\Delta| < \frac{0.15 + 0.25 |p-q|}{(npq)^{3/2}} + e^{-\frac{3}{2}\sqrt{npq}}.$$

15. n independent trials are made. The probability that an event A occurs in the i-th trial is p_i; $P_n(m)$ is the probability of the m-fold occurrence of the event A in the n trials. Prove that

a) $$\frac{P_n(1)}{P_n(0)} \geq \frac{P_n(2)}{P_n(1)} \geq \cdots \geq \frac{P_n(n)}{P_n(n-1)};$$

b) $P_n(m)$ at first increases and then decreases (provided $P_n(0)$ or $P_n(n)$ are not themselves maximum values).

16. Prove that for $x > 0$, the function $\int_x^\infty e^{-\frac{z^2}{2}} dz$ satisfies the inequalities

$$\frac{x}{1+x^2} e^{-\frac{1}{2}x^2} \leq \int_x^\infty e^{-\frac{1}{2}z^2} dz \leq \frac{1}{x} e^{-\frac{1}{2}x^2}.$$

CHAPTER III

MARKOV CHAINS

§ 17. Definition of a Markov Chain. Transition Matrix

An immediate generalization of the scheme of independent trials is that of a so-called *Markov chain,* which was first systematically investigated by the renowned Russian mathematician A. A. Markov. We shall confine ourselves to a presentation of the elements of the theory.

Let us imagine that a sequence of trials is performed in each of which one of k mutually exclusive events $A_1^{(s)}, A_2^{(s)}, \ldots, A_k^{(s)}$ may be realized (as in the preceding chapter, the superscript denotes the number of the trial). We shall say that the sequence of trials forms a *Markov chain*, or more precisely, a *simple* Markov chain, if *the conditional probability of occurrence of the event $A_i^{(s+1)}$ ($i = 1, 2, \ldots, k$) in the $(s + 1)$-st trial ($s = 1, 2, 3, \ldots$) given that a known event has occurred in the s-th trial depends only on which event has occurred in the s-th trial and is not affected by the further information as to which events have occurred in the earlier trials.*

A different terminology is often used in the theory of Markov chains. One speaks of a physical system S which at any time can be in one of the states $A_1, A_2, \ldots, A_k$ and which changes its state only at the times $t_1, t_2, \ldots, t_n, \ldots$. For a Markov chain, the probability of transition of the system into some state A_i ($i = 1, 2, \ldots, k$) at time τ ($t_s < \tau < t_{s+1}$) depends only on what the state of the system was at time t ($t_{s-1} < t < t_s$) and is unaffected by anything that may become known about its states at an earlier time.

We shall now discuss two illustrative examples.

Example 1. Imagine a particle lying on a line to move along this line under the influence of random impacts occurring at times $t_1, t_2, t_3, \ldots$. The particle may be at the points with integral coordinates $a, a + 1, a + 2, \ldots, b$; there are reflecting barriers at the points a and b. Each impact moves the particle to the right with probability p or to the left with probability $q = 1 - p$, as long as the particle does not reach one of the barriers. However, if the particle is at a barrier then the impact moves it one unit towards the interior of the interval between the barriers. We see that this example of random walk is a typical Markov chain. In exactly the same way, one may consider the cases in which the particle is 'absorbed' at one or at both of the barriers.

Example 2. In the Bohr model of the hydrogen atom, the electron may be found in one of certain admissible orbits. Let A_i denote the event consisting in the electron being found in the i-th orbit. Let us suppose further that a change in the state of the atom can only happen at times $t_1, t_2, t_3, \ldots$ (in reality, these times are random quantities). The probability of a transition from the i-th orbit to the j-th orbit at time t_s depends only on i and j (the difference $j - i$ depends on the amount by which the energy of the atom is changed at time t_s) and not on the past orbits of the electron.

The last example is a Markov chain with an infinite number of states (although in principle only); this example would be far closer to actual conditions if the time of transition of the system from a given state to a new state were to vary continuously.

We shall further confine our presentation to the simplest facts relating to *homogeneous Markov chains,* in which the conditional probability of occurrence of the event $A_j^{(s+1)}$ in the $(s + 1)$-st trial, given that the event $A_i^{(s)}$ has been realized in the s-th trial, is independent of the number of the trial. This probability is called the *probability of transition* and will be denoted by p_{ij}; in this notation, the first subscript always stands for the outcome of the preceding trial and the second subscript indicates the state into which the system passes in the following moment of time.

The complete probabilistic picture of the possible changes in state which are realizable by transition from one trial to the one directly following is given by the matrix

$$\pi_1 = \begin{pmatrix} p_{11} & p_{12} \cdots p_{1k} \\ p_{21} & p_{22} \cdots p_{2k} \\ \cdots \cdots \cdots \cdots \\ p_{k1} & p_{k2} \cdots p_{kk} \end{pmatrix}$$

constructed from the transition probabilities, and which is called the *matrix of transition probabilities.*

§ 17. Definition of Markov Chain. Transition Matrix

For illustrative purposes, we consider some examples.

Example 3. A system S under investigation may be found in the states A_1, A_2, and A_3; the transition from state to state takes place according to the scheme of a homogeneous Markov chain; the transition probabilities are given by the matrix

$$\begin{pmatrix} 1/2 & 1/6 & 1/3 \\ 1/2 & 0 & 1/2 \\ 1/3 & 1/3 & 1/3 \end{pmatrix}.$$

We see that if the system was in the state A_1, then after a one-step change in state it remains in the same state with probability 1/2, it passes into the state A_2 with probability 1/6, and passes into the state A_3 with probability 1/3. However, if the system was in the state A_2, then after transition it may exist, with equal probability, only in the states A_1 and A_3; it can not pass from A_2 into A_2. The last row of the matrix shows us that the system can pass from the state A_3 into any of the three possible states with the same probability 1/3.

Example 4. Let us write the transition matrix for the case of the random walk of a particle between two reflecting barriers. Such a random walk is described in the first example. If we denote by A_1 the event consisting in the particle being at the point with coordinate a, by A_2 its being at the point with coordinate $a+1$, ..., and by A_s ($s = b-a+1$) its being at the point with coordinate b, then the transition matrix will be as follows:

$$\begin{pmatrix} 0 & 1 & 0 & 0 \ldots 0 & 0 & 0 \\ q & 0 & p & 0 \ldots 0 & 0 & 0 \\ 0 & q & 0 & p \ldots 0 & 0 & 0 \\ \multicolumn{7}{c}{\dotfill} \\ 0 & 0 & 0 & 0 \ldots 0 & 1 & 0 \end{pmatrix}.$$

Example 5. Let us also write the transition matrix for the random walk of a particle between two absorbing barriers. The descriptions of the events as well as the other conditions of the preceding problem remain the same. The difference will merely lie in the fact that a particle that reaches the state A_1 or A_s will remain in that state with probability one. The transition matrix for this problem is the following:

$$\begin{pmatrix} 1 & 0 & 0 & 0 & \ldots & 0 \\ q & 0 & p & 0 & \ldots & 0 \\ 0 & q & 0 & p & \ldots & 0 \\ \multicolumn{6}{c}{\dotfill} \\ 0 & 0 & 0 & 0 & \ldots & 1 \end{pmatrix}.$$

Let us note the conditions that must be satisfied by the elements of a transition matrix. First of all, as probabilities, they should be nonnegative quantities, i.e., for all i and j,

$$0 \leq p_{ij} \leq 1.$$

Furthermore, since in leaving the state $A_i^{(s)}$ in the s-th trial, the system must pass into only one of the states $A_j^{(s+1)}$ in the $(s+1)$-st trial, it follows that

$$\sum_{j=1}^{k} p_{ij} = 1 \qquad (i = 1, 2, \ldots, k).$$

Thus, the sum of the elements of any row of the transition matrix is one.

Our first problem in the theory of Markov chains is the determination of the probability of transition from the state $A_i^{(s)}$ in the s-th trial into the state $A_j^{(s+n)}$ after n further steps. Let us denote this probability by the symbol $P_{ij}(n)$.

Consider any intermediate trial, numbered $s+m$. In this trial, one of the possible outcomes $A_r^{(s+m)}$ ($1 \leq r \leq k$) will occur. In terms of the notation just introduced, the probability of transition into this state is $P_{ir}(m)$. The probability of the transition from the state $A_r^{(s+m)}$ into the state $A_j^{(s+n)}$ is $P_{rj}(n-m)$. By the formula on total probability, we have:

$$P_{ij}(n) = \sum_{r=1}^{k} P_{ir}(m) \cdot P_{rj}(n-m). \tag{1}$$

Let π_n denote the matrix of transition probabilities after n steps:

$$\pi_n = \begin{pmatrix} P_{11}(n) & P_{12}(n) & \ldots & P_{1k}(n) \\ \multicolumn{4}{c}{\dotfill} \\ P_{k1}(n) & P_{k2}(n) & \ldots & P_{kk}(n) \end{pmatrix}.$$

§ 17. Definition of Markov Chain. Transition Matrix

From (1), it follows that the matrices π_n with different subscripts satisfy the relation

$$\pi_n = \pi_m \cdot \pi_{n-m} \qquad (0 < m < n).$$

In particular, for $n = 2$ we find:

$$\pi_2 = \pi_1 \cdot \pi_1 = \pi_1^2;$$

for $n = 3$,

$$\pi_3 = \pi_1 \cdot \pi_2 = \pi_2 \cdot \pi_1 = \pi_1^3;$$

and for any n whatever, we have

$$\pi_n = \pi_1^n.$$

We note a special case of formula (1): for $m = 1$

$$P_{ij}(n) = \sum_{r=1}^{k} p_{ir} P_{rj}(n-1).$$

Example 6. A simple computation shows that the two-step transition matrices for Examples 4 and 5 of this section are given by

$$\begin{pmatrix} q & 0 & p & 0 & 0 & \cdots & 0 & 0 & 0 \\ 0 & q+pq & 0 & p^2 & 0 & \cdots & 0 & 0 & 0 \\ q^2 & 0 & 2pq & 0 & p^2 & \cdots & 0 & 0 & 0 \\ \multicolumn{9}{c}{\dotfill} \\ 0 & 0 & 0 & 0 & 0 & \cdots & q & 0 & p \end{pmatrix}$$

for the random walk of a particle between reflecting barriers and by

$$\begin{pmatrix} 1 & 0 & 0 & 0 & 0 & \cdots & 0 \\ q & pq & 0 & p^2 & 0 & \cdots & 0 \\ q^2 & 0 & 2pq & 0 & p^2 & \cdots & 0 \\ \multicolumn{7}{c}{\dotfill} \\ 0 & 0 & 0 & 0 & 0 & \cdots & 1 \end{pmatrix}$$

for the random walk of a particle between absorbing barriers. In both cases $s \geq 5$.

In the case of reflecting barriers, it is intuitively clear that after a large number of steps the particle has a chance of appearing at any point between the reflecting barriers. However, in the case of absorbing barriers, the larger the number of steps the system has gone through, the larger is the probability that the particle will have been absorbed at one of the barriers.

§ 18. Classification of Possible States

The classification of states which is proposed here was almost simultaneously described by Kolmogorov for Markov chains with a denumerable number of states and by Doeblin for Markov chains with a finite number of states.

A state A_i is called *unessential*[1] if there exists a state A_j and a positive integer n such that $P_{ij}(n) > 0$ but $P_{ji}(m) = 0$ for all positive integers m. Thus, an unessential state has the property that it is possible with positive probability to get from it to a certain other state but impossible ever to get back again from this other state to the original unessential state. Consider the fifth example of the preceding section, namely, the random walk of a particle between absorbing barriers. It is easy to see that all of the states in this example with the exception of A_1 and A_s are unessential. In fact, no matter what state (distinct from A_1 and A_s) the particle may be in, both A_1 and A_s can be reached with positive probability after a finite number of steps, but once these states are reached there can be no return to any of the other states.

All states which are not unessential are called *essential*. From the definition, it follows that if A_i and A_j are essential states such that $P_{ij}(m) > 0$ for some positive integer m, then there exists a positive integer n such that the inequality $P_{ji}(n) > 0$ is also satisfied. If A_i and A_j are such that these two inequalities are satisfied for some m and n, then they are said to *communicate*. It is clear that if A_i communicates with A_j and A_j communicates with A_k, then A_i also communicates with A_k. Thus, all essential states can be decomposed into equivalence classes such that all the states belonging to a single class communicate but any belonging to distinct classes do not. All of the states in Examples 3 and 4 of the preceding section are essential and, in each instance, form a single class of states.

Since the equation $P_{ij}(m) = 0$ holds for arbitrary m if A_i is an essential state and A_j is an unessential one, we can conclude the following: If a system has reached one of the states of a certain class of essential states, then it can never leave this class. In Example 5 of § 17, there are two classes of essential states each consisting of a single element; one class is the state A_1, the other the state A_s.

---

[1] The term *transient* is also frequently used here. For other terminology, see the chapter on Markov chains in, for example, Feller's book, *An Introduction to Probability Theory and Its Applications*, 2nd ed. (New York, 1957). [*Trans.*]

§ 18. Classification of Possible States

Let us now consider more carefully the mechanism of transition from state to state within a single class. To do this, we take some essential state A_i and we denote by M_i the set of all positive integers m for which $P_{ii}(m) > 0$. This set cannot be empty, in view of the definition of essential state. It is immediately evident that if m and n belong to the set M_i then their sum $m + n$ also belongs to this set. Let d_i denote the greatest common divisor of all integers in the set M_i. Clearly, M_i merely consists of multiples of d_i.[2] The number d_i is called the *period of the state A_i*.

Let A_i and A_j be two states belonging to the same class. From the preceding, it follows that there exist integers m and n such that $P_{ij}(m) > 0$ and $P_{ji}(n) > 0$. Clearly, the number $m + n$ belongs to M_i and is therefore divisible by d_i. Let r be an arbitrary, sufficiently large positive integer; then rd_j belongs to M_j, so that $P_{jj}(rd_j) > 0$. But since

$$P_{ii}(m + rd_j + n) \geq P_{ij}(m) \cdot P_{jj}(rd_j) \cdot P_{ji}(n),$$

all numbers of the form $m + rd_j + n$ belong to the set M_i for sufficiently large values of r. Since, from the above, the number $m + n$ is divisible by d_i, it follows that rd_j must be divisible by d_i. But inasmuch as r is arbitrary, d_j must be divisible by d_i. By a similar argument, it can shown that d_i is divisible by d_j. It follows from this that $d_i = d_j$.

Thus, *all states of the same class of essential states have the same period;* we shall denote this period by d.

This result allows us to conclude the following: The inequalities $P_{ij}(m) > 0$ and $P_{ji}(n) > 0$ can hold for two states A_i and A_j belonging to the same class only if m and $-n$ are congruent modulo d.[3] Thus, if we select a specific state A_a from the class in question, then we can set up a correspondence between each state A_i of this class and a certain number $\beta(i)$ ($\beta(i) = 1, 2, \ldots, d$) such that the inequality $P_{ai}(n) > 0$ is possible only for those values of n which satisfy the congruence $n \equiv \beta(i) \pmod{d}$. We now collect all the states A_i which correspond to the same number β, thus forming a subclass S_β. As a result, the class of essential states is decomposed into d subclasses S_β. These subclasses have the property that with each transition of the system it is only possible to pass from a state belonging to the subclass S_β into one of the states of the subclass $S_{\beta+1}$. If $\beta = d$, the system passes into one of the states of the subclass S_1.

---

[2] It is easily shown that M_i contains all sufficiently large integral multiples of d_i.
[3] In other words, if the sum $m + n$ is divisible by d.

Let A_i belong to the subclass S_β and A_j to the subclass S_γ. From the preceding, it is clear that the probability $P_{ij}(n)$ can be distinct from zero only if $n \equiv \gamma - \beta \pmod{d}$. However, if n satisfies this congruence and is sufficiently large, then the inequality $P_{ij}(n) > 0$ indeed holds.

For illustrative purposes, we consider Example 4 of the preceding section. We have seen that all of the states of the system form a single class. Since for any i the state A_i can with positive probability be reached again in two steps (and no less than two steps), it follows that $d = 2$. Thus, all of the states of the system can be partitioned into two subclasses, S_1 and S_2. Let us assign all states with odd subscripts to the subclass S_1 and all states with even subscripts to the subclass S_2. Clearly, it is only possible to go, in one step, either from a state in the subclass S_1 to a state in the subclass S_2 or from a state in the subclass S_2 to one in the subclass S_1.

§ 19. A Theorem on Limiting Probabilities

THEOREM. *If for some value of $s > 0$ all elements of the transition matrix π_s are positive, then there exist constants p_j ($j = 1, 2, \ldots, k$) such that the limit relations*
$$\lim_{n \to \infty} P_{ij}(n) = p_j$$
hold independently of the subscript i.

Proof. The gist of the proof of this theorem is very simple: we first establish that as n increases the largest of the probabilities $P_{ij}(n)$ cannot increase and the smallest cannot decrease; we then show that the maximum of the difference $|P_{ij}(n) - P_{lj}(n)|$ ($i, l = 1, 2, \ldots, k$) approaches zero as $n \to \infty$. With this, the proof of the theorem is clearly complete. For by virtue of the limit theorem for bounded monotonic sequences, we conclude from the first two properties mentioned above of the probabilities $P_{ij}(n)$ that the following limits exist:

$$\lim_{n \to \infty} \min_{1 \leq i \leq k} P_{ij}(n) = \bar{p}_j$$

and

$$\lim_{n \to \infty} \max_{1 \leq i \leq k} P_{ij}(n) = \bar{\bar{p}}_j.$$

And since, in view of the third property mentioned, we have

$$\lim_{n \to \infty} \max_{1 \leq i, l \leq k} |P_{ij}(n) - P_{lj}(n)| = 0,$$

it follows that

§ 19. A Theorem on Limiting Probabilities

$$\bar{p}_j = \bar{\bar{p}}_j = p_j.$$

We now proceed to carry out the above plan. We observe, first of all, that for $n > 1$ the inequality

$$P_{ij}(n) = \sum_{l=1}^{k} p_{il} \, P_{lj}(n-1) \geq \min_{1 \leq l \leq k} P_{lj}(n-1) \sum_{l=1}^{k} p_{il} = \min_{1 \leq l \leq k} P_{lj}(n-1)$$

holds. This inequality is valid for every i and, in particular, for the value for which

$$P_{ij}(n) = \min_{1 \leq l \leq k} P_{lj}(n).$$

Thus, we have

$$\min_{1 \leq i \leq k} P_{ij}(n) \geq \min_{1 \leq i \leq k} P_{ij}(n-1).$$

In a similar way, it is easy to show that

$$\max_{1 \leq i \leq k} P_{ij}(n) \leq \max_{1 \leq i \leq k} P_{ij}(n-1).$$

We may assume that $n > s$ and therefore, by formula (1) of § 17, it is permissible to write:

$$P_{ij}(n) = \sum_{r=1}^{k} P_{ir}(s) \cdot P_{rj}(n-s).$$

Consider the difference

$$P_{ij}(n) - P_{lj}(n) = \sum_{r=1}^{k} P_{ir}(s) \, P_{rj}(n-s) - \sum_{r=1}^{k} P_{lr}(s) \cdot P_{rj}(n-s) \quad (1)$$

$$= \sum_{r=1}^{k} [P_{ir}(s) - P_{lr}(s)] \, P_{rj}(n-s).$$

Let us denote those differences $P_{ir}(s) - P_{lr}(s)$ that are positive by the symbol $\beta_{il}^{(r)}$ and those that are non-positive by $-\beta_{il}'^{(r)}$. Since

$$\sum_{r=1}^{k} P_{ir}(s) = \sum_{r=1}^{k} P_{lr}(s) = 1,$$

it follows that

$$\sum_{r=1}^{k}[P_{ir}(s)-P_{lr}(s)]=\sum_{(r)}\beta_{il}^{(r)}-\sum_{(r)}\beta_{il}^{\prime(r)}=0. \tag{2}$$

From this equality, we conclude that

$$h_{il}=\sum_{(r)}\beta_{il}^{(r)}=\sum_{(r)}\beta_{il}^{\prime(r)}.$$

Since, by assumption, $P_{ir}(s) > 0$ for all values of i and r ($i, r = 1, 2, 3, \ldots, k$), we have

$$\sum_{(r)}\beta_{il}^{(r)}<\sum_{r=1}^{k}P_{ir}(s)=1.$$

Thus
$$0 \leq h_{il} < 1.$$

Let
$$h = \max_{1 \leq i,\, l \leq k} h_{il}.$$

Since the number of possible outcomes is finite, the quantity h along with h_{il} satisfies the inequality

$$0 \leq h < 1. \tag{3}$$

From (1), we find that for any i and l ($i, l = 1, 2, \ldots, k$)

$$|P_{ij}(n)-P_{lj}(n)| = \left|\sum_{(r)}\beta_{il}^{(r)} P_{rj}(n-s) - \sum_{(r)}\beta_{il}^{\prime(r)} P_{rj}(n-s)\right|$$

$$\leq \left|\max_{1 \leq r \leq k} P_{rj}(n-s) \sum_{(r)}\beta_{il}^{(r)} - \min_{1 \leq r \leq k} P_{rj}(n-s) \sum_{(r)}\beta_{il}^{\prime(r)}\right|$$

$$\leq h \left|\max_{1 \leq r \leq k} P_{rj}(n-s) - \min_{1 \leq r \leq k} P_{rj}(n-s)\right|$$

$$= h \max_{1 \leq i,\, l \leq k} |P_{ij}(n-s) - P_{lj}(n-s)|$$

and, therefore, also that

$$\max_{1 \leq i,\, l \leq k} |P_{ij}(n)-P_{lj}(n)| \leq h \max_{1 \leq i,\, l \leq k} |P_{ij}(n-s) - P_{lj}(n-s)|.$$

By applying this inequality $\left[\dfrac{n}{s}\right]$ times, we find:

§ 19. A Theorem on Limiting Probabilities

$$\max_{1\leq i,\, l\leq k} |P_{ij}(n) - P_{lj}(n)| \leq$$
$$\leq h^{\left[\frac{n}{s}\right]} \max_{1\leq i,\, l\leq k} \left|P_{ij}\left(n - \left[\frac{n}{s}\right]s\right) - P_{lj}\left(n - \left[\frac{n}{s}\right]s\right)\right|.$$

Since we always have

$$|P_{ij}(m) - P_{lj}(m)| \leq 1$$

it follows that

$$\max_{1\leq i,\, l\leq k} |P_{ij}(n) - P_{lj}(n)| \leq h^{\left[\frac{n}{s}\right]}.$$

As $n \to \infty$, $\left[\frac{n}{s}\right] \to \infty$ also, and therefore, by (3), it follows from this that

$$\lim_{n\to\infty} \max_{1\leq i,\, l\leq k} |P_{ij}(n) - P_{lj}(n)| = 0.$$

The theorem proved also implies that

$$\sum_{j=1}^{k} p_j = 1.$$

In fact,

$$\sum_{j=1}^{k} p_j = \lim_{n\to\infty} \sum_{j=1}^{k} P_{ij}(n) = \lim_{n\to\infty} 1 = 1.$$

Thus, one may interpret the probability p_j as the probability of occurrence of the outcome $A_j^{(n)}$ in the n-th trial when n is large.

The physical significance of this theorem is clear: The probability that the system is in the state A_j is practically independent of what state it was in in the remote past.

The above theorem was first proved by the creator of the concept of chain dependence, Markov; of the so-called ergodic theorems that play such an important role in modern physics, it was the first to be proved rigorously.

It can be shown that the ergodic theorem holds if the possible states constitute a single class of essential states.

§ 20. Generalization of the DeMoivre-Laplace Theorem to a Sequence of Chain-Dependent Trials

We now focus our attention on the consideration of a sequence of trials in each of which the event E may or may not occur. In so doing, we shall assume that the trials are dependent and connected in a simple Markov chain. Thus, if the event E has occurred in the k-th trial, the probability of the event E occurring again in the $(k+1)$-st trial is α; the probability of the event E occurring in the $(k+1)$-st trial, given that the event $\bar{E}$ has occurred in the k-th trial, is β. Thus, the transition probabilities in this case are given by the matrix

$$\begin{pmatrix} \alpha & 1-\alpha \\ \beta & 1-\beta \end{pmatrix}.$$

In the following, we shall assume that both α and β are distinct from 0 and 1, since these cases are of no real interest. Clearly, the scheme in question is the natural generalization of the scheme of independent trials, first considered by Bernoulli, which we investigated in the preceding chapter.

We should note that the assignment of the transition matrix does not give us a complete characterization of the trials, since there is obviously no trial preceding the first, and consequently the probabilities of occurrence of the events E and $\bar{E}$ in the first trial are unknown. Therefore, let p_1 denote the probability of occurrence of the event E in the first trial and $q_1 = 1 - p_1$ the probability of occurrence of the event $\bar{E}$ in the first trial.

We shall first solve the following two problems: 1) Find the probability that the event E will occur in the k-th trial; 2) Find the probability of occurrence of the event E in the k-th trial if the event E has occurred in the i-th trial $(i < k)$.

Let p_k denote the probability of occurrence of the event E in the k-th trial, and set $q_k = 1 - p_k$. Clearly, the event E can occur in the k-th trial in two mutually exclusive ways: the event E occurs in the $(k-1)$-st trial and then it occurs again in the next trial, or the event E occurs in the $(k-1)$-st trial and the event $\bar{E}$ occurs in the next one. By the formula on total probability, we find that

$$p_k = p_{k-1}\alpha + q_{k-1}\beta.$$

§ 20. Generalization of DeMoivre-Laplace Theorem

Since $q_{k-1} = 1 - p_{k-1}$, we obtain, setting $\delta = a - \beta$,

$$p_k = p_{k-1}\delta + \beta.$$

In particular, for $k = 2$,

$$p_2 = p_1\delta + \beta.$$

For $k = 3$,

$$p_3 = p_2\delta + \beta = p_1\delta^2 + \beta(1 + \delta).$$

We can easily show that for $k > 1$

$$p_k = p_1\delta^{k-1} + \beta(1 + \delta + \ldots + \delta^{k-2}) =$$

$$= \left(p_1 - \frac{\beta}{1-\delta}\right)\delta^{k-1} + \frac{\beta}{1-\delta}. \tag{1}$$

Because of our assumptions concerning a and β, the quantity δ satisfies the inequality $|\delta| < 1$. From the preceding expression, it thus follows that

$$p_k \to \frac{\beta}{1-\delta}$$

as $k \to \infty$.

It is interesting to note that the limiting value of p_k is independent of the probability p_1.

Since the quantity $\beta/(1-\delta)$ plays the role of a "limiting probability," it is natural to introduce the notation

$$p = \frac{\beta}{1-\delta} = \frac{\beta}{1-a+\beta}, \qquad q = 1 - p = \frac{1-a}{1-\delta}.$$

In this notation

$$p_k = p + (p_1 - p)\delta^{k-1}. \tag{1'}$$

Let us now denote by $p_j^{(i)}$ the probability of the event E in the j-th trial, given that it has occurred in the i-th trial. Then by proceeding along the same lines as above, we may satisfy ourselves that the probability $p_j^{(i)}$ satisfies the difference equation

$$p_j^{(i)} = p_{j-1}^{(i)}\delta + \beta$$

for all $j > i + 1$. But $p_{i+1}^{(i)} = a$, and therefore, by the procedure already used, we find that

$$p_j^{(i)} = a\delta^{j-i-1} + \beta(1+\delta+\ldots+\delta^{j-i-2}) =$$

$$= \frac{\beta}{1-\delta} + \frac{1-a}{1-\delta}\delta^{j-i} \tag{2}$$

or

$$p_j^{(i)} = p + q\delta^{j-i}. \tag{2'}$$

Let us now proceed to find the m-fold occurrence of the event E in n trials. With this aim, we break up the required probability, which as before will be denoted by $P_n(m)$, into four terms

$$P_n(m) = P_n(m, EE) + P_n(m, E\bar{E}) + P_n(m, \bar{E}E) + P_n(m, \bar{E}\bar{E}).$$

The first term stands for the probability of the m-fold occurrence of event E in n trials under the condition that the event E occurs in the first and last trials. The meaning of the remaining symbols is now evident without further explanation. To compute $P_n(m, EE)$, we first consider the following combination of outcomes of the trials:

In the first r_1 trials, the event E has occurred;
" " next s_1 " " " $\bar{E}$ " " ;
" " " r_2 " " " E " " ;
.... ;
" " " s_{k-1} " " " $\bar{E}$ " " ;
" " " r_k " " " E " " .

As is easily seen, the probability of such an outcome is

$$p_1 a^{r_1-1}(1-a)(1-\beta)^{s_1-1}\beta\ldots\beta a^{r_k-1} =$$

$$= p_1 a^{r_1+\cdots+r_k-k}(1-a)^{k-1}(1-\beta)^{s_1+\cdots+s_{k-1}-k+1}\beta^{k-1}.$$

But since

$$\sum_{i=1}^{k} r_i = m, \quad \sum_{i=1}^{k-1} s_i = n-m,$$

it follows that this probability is equal to

$$p_1 a^{m-k}(1-a)^{k-1}(1-\beta)^{n-m-k+1}\beta^{k-1}.$$

§ 20. Generalization of DeMoivre-Laplace Theorem

We see that it merely depends on n, m, and k and not on the values of r_j and s_j. Since the number m can be expressed as a sum of k positive integers in $\binom{m-1}{k-1}$ ways and the number $n-m$ as a sum of $k-1$ positive integers in $\binom{n-m-1}{k-2}$ ways, the probability of the m-fold occurrence of event E in which the event E appears in k groups and the event $\bar{E}$ in $k-1$ groups is

$$\binom{m-1}{k-1}\binom{n-m-1}{k-2} p_1 a^{m-k}(1-a)^{k-1}(1-\beta)^{n-m-k+1}\beta^{k-1}.$$

Since k may take on any value from 2 up to m, we find, for $P_n(m, EE)$:

$$p_1 \sum_{k=2}^{m} \binom{m-1}{k-1} a^{m-k}(1-a)^{k-1} \binom{n-m-1}{k-2}(1-\beta)^{n-m-k+1}\beta^{k-1}.$$

Similarly, we find, for the probabilities $P_n(m, E\bar{E})$, $P_n(m, \bar{E}E)$, and $P_n(m, \bar{E}\bar{E})$, respectively:

$$p_1 \sum_{k=1}^{m} \binom{m-1}{k-1} a^{m-k}(1-a)^{k} \binom{n-m-1}{k-1}(1-\beta)^{n-m-k}\beta^{k-1},$$

$$q_1 \sum_{k=1}^{m} \binom{m-1}{k-1} a^{m-k}(1-a)^{k-1} \binom{n-m-1}{k-1}(1-\beta)^{n-m-k}\beta^{k},$$

$$q_1 \sum_{k=2}^{m} \binom{m-1}{k-1} a^{m-k+1}(1-a)^{k-1} \binom{n-m-1}{k-1}(1-\beta)^{n-m-k}\beta^{k-1}.$$

To estimate each of these four probabilities, we consider the expression

$$A_n(m) = \sum_{k=1}^{m} \binom{m}{k} a^{m-k}(1-a)^{k} \binom{n-m}{k}(1-\beta)^{n-m-k}\beta^{k}$$

and introduce quantities z, u, and v given by

$$m = np + z\sqrt{\frac{ma(1-a)+(n-m)\beta(1-\beta)}{(1-a+\beta)^2}}, \qquad (3)$$

$$k = m(1-a) + u\sqrt{ma(1-a)},$$

and

$$k = \beta(n-m) + v\sqrt{(n-m)\beta(1-\beta)}.$$

In carrying out our computation, we shall assume that

$$u = o(m^{1/6}), \qquad z = o(n^{\gamma}),$$

where γ is some number in the range $0 < \gamma < 1/6$.

We split up the quantity $A_n(m)$ into the sum of three terms

$$A_n(m) = \Sigma_1 + \Sigma_2 + \Sigma_3,$$

where, in the first sum, k goes from

$$1 \quad \text{to} \quad m(1-a) - u_1\sqrt{ma(1-a)}$$

and in the second sum from

$$m(1-a) - u_1\sqrt{ma(1-a)} \quad \text{to} \quad m(1-a) + u_1\sqrt{ma(1-a)}$$

and in the third sum from

$$m(1-a) + u_1\sqrt{ma(1-a)} \quad \text{to} \quad m.$$

Let us begin our computation with the middle sum.

By repeating verbatim the reasoning that was used in the proof of the Local DeMoivre-Laplace Theorem, we find that

$$\binom{m}{k} a^{m-k}(1-a)^k = \frac{1}{\sqrt{2\pi ma(1-a)}} \exp(-u^2/2) \cdot (1+\omega_n'),$$

$$\binom{n-m}{k} \beta^k(1-\beta)^{n-m-k} = \frac{1}{\sqrt{2\pi(n-m)\beta(1-\beta)}} \exp(-v^2/2) \cdot (1+\omega_n'').$$

The quantities ω_n' and ω_n'' approach zero uniformly within chosen bounds.

Thus,

$$\Sigma_2 = \frac{1}{2\pi\sqrt{m(n-m)a\beta(1-\beta)(1-a)}} \cdot \sum_{u=-u_1}^{u_1} \exp(-[u^2+v^2]/2) \cdot (1+\omega_n')(1+\omega_n'').$$

By the DeMoivre-Laplace Integral Theorem, we have

$$\Sigma_2 = \frac{1}{2\pi\sqrt{(n-m)\beta(1-\beta)}} \int_{-u_1}^{u_1} \exp(-[u^2+v^2]/2) \, du \cdot (1+\omega_n).$$

§ 20. Generalization of DeMoivre-Laplace Theorem

Since u_1 tends to ∞ as $n \to \infty$, we may write

$$\Sigma_2 = \frac{1}{2\pi\sqrt{(n-m)\beta(1-\beta)}} \int \exp(-[u^2+v^2]/2) \, du \cdot (1+\omega_n).$$

But u and v are connected by the equation

$$m(1-a) + u\sqrt{ma(1-a)} = \beta(n-m) + v\sqrt{(n-m)\beta(1-\beta)}.$$

We now substitute in this the value of m from (3). After obvious simplifications, we find that

$$z\sqrt{ma(1-a) + (n-m)\beta(1-\beta)}$$
$$+ u\sqrt{ma(1-a)} = v\sqrt{(n-m)\beta(1-\beta)}.$$

Hence, v is equal to

$$\frac{1}{\sqrt{(n-m)\beta(1-\beta)}} [z\sqrt{ma(1-a) + (n-m)\beta(1-\beta)} + u\sqrt{ma(1-a)}].$$

Thus

$$u^2 + v^2 = z^2 + \frac{ma(1-a) + (n-m)\beta(1-\beta)}{(n-m)\beta(1-\beta)}$$

$$\left[u + z\sqrt{\frac{ma(1-a)}{ma(1-a) + (n-m)\beta(1-\beta)}}\right]^2$$

and therefore

$$\Sigma_2 = \frac{1}{\sqrt{2\pi[ma(1-a) + (n-m)\beta(1-\beta)]}} \exp(-z^2/2) \cdot (1+\bar{\omega}_n).$$

Using (3), we now note that

$$ma(1-a) + (n-m)\beta(1-\beta) = npa(1-a) + nq\beta(1-\beta) + O(z\sqrt{n})$$
$$= npq(1+a-\beta)(1-a+\beta) + O(z\sqrt{n}).$$

Thus, asymptotically,

$$ma(1-a) + (n-m)\beta(1-\beta) = npq(1+a-\beta)(1-a+\beta),$$

and

$$\Sigma_2 = \frac{1}{\sqrt{2\pi npq(1+\alpha-\beta)(1-\alpha+\beta)}} \exp(-z^2/2) \cdot (1+\bar{\omega}_n').$$

To estimate the sum Σ_1, we introduce the quantity

$$c_i = \binom{m}{i} \alpha^{m-i}(1-\alpha)^i \binom{n-m}{i}(1-\beta)^{n-m-i}\beta^i.$$

We note that the ratio

$$\frac{c_i}{c_{i+1}} = \frac{(i+1)^2}{(m-i)(n-m-i)} \cdot \frac{(1-\beta)\alpha}{\beta(1-\alpha)}$$

increases with increasing i and remains less than one for values of i that are not too large. Let

$$j = m(1-\alpha) - c_1\sqrt{m\alpha(1-\alpha)},$$

$$d_j = c_j, \qquad d_{j-1} = c_{j-1}, \qquad d_{j-1}/d_j = \kappa,$$

and for the remaining values of i

$$d_i = d_j \kappa^{j-i}.$$

Clearly,

$$\Sigma_1 < d_1 + d_2 + \ldots + d_j < d_j \frac{1}{1-\kappa}.$$

Since, by the computation carried out before,

$$d_j = \frac{1}{2\pi\sqrt{m(n-m)\beta(1-\beta)\alpha(1-\alpha)}} \cdot$$
$$\cdot \exp(-[u_1^2+v_1^2]/2) \cdot (1+\omega_n')(1+\omega_n''),$$

and since for sufficiently large values of n,

$$\kappa < \frac{1}{2},$$

it follows that

$$\Sigma_1 = o(1).$$

§ 20. Generalization of DeMoivre-Laplace Theorem

In an analogous way, we can show that

$$\Sigma_3 = o(1).$$

As a result, we find that the main part of $A_n(m)$ is Σ_2.

Comparing $A_n(m)$ with the required probabilities and setting

$$Q = \frac{1}{\sqrt{2\pi npq(1+\alpha-\beta)(1-\alpha+\beta)}},$$

we obtain,

$$P_n(m, EE) = p_1\beta Q\, e^{-z^2/2}(1+\bar{\omega}_n'),$$

$$P_n(m, E\bar{E}) = p_1(1-\alpha)Qe^{-z^2/2}(1+\bar{\omega}_n'),$$

$$P_n(m, \bar{E}E) = q_1\beta Q\, e^{-z^2/2}(1+\bar{\omega}_n'),$$

$$P_n(m, \bar{E}\bar{E}) = q_1(1-\alpha)Qe^{-z^2/2}(1+\bar{\omega}_n').$$

Hence, we conclude that

$$P_n(m) = \frac{1}{\sqrt{2\pi npq(1+\alpha-\beta)/(1-\alpha+\beta)}}\, e^{-z^2/2}(1+\bar{\omega}_n').$$

Thus, the local theorem is proved.

We note that if the transition probabilities satisfy the equation

$$\alpha = \beta,$$

then the local theorem assumes the same form as in the case of independent trials.

Proceeding in the usual way and making use of the local theorem, we may derive an integral limit theorem: for any z_1 and z_2

$$\mathsf{P}\left\{z_1 \leqq \frac{m-np}{\sqrt{npq(1+\alpha-\beta)/(1-\alpha+\beta)}} < z_2\right\} =$$

$$= \frac{1}{\sqrt{2\pi}} \int_{z_1}^{z_2} e^{-z^2/2}\, dz\, (1+\omega_n).$$

The quantity ω_n approaches zero uniformly in z_1 and z_2 as n tends to infinity.

EXERCISES

1. Given the matrix of transition probabilities

$$\pi_1 = \begin{pmatrix} \frac{1}{3} & \frac{1}{4} & \frac{5}{12} \\ \frac{1}{3} & \frac{1}{4} & \frac{5}{12} \\ \frac{1}{3} & \frac{1}{4} & \frac{5}{12} \end{pmatrix}.$$

How many states are there? Find the two-step transition probabilities.

2. An electron may be found in one of a denumerable number of orbits depending on the available energy. The transition from the i-th orbit to the j-th orbit takes place in one second with probability $c_i e^{-a|i-j|}$. Find a) the transition probabilities for two seconds, b) the constant c_i.

3. Given the matrix of transition probabilities

$$\pi_1 = \begin{pmatrix} 0 & \frac{1}{2} & \frac{1}{2} \\ \frac{1}{2} & 0 & \frac{1}{2} \\ \frac{1}{2} & \frac{1}{2} & 0 \end{pmatrix}.$$

Is the Ergodic Theorem of Markov applicable in this case? If so, find the limiting probabilities.

CHAPTER IV

RANDOM VARIABLES AND DISTRIBUTION FUNCTIONS

§ 21. Fundamental Properties of Distribution Functions

One of the fundamental concepts of probability theory is that of a random variable. Before proceeding to give the formal definition of this concept, we shall pause to consider some examples.

The number of cosmic particles that strike a specific portion of the earth's surface in a given interval of time is subject to considerable fluctuations depending on many chance factors.

The number of calls received from subscribers at a telephone exchange in a given interval of time is also a random variable that takes on different values depending on chance factors.

The amount by which the point of impact of a shell deviates from the center of the target is determined by a wide variety of causes of a random nature. Thus, in the theory of firing, one is compelled to regard the phenomenon of shell dispersion about the target as a random process and the deviations from the center of the target as random variables.

The velocity of a gas molecule does not remain constant but changes depending on the collisions it has with other molecules. There are a great many such collisions, even in a short interval of time. A knowledge of the velocity of a molecule at a given instant does not enable us to specify its value with complete definiteness after, say, 0.01 or 0.001 seconds. The variation of the velocity of a molecule is thus of a random character.

The examples discussed show quite clearly that we have to deal with random variables in the most varied fields of science and engineering. The natural as well as important question arises of formulating methods for studying random variables.

Despite the complete heterogeneity of the specific content of the examples given, from the mathematical standpoint they all present the same picture. That is, in each example, we deal with a variable which characterizes each of the phenomena concerned. Each of these variables is capable of taking on various values, under the influence of chance. To specify in advance which value the variable will take on is impossible, since it changes in random fashion from experiment to experiment.

Thus, in order to know a random variable, it is first of all necessary to know which values it can assume. However, a mere listing of the values of a random variable is still an insufficient basis on which to make any important deductions. In fact, if in the third example, we consider the gas for different temperatures, the possible values of the molecular velocities will remain the same, although the states of the gas will be different. Thus, to assign a random variable, it is not only necessary to know what values it can take on but also, how often—i.e., with what probability—it can assume these values.

The variety of random variables is very large. The number of values they take on may be finite, denumerable, or non-denumerable; these values may be distributed discretely or continuously in some interval, or, without filling out an interval, they may be everywhere dense. In order to assign probabilities to the values of random variables so different in nature and yet assign them in a consistent way, we introduce into the theory of probability the concept of *a distribution function of a random variable*.

Let ξ be a random variable and x an arbitrary real number. The probability that ξ will assume a value less than x is called the *distribution function* of the random variable ξ:

$$F(x) = \mathsf{P}\{\xi < x\}.$$

Let us agree that henceforth we shall, as a rule, denote random variables by *Greek* letters and the values which they take on by *lower-case italic* letters.

Let us summarize the above: A *random variable* is a variable quantity whose values depend on chance and for which there exists a distribution function.[1]

Let us consider some examples of distribution functions.

Example 1. Let μ denote the number of occurrences of an event A in a sequence of n independent trials in each of which the probability of its occurrence is the same and equal to p. Depending on chance, μ may

---

[1] A formal definition of a random variable will be given on page 148.

§ 21. Fundamental Properties of Distribution Functions

assume any integral value between 0 and n, inclusive. According to the results of Chapter II, we have:

$$P_n(m) = \mathsf{P}\{\mu = m\} = \binom{n}{m} p^m q^{n-m}.$$

The distribution function for the variable μ is defined as follows:

$$F(x) = \begin{cases} 0 & \text{for } x \leq 0, \\ \sum_{k<x} P_n(k) & \text{for } 0 < x \leq n, \\ 1 & \text{for } x > n. \end{cases}$$

The distribution function is a step function with jumps at the points $x = 0, 1, 2, \ldots, n$; the height of the jump at the point $x = k$ is $P_n(k)$.

The example considered shows that the so-called Bernoulli scheme can be included in the general theory of random variables.

Example 2. Let the random variable ξ take on the values $0, 1, 2, \ldots$ with the probabilities

$$p_n = \mathsf{P}\{\xi = n\} = \frac{\lambda^n e^{-\lambda}}{n!} \quad (n = 0, 1, 2, \ldots),$$

where λ is a positive constant. The distribution function for the variable ξ, which resembles a staircase with an infinite number of steps, has jumps at all the non-negative integral points. The height of the jump at the point $x = n$ is equal to p_n; for $x \leq 0$, we have $F(x) = 0$. The random variable in this example is said to be distributed according to the *Poisson law*.

Example 3. We shall say that a random variable has a *normal*, or *Gaussian*, distribution if its distribution function is given by

$$\Phi(x) = C \int_{-\infty}^{x} e^{-\frac{(z-a)^2}{2\sigma^2}} dz,$$

where C and σ are positive constants and a is an arbitrary constant. Later on, we shall establish a connection between the constants σ and C, and we shall explain what significance the parameters a and σ have in the theory of probability. Normally distributed random variables play an especially important role in probability theory and its applications; in the sequel, we shall have ample occasion to convince ourselves of this.

IV. Random Variables and Distribution Functions

We note that while the random variables in our first two examples could only assume a finite or a denumerable number of values (they are *discrete* variables), a random variable which is distributed according to the normal law may take on the values in any interval. In fact, as we shall see below, the probability that a normally distributed random variable takes on values in the interval $x_1 \leq \xi < x_2$ is

$$\Phi(x_2) - \Phi(x_1) = C \int_{x_1}^{x_2} e^{-\frac{(z-a)^2}{2\sigma^2}} \, dz \, ,$$

and therefore, for any distinct x_1 and x_2, this probability is positive.

With these preliminary remarks of an intuitive nature, we now proceed to a strictly formal presentation of the concept of random variable.

In defining a random variable, we start from a set of elementary events U, as we did in defining the concept of random event. We put into correspondence with each elementary event e a certain number

$$\xi = f(e).$$

We say that *ξ is a random variable* if *the function $f(e)$ is measurable with respect to the probability measure defined on the set U in question*. In other words, for every Borel-measurable set A_ξ of values of ξ, the set A_e of those events such that $f(e) \subset A_\xi$ is required to be in the set of random events F and, therefore, the probability

$$\mathsf{P}\{\xi \subset A_\xi\} = \mathsf{P}\{A_e\}$$

be defined for it. In particular, if the set A_ξ consists of the points on the half-line $\xi < x$, then the probability $\mathsf{P}\{A_e\}$ is a function of x, namely, the function

$$\mathsf{P}\{\xi < x\} = \mathsf{P}\{A_e\} = F(x) \, ,$$

which we have called the distribution function of the random variable ξ.

Example 4. Consider a sequence of n independent trials in each of which the probability of occurrence of an event A is constant and equal to p. In this example, the elementary events consist of sequences of occurrences and non-occurrences of the event A in the n trials. Thus, one of the elementary events will be the occurrence of the event A in every trial. It is easily shown that there are 2^n elementary events in all.

§ 21. Fundamental Properties of Distribution Functions

We now define a function $\mu = f(e)$ of the elementary event e as follows: It is equal to the number of occurrences of the event A in the elementary event e. By the results of Chapter II, we have:

$$P\{\mu = k\} = P_n(k) = \binom{n}{k} p^k q^{n-k}.$$

The measurability of the function $\mu = f(e)$ in the field of probabilities is immediately evident. From this and the definition, we conclude that μ is a random variable.

Example 5. Three observations are made of the position of a molecule lying on a straight line. The set of elementary events consists of the points in three-dimensional euclidean space R_3. The set of random events F consists of all possible Borel sets in the space R_3.

For every random event A, the probability $P\{A\}$ is defined by means of the relation

$$P\{A\} = \frac{1}{(\sigma\sqrt{2\pi})^3} \int\!\!\int\!\!\int_A e^{-\frac{1}{2\sigma^2}[(x_1-a)^2+(x_2-a)^2+(x_3-a)^2]} dx_1\, dx_2\, dx_3.$$

Now consider the function $\xi = f(e)$ of the elementary event $e = (x_1, x_2, x_3)$ defined by means of the equation

$$\xi = \frac{1}{3}(x_1 + x_2 + x_3).$$

This function is measurable with respect to the probability that has been introduced and hence ξ is a random variable. Its distribution function is given by

$$F(x) = P\{\xi < x\} = \frac{1}{(\sigma\sqrt{2\pi})^3} \int\!\!\int\!\!\int_{x_1+x_2+x_3<3x} e^{-\frac{1}{2\sigma^2}\sum_{k=1}^{3}(x_k-a)^2} dx_1\, dx_2\, dx_3$$

$$= \frac{1}{\sigma\sqrt{\frac{2}{3}\pi}} \int_{-\infty}^{x} e^{-\frac{3(z-a)^2}{2\sigma^2}} dz.$$

From the point of view just elaborated, any operation on random variables can be reduced to familiar operations on functions. Thus, if

ξ_1 and ξ_2 are random variables, i.e., if

$$\xi_1 = f_1(e) \text{ and } \xi_2 = f_2(e)$$

are measurable functions with respect to the probability defined, then any Borel function of these variables is also a random variable. For example, the quantity

$$\zeta = \xi_1 + \xi_2$$

is measurable with respect to the probability defined and is therefore a random variable.

In § 24, we shall elaborate on the remark just made, and we shall deduce a number of results important both for the theory and its applications. In particular, a formula will be derived in that section for the distribution function of a sum based on the distribution functions of the summands.

It is possible to determine the probability of the inequality $x_1 \leqq \xi < x_2$ for any values of x_1 and x_2 by means of the distribution function of the random variable ξ. In fact, if we denote by A the event consisting in ξ assuming a value less than x_2, by B the event that $\xi < x_1$, and finally, by C the event that $x_1 \leqq \xi < x_2$, then the following relation obviously holds:

$$A = B + C.$$

Since the events B and C are mutually exclusive, we have

$$\mathsf{P}(A) = \mathsf{P}(B) + \mathsf{P}(C).$$

But

$$\mathsf{P}(A) = F(x_2), \quad \mathsf{P}(B) = F(x_1), \quad \mathsf{P}(C) = \mathsf{P}\{x_1 \leqq \xi < x_2\},$$

and therefore

$$\mathsf{P}\{x_1 \leqq \xi < x_2\} = F(x_2) - F(x_1). \tag{1}$$

Since by definition a probability is a non-negative number, it follows from equation (1) that the inequality

$$F(x_2) \geqq F(x_1)$$

holds for any values of x_1 and x_2 ($x_2 > x_1$), i.e., *the distribution function of every random variable is a non-decreasing function.*

§ 21. Fundamental Properties of Distribution Functions

It is further evident that, for any x, the distribution function $F(x)$ satisfies the inequality

$$0 \leq F(x) \leq 1. \tag{2}$$

We shall say that a distribution function $F(x)$ has a *jump* at the point $x = x_0$ if

$$F(x_0 + 0) - F(x_0 - 0) = C_0 > 0.$$

A distribution function can have at most a denumerable number of jumps. In fact, a distribution function can have at most one jump of magnitude greater than 1/2, at most three jumps of magnitude between 1/4 and 1/2 ($1/4 < C_0 \leq 1/2$), and, in general, at most $2^n - 1$ jumps of magnitude between 2^{-n} and 2^{1-n} (or equal to the latter value). It is perfectly clear that the number of jumps can be enumerated by arranging them according to size, starting with the largest and repeating equal values as many times as the function $F(x)$ has jumps of that size.

We shall now establish a few additional general properties of distribution functions. Let us define $F(-\infty)$ and $F(+\infty)$ by the relations

$$F(-\infty) = \lim_{n \to +\infty} F(-n), \quad F(+\infty) = \lim_{n \to \infty} F(+n);$$

we shall prove that

$$F(-\infty) = 0, \quad F(+\infty) = 1.$$

In fact, since the inequality $\xi < \infty$ is certain, it follows that

$$\mathsf{P}\{\xi < +\infty\} = 1.$$

Let Q_k denote the event $k - 1 \leq \xi < k$. Because the event $\xi < \infty$ is equivalent to the sum of the events Q_k, from the extended addition axiom we have that

$$\mathsf{P}\{\xi < +\infty\} = \sum_{k=-\infty}^{\infty} \mathsf{P}\{Q_k\}.$$

Therefore, as $n \to \infty$

$$\sum_{k=1-n}^{n} \mathsf{P}\{Q_k\} = \sum_{k=1-n}^{n} [F(k) - F(k-1)] = F(n) - F(-n) \to 1.$$

Taking into account inequality (2), we conclude from this that, as $n \to \infty$,

$$F(-n) \to 0, \quad F(+n) \to 1.$$

Distribution functions are continuous from the left.

Let us choose some increasing sequence $x_0 < x_1 < x_2 < \ldots < x_n < \ldots$ which converges to x.

Denote by A_n the event $\{x_n \leq \xi < x\}$. Then it is clear that $A_i \subset A_j$ for $i > j$, and that the product of all the events A_n is the impossible event. By the axiom of continuity, we must have:

$$\lim_{n\to\infty} \mathsf{P}(A_n) = \lim_{n\to\infty} \{F(x) - F(x_n)\} = F(x) - \lim_{n\to\infty} F(x_n)$$
$$= F(x) - F(x-0) = 0,$$

Q.E.D.

In an analogous way, we can show that

$$\mathsf{P}\{\xi \leq x\} = F(x+0).$$

We thus see that *every distribution function is a non-decreasing function which is continuous from the left and satisfies the conditions* $F(-\infty) = 0$ *and* $F(\infty) = 1$. The converse is also true: *Every function satisfying the conditions just stated can be regarded as the distribution function of some random variable.*

We note that whereas each random variable determines its distribution function uniquely, there exist random variables as distinct as we please that have the same distribution function. Thus, if ξ assumes the two values -1 and 1 each with a probability of $1/2$ and if $\eta = -\xi$, then clearly ξ is always different from η. Nevertheless, both of these random variables have the same distribution function

$$F(x) = \begin{cases} 0 & \text{for} \quad x \leq -1, \\ \frac{1}{2} & \text{for} \quad -1 < x \leq 1, \\ 1 & \text{for} \quad x > 1. \end{cases}$$

§ 22. Continuous and Discrete Distributions

Sometimes the behavior of a random variable is characterized not by the assignment of its distribution function but in some other way. Any such characterization is referred to as a *distribution law* of the random variable, provided the distribution function can be obtained from it by some prescribed rule. Thus, the interval function $P\{x_1, x_2\}$, representing the probability of the inequality $x_1 \leq \xi < x_2$, is such a distribution law.

§ 22. CONTINUOUS AND DISCRETE DISTRIBUTIONS

In fact, knowing $P\{x_1, x_2\}$, we can find the distribution function by using the formula

$$F(x) = P\{-\infty, x\}.$$

We already know that for any x_1 and x_2 the function $P\{x_1, x_2\}$ can be found from $F(x)$ by using

$$P\{x_1, x_2\} = F(x_2) - F(x_1).$$

It is often useful to take as a distribution law the set function $P\{E\}$ defined on all Borel sets and representing the probability that the random variable assumes a value belonging to the set E. By the extended addition axiom, the probability $P(E)$ is a completely additive set function, i.e., for any set E which is the sum of a finite or denumerable number of disjoint sets E_k, we have:

$$P(E) = \sum P\{E_k\}.$$

Of the various random variables possible, we first single out those that may only take on either a finite or a denumerable number of values. Such variables will be called *discrete*. For a complete probabilistic characterization of a discrete random variable ξ taking on the values $x_1, x_2, x_3, \ldots$ with positive probability, it suffices to know the probabilities $p_k = \mathsf{P}\{\xi = x_k\}$.[2] It is evident that we can determine the distribution function $F(x)$ in terms of the probabilities p_k by means of the relation

$$F(x) = \sum p_k,$$

in which the summation extends over all values of the subscript for which $x_k < x$.

The distribution function $F(x)$ of an arbitrary discrete variable is discontinuous and increases by jumps at the values of x which are the possible values for ξ. The size of the jump at the point x, as we showed earlier, is equal to the difference $F(x + 0) - F(x)$.

If two of the possible values of the variable ξ are separated by an interval in which no other possible values of ξ appear, then the distribution function $F(x)$ is constant in this interval. If the number of possible values of ξ is finite, say n, then the distribution function $F(x)$ is a step function which is constant in each of $n + 1$ intervals. If, however, the

---

[2] These, and only these, values x_n will be called *possible values* of the discrete random variable ξ.

number of possible values of ξ is denumerable, these possible values may be everywhere dense, so that there may or may not be any interval in which the distribution function of the discrete random variable is constant. For example, let the possible values of ξ be all of the rational numbers. Suppose that these numbers are ordered in some fashion: $r_1, r_2, \ldots$ and that the probabilities $\mathsf{P}\{\xi = r_k\} = p_k$ are defined by means of the relation $p_k = 2^{-k}$. In this example, every rational point is a point of discontinuity of the distribution function.

As another important class of random variables, we single out those random variables for which there exists a non-negative function $p(x)$ satisfying the equation

$$F(x) = \int_{-\infty}^{x} p(z)\, dz$$

for arbitrary values of x. A random variable possessing this property is called *continuous*; the function $p(x)$ is called the *probability density function*.

Note that a probability density function possesses the following properties:

1) $p(x) \geq 0$;
2) For any x_1 and x_2, it satisfies the relation

$$\mathsf{P}\{x_1 \leq \xi < x_2\} = \int_{x_1}^{x_2} p(x)\, dx;$$

in particular, if $p(x)$ is continuous at x, then $\mathsf{P}\{x \leq \xi < x+dx\} = p(x)dx$ up to infinitesimals of higher order;

3) $\int_{-\infty}^{\infty} p(x)dx = 1$.

Examples of continuous random variables are those which are distributed according to the normal or the uniform[3] law.

Example. Let us examine the normal distribution more closely. Its probability density function is given by

$$p(x) = C \cdot e^{-\frac{(x-a)^2}{2\sigma^2}}.$$

---

[3] This law refers to a distribution function which varies linearly from 0 up to 1 in some interval (a, b) and which is zero to the left of the point a and one to the right of b.

§ 22. CONTINUOUS AND DISCRETE DISTRIBUTIONS

The constant C can be determined by starting from Property 3. In fact,

$$C\int e^{-\frac{(x-a)^2}{2\sigma^2}}\,dx = 1.$$

By the change of variables $(x-a)/\sigma = z$, this relation is expressible in the form

$$C\sigma\int e^{-\frac{z^2}{2}}\,dz = 1.$$

The integral appearing in the left-hand member of this equation is known as the *Poisson integral* and has the value

$$\int e^{-\frac{z^2}{2}}\,dz = \sqrt{2\pi}.$$

Thus, we find:

$$C = \frac{1}{\sigma\sqrt{2\pi}}$$

and this implies that

$$p(x) = \frac{1}{\sigma\sqrt{2\pi}} e^{-\frac{(x-a)^2}{2\sigma^2}}$$

for the normal distribution.

The function $p(x)$ attains a maximum value at $x = a$ and has points of inflection at $x = a \pm \sigma$; its graph is asymptotic to the x-axis as $x \to \pm\infty$. To illustrate the effect of the parameter σ on the shape of the graph of the normal density function, we have plotted the graph of $p(x)$ in Fig. 14 for $a = 0$ and σ^2 equal to a) 1/4, b) 1, and c) 4. We see that the smaller the value of σ, the larger the maximum value of $p(x)$ and the steeper the curve. This means, in particular, that for a normally distributed random variable (with parameter $a = 0$) the probability of falling in an interval $(-a, a)$ is greater when the value of σ is smaller. Therefore, we may look upon σ as an index of the dispersion of the values of the variable ξ. For $a \neq 0$, the density curves have the same shape; they are merely translated to the right $(a > 0)$ or to the left $(a < 0)$ depending on the sign of the parameter a.

156   IV. RANDOM VARIABLES AND DISTRIBUTION FUNCTIONS

Of course, there exist still other random variables in addition to those that are discrete or continuous. Besides those that behave like continuous variables in some intervals and like discrete variables in others, there are variables that are neither discrete nor continuous in any interval. To this category belong the random variables, for example, whose distribution functions are continuous but which at the same time only increase at points of a set of Lebesgue measure zero. As an example of such a random variable, we cite the one having the well-known Cantor ternary function as its distribution function. Let us recall how this

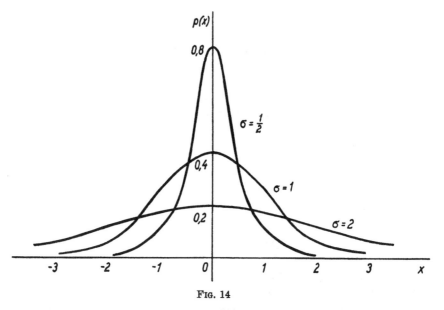

FIG. 14

function is constructed. The variable ξ takes on only values between 0 and 1. Therefore, its distribution function satisfies the conditions

$$F(x) = 0 \text{ for } x \leq 0, \quad F(x) = 1 \text{ for } x > 1.$$

Within the interval $(0, 1)$, ξ assumes values only in the first third and last third of this interval, in each with probability 1/2. Thus,

$$F(x) = \frac{1}{2} \quad \text{for} \quad \frac{1}{3} < x \leq \frac{2}{3}.$$

In the intervals $(0, 1/3)$ and $(2/3, 1)$, ξ can, again, only assume values in the first third and last third of each of these intervals, with probability

1/4 in each sub-interval. This defines the values of $F(x)$ in two more intervals:

$$F(x) = \frac{1}{4} \quad \text{for} \quad \frac{1}{9} < x \leq \frac{2}{9},$$

$$F(x) = \frac{3}{4} \quad \text{for} \quad \frac{7}{9} < x \leq \frac{8}{9}.$$

This construction is repeated in each of the remaining intervals and the process is continued to infinity. The resulting function $F(x)$ turns out to be defined on a denumerable set of intervals, and undefined at the points separating these intervals constituting a nowhere-dense perfect set of measure zero. On this set, we define the values of the function by continuity. The variable ξ having the distribution function thus defined is not discrete, since its distribution function is continuous; but at the same time, ξ is not continuous, because its distribution function is not the integral of its derivative.

All of the definitions that we have introduced are readily carried over to the case of conditional probability. Thus, for example, if the event B is such that $\mathsf{P}\{B\} > 0$, then the function $F(x/B) = \mathsf{P}\{\xi < x/B\}$ will be called the *conditional distribution function* of the random variable ξ under the condition B. It is evident that $F(x/B)$ has all the properties of an ordinary distribution function.

§ 23. Multi-Dimensional Distribution Functions

In the sequel, we shall need in addition to the concept of a random variable the concept of a random vector or, as it is often called, a multi-dimensional random variable.

If a correspondence

$$\xi_1 = f_1(e), \quad \xi_2 = f_2(e), \ldots, \quad \xi_n = f_n(e)$$

is set up between each elementary event e and a point of n-dimensional euclidean space such that each function $f_k(e)$ ($k = 1, 2, \ldots, n$) is measurable with respect to the probability defined on the set of random events F, then the set of numbers $(\xi_1, \xi_2, \ldots, \xi_n)$ is called an *n-dimensional random variable*.

The function
$$F(x_1, x_2, \ldots, x_n) = \mathsf{P}\{\xi_1 < x_1, \xi_2 < x_2, \ldots, \xi_n < x_n\}$$
is called the *n-dimensional distribution function of the random vector* $(\xi_1, \xi_2, \ldots, \xi_n)$.

In the following, we shall resort to geometrical language, and we begin by regarding the quantities $\xi_1, \xi_2, \ldots, \xi_n$ as coordinates of a point in n-dimensional euclidean space. Clearly, the position of a point $(\xi_1, \xi_2, \ldots, \xi_n)$ depends on chance, and the function $F(x_1, x_2, \ldots, x_n)$, under this interpretation, yields the probability that the point $(\xi_1, \xi_2, \ldots, \xi_n)$ fall within the n-dimensional parallelepiped $\xi_1 < x_1, \xi_2 < x_2, \ldots, \xi_n < x_n$ whose edges are parallel to the coordinate axes.

By means of the distribution function, it is easy to compute the probability that the point $(\xi_1, \xi_2, \ldots, \xi_n)$ fall inside the parallelepiped
$$a_i \leq \xi_i < b_i, \qquad (i = 1, 2, \ldots, n)$$
where a_i and b_i are arbitrary constants. It is not difficult to show that

$$\mathsf{P}\{a_1 \leq \xi_1 < b_1,\ a_2 \leq \xi_2 < b_2, \ldots,\ a_n \leq \xi_n < b_n\} \qquad (1)$$
$$= F(b_1, b_2, \ldots, b_n) - \sum_{i=1}^n p_i + \sum_{i<j} p_{ij} \mp \cdots +$$
$$+ (-1)^n F(a_1, a_2, \ldots, a_n),$$

where $p_{ij\ldots k}$ denotes the value of the function $F(c_1, c_2, \ldots, c_n)$ for $c_i = a_i, c_j = a_j, \ldots, c_k = a_k$ and the remaining c_s equal to b_s. In particular, we note that $F(x_1, \ldots, x_{k-1}, +\infty, x_{k+1}, \ldots, x_n)$ gives the probability that the following system of inequalities is satisfied:
$$\xi_1 < x_1,\ \xi_2 < x_2, \ldots, \xi_{k-1} < x_{k-1},\ \xi_{k+1} < x_{k+1}, \ldots, \xi_n < x_n.$$

Since by the extended addition axiom for probabilities,
$$\mathsf{P}\{\xi_1 < x_1, \ldots, \xi_{k-1} < x_{k-1}, \xi_{k+1} < x_{k+1}, \ldots, \xi_n < x_n\}$$
$$= \sum_{s=-\infty}^{\infty} \mathsf{P}\{\xi_1 < x_1, \ldots, \xi_{k-1} < x_{k-1}, s \leq \xi_k < s+1, \xi_{k+1} < x_{k+1}, \ldots, \xi_n < x_n\}$$
$$= F(x_1, \ldots, x_{k-1}, \infty, x_{k+1}, \ldots, x_n)$$

it follows that $F(x_1, \ldots, x_{k-1}, +\infty, x_{k+1}, \ldots, x_n)$ is the distribution function for the $(n-1)$-dimensional random variable
$$(\xi_1, \xi_2, \ldots, \xi_{k-1}, \xi_{k+1}, \ldots, \xi_n).$$

§ 23. MULTI-DIMENSIONAL DISTRIBUTION FUNCTIONS

By then continuing this process, we can determine the k-dimensional distribution function of any group of k quantities $\xi_{i_1}, \xi_{i_2}, \ldots, \xi_{i_k}$ ($i_1 < i_2 < \ldots < i_k$) from the formula:

$$F_k(x_{i_1}, x_{i_2}, \ldots, x_{i_k}) = \mathsf{P}\{\xi_{i_1} < x_{i_1}, \ldots, \xi_{i_k} < x_{i_k}\} = F(c_1, c_2, \ldots, c_n),$$

where $c_s = x_s$ if $s = i_r$ ($1 \leq r \leq k$) and $c_s = \infty$ in all other cases. In particular, the distribution function for the random variable ξ_k is given by

$$F_1(x_k) = F(c_1, c_2, \ldots, c_n),$$

where, for all $i \neq k$, $c_i = \infty$ and $c_k = x$.

The behavior of a multi-dimensional random variable, like that of a one-dimensional random variable, may be characterized by means other than its distribution function, say, by a non-negative completely additive set function $\Phi\{E\}$ defined on arbitrary Borel sets in n-dimensional space. This function is defined as the probability that the point $(\xi_1, \xi_2, \ldots, \xi_n)$ lies in the set E. This method of characterizing an n-dimensional random variable probabilistically must be deemed the most natural one and, from a theoretical point of view, the most successful one.

Let us consider some examples.

Example 1. A random vector $(\xi_1, \xi_2, \ldots, \xi_n)$ is said to be *uniformly distributed* in the parallelepiped $a_i \leq \xi_i < b_i$ ($1 \leq i \leq n$) if the probability of the point $(\xi_1, \xi_2, \ldots, \xi_n)$ lying in an arbitrary region interior to the parallelepiped is proportional to its volume and if the probability of its falling inside the parallelepiped is one.

The distribution function of the required variable is given by

$$F(x_1, \ldots, x_n) = \begin{cases} 0 & \text{if } x_i \leq a_i \text{ for at least one } i, \\ \prod_{i=1}^{n} \frac{c_i - a_i}{b_i - a_i}, & \text{where } c_i = x_i \text{ if } a_i \leq x_i \leq b_i, \\ & \text{and } c_i = b_i \text{ if } x_i > b_i. \end{cases}$$

Example 2. A two-dimensional random variable (ξ_1, ξ_2) is distributed *normally* if its distribution function is given by

$$F(x, y) = C \int_{-\infty}^{x} \int_{-\infty}^{y} e^{-Q(u, v)} \, du \, dv;$$

here, $Q(x, y)$ is a positive-definite quadratic form of $x - a$ and $y - b$, where a and b are constants.

IV. Random Variables and Distribution Functions

It is well known that a positive-definite quadratic form of $x-a$ and $y-b$ can be written in the form

$$Q(x, y) = \frac{(x-a)^2}{2A^2} - r\frac{(x-a)(y-b)}{AB} + \frac{(y-b)^2}{2B^2},$$

where A and B are positive quantities, r, a, and b are real numbers, and r satisfies the condition $|r| \leq 1$.

It is readily apparent that for $r^2 \neq 1$, each of the random variables ξ_1 and ξ_2 is subject to a one-dimensional normal law. In fact,

$$F_1(x_1) = \mathsf{P}\{\xi_1 < x_1\} = F(x_1, +\infty) = C \int\limits_{-\infty}^{x_1} \int e^{-Q(x,y)} \, dx \, dy =$$

$$= C \int\limits_{-\infty}^{x_1} e^{-\frac{(x-a)^2}{2A^2}(1-r^2)} \int e^{-\frac{1}{2}\left[\frac{y-b}{B} - \frac{r(x-a)}{A}\right]^2} dy \, dx.$$

Since

$$\int e^{-\frac{1}{2}\left[\frac{y-b}{B} - r\frac{x-a}{A}\right]^2} dy = B\sqrt{2\pi},$$

it follows that

$$F_1(x_1) = BC\sqrt{2\pi} \int\limits_{-\infty}^{x_1} e^{-\frac{(x-a)^2}{2A^2}(1-r^2)} dx. \qquad (2)$$

The constant C is expressible in terms of A, B, and r. This dependence can be found from the condition $F_1(+\infty) = 1$. We have:

$$1 = BC\sqrt{2\pi} \int e^{-\frac{(x-a)^2}{2A^2}(1-r^2)} dx = \frac{ABC\sqrt{2\pi}}{\sqrt{1-r^2}} \int e^{-\frac{z^2}{2}} dz = \frac{2ABC\pi}{\sqrt{1-r^2}}.$$

Hence,

$$C = \frac{\sqrt{1-r^2}}{2\pi AB}.$$

If $r^2 \neq 1$, then we set

$$A = \sigma_1\sqrt{1-r^2}, \qquad B = \sigma_2\sqrt{1-r^2}.$$

In these new variables the two-dimensional normal law takes the following form:

$$F(x_1, x_2) = \frac{1}{2\pi\sigma_1\sigma_2\sqrt{1-r^2}}$$

$$\int\limits_{-\infty}^{x_1}\int\limits_{-\infty}^{x_2} e^{-\frac{1}{2(1-r^2)}\left[\frac{(x-a)^2}{\sigma_1^2} - 2r\frac{(x-a)(y-b)}{\sigma_1\sigma_2} + \frac{(y-b)^2}{\sigma_2^2}\right]} dx \, dy.$$

§ 23. Multi-Dimensional Distribution Functions

The theoretic-probabilistic meaning of the parameters appearing in this formula will be explained in the following chapter.

If $r^2 = 1$, equation (2) ceases to have meaning. In this case, ξ_1 is a linear function of ξ_2.

We may establish a number of properties of multi-dimensional distribution functions in the same way as for the one-dimensional case. Here, we shall merely formulate them, leaving their justification to the reader. A distribution function

1) is a non-decreasing function of each of its arguments,
2) is continuous from the left in each of its arguments, and
3) satisfies the relations

$$F(+\infty, +\infty, \ldots, +\infty) = 1,$$
$$\lim_{x_k \to -\infty} F(x_1, x_2, \ldots, x_n) = 0 \qquad (1 \leq k \leq n)$$

for arbitrary values of the other arguments.

In the one-dimensional case, we have seen that the above properties are necessary and sufficient conditions for the function $F(x)$ to be the distribution function of some random variable. In the n-dimensional case, these properties no longer turn out to be sufficient. In order for the function $F(x_1, \ldots, x_n)$ to be a distribution function, it is necessary to have the following property, in addition to the three properties above:

4) For any a_i and b_i ($i = 1, 2, \ldots, n$) the expression (1) is non-negative.

That this condition may not be satisfied even if the function $F(x_1, \ldots, x_n)$ obeys conditions 1)-3) is shown by the following example. Let

$$F(x, y) = \begin{cases} 0 \text{ if } x \leq 0 \text{ or } x + y \leq 1 \text{ or } y \leq 0, \\ 1 \text{ in the rest of the plane.} \end{cases}$$

This function satisfies conditions 1)-3) but

$$F(1, 1) - F\left(1, \frac{1}{2}\right) - F\left(\frac{1}{2}, 1\right) + F\left(\frac{1}{2}, \frac{1}{2}\right) = -1, \qquad (3)$$

and therefore the fourth condition is not satisfied.

The function $F(x, y)$ cannot be a distribution function since, according to the relation (1), the expression (3) should be the probability that the point (ξ_1, ξ_2) fall in the square $1/2 \leq \xi_1 < 1$, $1/2 \leq \xi_2 < 1$.

If there exists a function $p(x_1, \ldots, x_n)$ such that the equation

$$F(x_1, x_2, \ldots, x_n) = \int_{-\infty}^{x_1} \int_{-\infty}^{x_2} \cdots \int_{-\infty}^{x_n} p(z_1, z_2, \ldots, z_n)\, dz_n \ldots dz_2\, dz_1$$

holds for any values of $x_1, x_2, \ldots, x_n$, then this function is called the *probability density function* of the random vector $(\xi_1, \xi_2, \ldots, \xi_n)$. It is easy to see that a density function possesses the following properties:

1) $p(x_1, x_2, \ldots, x_n) \geq 0$.
2) The probability that the point $(\xi_1, \xi_2, \ldots, \xi_n)$ will fall in some region R is equal to

$$\int_R \cdots \int p(x_1, \ldots, x_n)\, dx_n \ldots dx_1.$$

In particular, if the function $p(x_1, x_2, \ldots, x_n)$ is continuous at the point $(x_1, \ldots, x_n)$, then the probability that the point $(\xi_1, \xi_2, \ldots, \xi_n)$ will fall in the parallelepiped $x_k \leq \xi_k < x_k + dx_k$ $(k = 1, 2, \ldots, n)$ is, up to infinitesimals of higher order,

$$p(x_1, x_2, \ldots, x_n)\, dx_1\, dx_2 \ldots dx_n.$$

Example 3. As an example of an n-dimensional random variable having a density function, we can take a variable which is uniformly distributed in some n-dimensional region R. If V denotes the n-dimensional volume of the region R, then the density function is given by

$$p(x_1, x_2, \ldots, x_n) = \begin{cases} 0 & \text{if } (x_1, x_2, \ldots, x_n) \notin R, \\ \dfrac{1}{V} & \text{if } (x_1, x_2, \ldots, x_n) \in R. \end{cases}$$

Example 4. The density corresponding to the two-dimensional normal law is given by the expression

$$p(x, y) = \frac{1}{2\pi \sigma_1 \sigma_2 \sqrt{1-r^2}} e^{-\frac{1}{2(1-r^2)}\left[\frac{(x-a)^2}{\sigma_1^2} - 2r\frac{(x-a)(y-b)}{\sigma_1 \sigma_2} + \frac{(y-b)^2}{\sigma_2^2}\right]}$$

We observe that the normal density function has a constant value on the ellipses

§ 23. MULTI-DIMENSIONAL DISTRIBUTION FUNCTIONS

$$\frac{(x-a)^2}{\sigma_1^2} - 2r\frac{(x-a)(y-b)}{\sigma_1\sigma_2} + \frac{(y-b)^2}{\sigma_2^2} = \lambda^2, \tag{4}$$

where λ is a constant; for this reason, the ellipses (4) are referred to as the *ellipses of equal probability*.

Let us find the probability that the point (ξ_1, ξ_2) falls inside the ellipse (4). By the definition of a density function, we have

$$P(\lambda) = \iint_{R(\lambda)} p(x, y)\, dx\, dy, \tag{5}$$

where $R(\lambda)$ denotes the region bounded by the ellipse (4). To evaluate this integral, we introduce polar coordinates

$$x - a = \varrho \cos\theta,$$
$$y - b = \varrho \sin\theta.$$

The resultant integral in (5) takes the form

$$P(\lambda) = \frac{1}{2\pi\sigma_1\sigma_2\sqrt{1-r^2}} \int_0^{2\pi} \int_0^{\lambda/s\sqrt{1-r^2}} e^{-\frac{\varrho^2 s^2}{2}} \varrho\, d\varrho\, d\theta,$$

where for brevity we have set

$$s^2 = \frac{1}{1-r^2}\left[\frac{\cos^2\theta}{\sigma_1^2} - 2r\frac{\cos\theta\sin\theta}{\sigma_1\sigma_2} + \frac{\sin^2\theta}{\sigma_2^2}\right].$$

The integration with respect to ϱ yields:

$$P(\lambda) = \frac{1 - e^{-\frac{\lambda^2}{2(1-r^2)}}}{2\pi\sigma_1\sigma_2\sqrt{1-r^2}} \int_0^{2\pi} \frac{d\theta}{s^2}.$$

The integration with respect to θ can be carried out by the standard techniques for integrating trigonometric functions; but there is no need for this, since it can be carried out automatically with the help of probabilistic reasoning. In fact,

$$P(+\infty) = 1 = \frac{1}{2\pi\sigma_1\sigma_2\sqrt{1-r^2}} \int_0^{2\pi} \frac{d\theta}{s^2}.$$

Hence,
$$\int_0^{2\pi} \frac{d\theta}{s^2} = 2\pi\, \sigma_1\, \sigma_2 \sqrt{1-r^2},$$

and therefore
$$P(\lambda) = 1 - e^{-\frac{\lambda^2}{2(1-r^2)}}.$$

The normal distribution plays an exceptionally large role in various applied problems. The distributions of many random variables of practical importance prove to obey the normal distribution law. Thus, for example, a tremendous amount of artillery practice, carried out under diverse conditions, has shown that when a single gun is fired at a given target, the dispersion of the shells in the plane obeys the normal law. In Chapter VIII, we shall see that this "universality" of the normal law is explained by the fact that every random variable which is the sum of a very large number of independent random variables each having only a negligible effect on the sum is distributed almost according to the normal law.

That most important concept of probability theory—the independence of events—also retains its significance for random variables. Let $\xi_{i_1}, \xi_{i_2}, \ldots, \xi_{i_k}$ ($i_1 < i_2 < \ldots < i_k$) be any from among the random variables $\xi_1, \xi_2, \ldots, \xi_n$. Then, in accordance with the definition of independence of events, we shall say that the *random variables* $\xi_1, \xi_2, \ldots, \xi_n$ *are independent* if the relation

$$\mathsf{P}\{\xi_{i_1} < x_{i_1}, \xi_{i_2} < x_{i_2}, \ldots, \xi_{i_k} < x_{i_k}\} =$$
$$= \mathsf{P}\{\xi_{i_1} < x_{i_1}\}\mathsf{P}\{\xi_{i_2} < x_{i_2}\} \ldots \mathsf{P}\{\xi_{i_k} < x_{i_k}\}$$

holds for arbitrary values of $x_{i_1}, x_{i_2}, \ldots, x_{i_k}$ and any k ($1 \leq k \leq n$). In particular, the relation

$$\mathsf{P}\{\xi_1 < x_1, \xi_2 < x_2, \ldots, \xi_n < x_n\} =$$
$$= \mathsf{P}\{\xi_1 < x_1\}\mathsf{P}\{\xi_2 < x_2\} \ldots \mathsf{P}\{\xi_n < x_n\}$$

can be satisfied for arbitrary values of $x_1, x_2, \ldots, x_n$ or, in terms of distribution functions, the relation

$$F(x_1, x_2, \ldots, x_n) = F_1(x_1) \cdot F_2(x_2) \ldots F_n(x_n),$$

where $F_k(x_k)$ denotes the distribution function of the variable ξ_k.

§ 23. MULTI-DIMENSIONAL DISTRIBUTION FUNCTIONS

It is easy to see that the converse proposition is also true: If the distribution function of the system of random variables $\xi_1, \xi_2, \ldots, \xi_n$ is given by

$$F(x_1, x_2, \ldots, x_n) = F_1(x_1) F_2(x_2) \ldots F_n(x_n),$$

where the functions $F_k(x_k)$ satisfy the relations

$$F_k(+\infty) = 1 \quad (k = 1, 2, \ldots, n),$$

then the quantities $\xi_1, \xi_2, \ldots, \xi_n$ are independent, and the functions $F_1(x_1), F_2(x_2), \ldots, F_n(x_n)$ are their distribution functions.

The proof of this theorem will be left to the reader.

If the independent random variables $\xi_1, \xi_2, \ldots, \xi_n$ have density functions $p_1(x), p_2(x), \ldots, p_n(x)$, then the n-dimensional random variable $(\xi_1, \xi_2, \ldots, \xi_n)$ has a density function given by

$$p(x_1, x_2, \ldots, x_n) = p_1(x_1) p_2(x_2) \ldots p_n(x_n).$$

Example 5. Let us consider an n-dimensional random variable whose components $\xi_1, \xi_2, \ldots, \xi_n$ are mutually independent random variables distributed according to the normal law; let their distribution functions be

$$F_k(x_k) = \frac{1}{\sigma_k \sqrt{2\pi}} \int_{-\infty}^{x_k} e^{-\frac{(z-a_k)^2}{2\sigma_k^2}} dz.$$

In this example, the distribution function of the variable $(\xi_1, \xi_2, \ldots, \xi_n)$ is

$$F(x_1, x_2, \ldots, x_n) = (2\pi)^{-\frac{n}{2}} \prod_{k=1}^{n} \sigma_k^{-1} \int_{-\infty}^{x_k} e^{-\frac{(z-a_k)^2}{2\sigma_k^2}} dz.$$

The n-dimensional density function is given by

$$p(x_1, x_2, \ldots, x_n) = \frac{(2\pi)^{-\frac{n}{2}}}{\sigma_1 \sigma_2 \ldots \sigma_n} e^{-\frac{1}{2} \sum_{k=1}^{n} \frac{(x_k - a_k)^2}{\sigma_k^2}}. \tag{6}$$

For $n = 2$, this expression becomes

$$p(x_1, x_2) = \frac{1}{2\pi \sigma_1 \sigma_2} e^{-\frac{(x_1-a_1)^2}{2\sigma_1^2} - \frac{(x_2-a_2)^2}{2\sigma_2^2}}.$$

A comparison between this function and the density function for the two-dimensional normal law in Example 4 shows that when the random variables ξ_1, ξ_2 are independent, the parameter r is equal to zero.

For $n = 3$, formula (6) may be interpreted as the probability density function for the components ξ_1, ξ_2, ξ_3 of the velocity of a molecule along the coordinate axes (the Maxwell distribution) provided it is assumed that

$$\sigma_1^2 = \sigma_2^2 = \sigma_3^2 = \frac{1}{hm},$$

where m is the mass of a molecule and h is a constant.

§ 24. Functions of Random Variables

The knowledge that we have acquired about distribution functions allows us to set about solving the following problem: Given the distribution function $F(x_1, x_2, \ldots, x_n)$ of the aggregate of random variables $\xi_1, \xi_2, \ldots, \xi_n$, to determine the distribution function $\Phi(y_1, y_2, \ldots, y_k)$ of the variables $\eta_1 = f_1(\xi_1, \xi_2, \ldots, \xi_n)$, $\eta_2 = f_2(\xi_1, \xi_2, \ldots, \xi_n)$, $\ldots$, $\eta_k = f_k(\xi_1, \xi_2, \ldots, \xi_n)$.

This problem can be solved very simply in the general case but requires extension of the concept of an integral. In order not to be sidetracked by purely analytical questions, we shall confine ourselves to the most important special cases—discrete and continuous random variables. In the following section, the definition and the fundamental properties of a Stieltjes integral will be presented; we shall there give the general form of the most important results of the present section.

Let us first consider the case where the n-dimensional vector $(\xi_1, \ldots, \xi_n)$ possesses a probability density function $p(x_1, x_2, \ldots, x_n)$. From the foregoing, it is apparent that the desired distribution function can be defined by the relation

$$\Phi(y_1, y_2, \ldots, y_k) = \int \ldots \int_D p(x_1, x_2, \ldots, x_n)\, dx_1\, dx_2 \ldots dx_n,$$

where the region of integration D is defined by the inequalities

$$f_i(x_1, x_2, \ldots, x_n) < y_i \quad (i = 1, 2, \ldots, k).$$

In the case of discrete random variables, the solution is given by means of an n-fold sum, also extended over the region D.

§ 24. Functions of Random Variables

These general observations on the solution of the general problem posed above will now be applied to several important special cases.

The distribution function of a sum. Let it be required to find the distribution function of the sum

$$\eta = \xi_1 + \xi_2 + \cdots + \xi_n,$$

if $p(x_1, x_2, \ldots, x_n)$ is the probability density function of the vector $(\xi_1, \xi_2, \ldots, \xi_n)$. The required function is equal to the probability that the point $(\xi_1, \xi_2, \ldots, \xi_n)$ falls in the half-space $\xi_1 + \xi_2 + \ldots + \xi_n < y$, and therefore

$$\Phi(y) = \int \cdots \int_{\Sigma x_k < y} p(x_1, x_2, \ldots, x_n)\, dx_1\, dx_2 \ldots dx_n.$$

Let us consider the case $n = 2$ in more detail. In this case, the preceding formula becomes:

$$\Phi(y) = \int\int_{x_1+x_2<y} p(x_1, x_2)\, dx_1\, dx_2 = \int \int_{-\infty}^{y-x_1} p(x_1, x_2)\, dx_2\, dx_1. \quad (1)$$

If the variables ξ_1 and ξ_2 are independent, then $p(x_1, x_2) = p_1(x_1) p_2(x_2)$, and equation (1) is expressible in the form:

$$\Phi(y) = \int dx_1 \int_{-\infty}^{y-x_1} p_1(x_1)\, p_2(x_2)\, dx_2 = \int dx_1 \int_{-\infty}^{y} p_1(x_1)\, p_2(z - x_1)\, dz$$

$$= \int_{-\infty}^{y} dz \left\{ \int p_1(x_1)\, p_2(z - x_1)\, dx_1 \right\}. \quad (2)$$

In the general case, formula (1) yields

$$\Phi(y) = \int_{-\infty}^{y} dx_1 \int p(z, x_1 - z)\, dz. \quad (3)$$

The last result shows that if the random vector formed from the variables in a sum has a probability density function, then the sum of the variables also has a density function. In the case of independent summands, this density function is expressible in the form:

$$p(y) = \int p_1(z)\, p_2(y - z)\, dz. \quad (4)$$

Let us consider some examples.

IV. RANDOM VARIABLES AND DISTRIBUTION FUNCTIONS

Example 1. Let ξ_1 and ξ_2 be independent and uniformly distributed in the interval (a, b). Find the density function for the sum $\eta = \xi_1 + \xi_2$.

The probability density functions of ξ_1 and ξ_2 are given by

$$p_1(x) = p_2(x) = \begin{cases} 0 & \text{if } x \leq a \text{ or } x > b, \\ \dfrac{1}{b-a} & \text{if } a < x \leq b. \end{cases}$$

By equation (4), we find that

$$p_\eta(y) = \int_a^b p_1(z) p_2(y-z) \, dz = \frac{1}{b-a} \int_a^b p_2(y-z) \, dz.$$

From the fact that

$$y - z < 2a - z < a$$

for $y < 2a$ and

$$y - z > 2b - z > b$$

for $y > 2b$, we conclude that

$$p_\eta(y) = 0$$

for $y < 2a$ and $y > 2b$. Now let $2a < y < 2b$. The integrand is different from zero only for those values of z that satisfy the inequality

$$a < y - z < b$$

or, what is the same, the inequality

$$y - b < z < y - a.$$

Since $y > 2a$, it follows that $y - a > a$. Obviously, $y - a \leq b$ for $y \leq a + b$. Therefore, if $2a < y \leq a + b$, we have:

$$p_\eta(y) = \int_a^{y-a} \frac{dz}{(b-a)^2} = \frac{y - 2a}{(b-a)^2}.$$

In an analogous way, we find for $a + b < y \leq 2b$:

$$p_\eta(y) = \int_{y-b}^b \frac{dz}{(b-a)^2} = \frac{2b - y}{(b-a)^2}.$$

§ 24. FUNCTIONS OF RANDOM VARIABLES

Collecting the results obtained, we find:

$$p_\eta(y) = \begin{cases} 0 & \text{for } y \leq 2a \text{ and } y > 2b, \\ \dfrac{y-2a}{(b-a)^2} & \text{for } 2a < y \leq a+b, \\ \dfrac{2b-y}{(b-a)^2} & \text{for } a+b < y \leq 2b. \end{cases} \quad (5)$$

The function $p_\eta(y)$ is called the *Simpson*, or *triangular*, distribution law.

The computations in the last example can be considerably simplified if we make use of geometrical reasoning. Let us, as usual, represent ξ_1 and ξ_2 as rectangular coordinates in the plane. Then the probability of

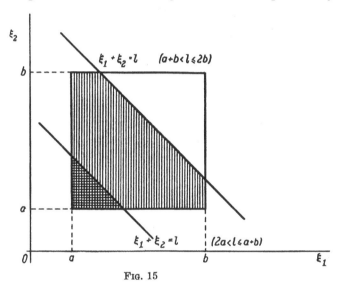

FIG. 15

the inequality $\xi_1 + \xi_2 < y$ for $2a < y \leq a+b$ is equal to the probability that η will correspond to a point in the doubly-shaded triangle (Fig. 15). As is easily shown, this probability is given by

$$F_\eta(y) = \frac{(y-2a)^2}{2(a-b)^2}.$$

For $a+b < y \leq 2b$, the probability of the inequality $\xi_1 + \xi_2 < y$ is the probability of the point lying in the entire shaded region of the figure.

This probability is given by
$$F_\eta(y) = 1 - \frac{(2b-y)^2}{2(b-a)^2}.$$

Differentiation with respect to y then yields formula (5).

In connection with the example considered, it is interesting to note the following.

Certain general questions in Geometry made it necessary for Lobachevsky to solve the following problem: Given a group of n mutually independent random variables $\xi_1, \xi_2, \ldots, \xi_n$ (the errors of observation), to find the probability distribution of their arithmetic mean.

He solved this problem only for the case where all of the errors were distributed *uniformly* in the interval $(-1, 1)$. It turns out in this case that the probability that the error in the arithmetic mean lies between the values of $-x$ and x is

$$P_n(x) = 1 - \frac{1}{2^{n-1}} \sum (-1)^r \frac{[n-nx-2r]^r}{r!(n-r)!},$$

where the summation is extended over all the integers r from $r = 0$ to $r = [(n-nx)/2]$.

Example 2. A two-dimensional random variable (ξ_1, ξ_2) is distributed according to the normal law

$$p(x, y) = \frac{1}{2\pi\sigma_1\sigma_2\sqrt{1-r^2}} \times$$
$$\times \exp\left\{-\frac{1}{2(1-r^2)}\left(\frac{(x-a^2)}{\sigma_1^2} - 2r\frac{(x-a)(y-b)}{\sigma_1\sigma_2} + \frac{(y-b)^2}{\sigma_2^2}\right)\right\}.$$

Find the distribution function for the sum $\eta = \xi_1 + \xi_2$.

According to formula (3),

$$p_\eta(y) = \frac{1}{2\pi\sigma_1\sigma_2\sqrt{1-r^2}} \times$$
$$\times \int \exp\left\{-\frac{1}{2(1-r^2)}\left(\frac{(z-a)^2}{\sigma_1^2} - 2r\frac{(z-a)(y-z-b)}{\sigma_1\sigma_2} + \frac{(y-z-b)^2}{\sigma_2^2}\right)\right\} dz.$$

For brevity, denote $y - a - b$ and $z - a$ by v and u, respectively; then

$$p_\eta(y) = \frac{1}{2\pi\sigma_1\sigma_2\sqrt{1-r^2}} \times$$
$$\times \int \exp\left\{-\frac{1}{2(1-r^2)}\left(\frac{u^2}{\sigma_1^2} - 2r\frac{u(v-u)}{\sigma_1\sigma_2} + \frac{(v-u)^2}{\sigma_2^2}\right)\right\} du.$$

§ 24. Functions of Random Variables

Since

$$\frac{u^2}{\sigma_1^2} - 2r\frac{u(v-u)}{\sigma_1\sigma_2} + \frac{(v-u)^2}{\sigma_2^2} = u^2\frac{\sigma_1^2 + 2r\sigma_1\sigma_2 + \sigma_2^2}{\sigma_1^2\sigma_2^2} - 2uv\frac{\sigma_1 + r\sigma_2}{\sigma_1\sigma_2^2} + \frac{v^2}{\sigma_2^2} =$$

$$= \left[u\frac{\sqrt{\sigma_1^2 + 2r\sigma_1\sigma_2 + \sigma_2^2}}{\sigma_1\sigma_2} - \frac{v}{\sigma_2}\frac{\sigma_1 + r\sigma_2}{\sqrt{\sigma_1^2 + 2r\sigma_1\sigma_2 + \sigma_2^2}}\right]^2 + \frac{v^2}{\sigma_2^2}\left(1 - \frac{(\sigma_1 + r\sigma_2)^2}{\sigma_1^2 + 2r\sigma_1\sigma_2 + \sigma_2^2}\right)$$

$$= \left[u\frac{\sqrt{\sigma_1^2 + 2r\sigma_1\sigma_2 + \sigma_2^2}}{\sigma_1\sigma_2} - \frac{v}{\sigma_2}\frac{\sigma_1 + r\sigma_2}{\sqrt{\sigma_1^2 + 2r\sigma_1\sigma_2 + \sigma_2^2}}\right]^2 + \frac{v^2(1-r^2)}{\sigma_1^2 + 2r\sigma_1\sigma_2 + \sigma_2^2},$$

by making the change of variables

$$t = \frac{1}{\sqrt{1-r^2}}\left[u\frac{\sqrt{\sigma_1^2 + 2r\sigma_1\sigma_2 + \sigma_2^2}}{\sigma_1\sigma_2} - \frac{v}{\sigma_2}\frac{\sigma_1 + r\sigma_2}{\sqrt{\sigma_1^2 + 2r\sigma_1\sigma_2 + \sigma_2^2}}\right],$$

we can reduce the expression for $p_\eta(y)$ to the form

$$p_\eta(y) = \frac{\exp\left\{-\frac{v^2}{2(\sigma_1^2 + 2r\sigma_1\sigma_2 + \sigma_2^2)}\right\}}{2\pi\sqrt{\sigma_1^2 + 2r\sigma_1\sigma_2 + \sigma_2^2}} \int e^{-\frac{t^2}{2}} dt.$$

Because

$$v = y - a - b \quad \text{and} \quad \int e^{-\frac{t^2}{2}} dt = \sqrt{2\pi},$$

it follows that

$$p_\eta(y) = \frac{1}{\sqrt{2\pi(\sigma_1^2 + 2r\sigma_1\sigma_2 + \sigma_2^2)}} e^{-\frac{(y-a-b)^2}{2(\sigma_1^2 + 2r\sigma_1\sigma_2 + \sigma_2^2)}}.$$

In particular, if the random variables ξ_1 and ξ_2 are independent, then $r = 0$ and the expression for $p_\eta(y)$ takes the form

$$p_\eta(y) = \frac{1}{\sqrt{2\pi(\sigma_1^2 + \sigma_2^2)}} e^{-\frac{(y-a-b)^2}{2(\sigma_1^2 + \sigma_2^2)}}.$$

We have thus obtained the following result: *The sum of the components of a normally distributed random vector is itself distributed according to the normal law.*

It is interesting to note that when the summands are independent, the converse proposition is also true (Cramér's Theorem): *If the sum of two independent random variables is distributed according to the normal law, then each of the summands is distributed normally.* We shall not pause to prove this theorem, since it requires more complex mathematical techniques.

Example 3. The χ^2 distribution. Let $\xi_1, \xi_2, \ldots, \xi_n$ be independent random variables all distributed according to the same normal law, with parameters a and σ.

The distribution function of the quantity

$$\chi^2 = \frac{1}{\sigma^2} \sum_{k=1}^{n} (\xi_k - a)^2$$

is called the χ^2 *distribution*.

This distribution plays an important role in various statistical problems.

We shall now compute the distribution function of the variable $\zeta = \chi/\sqrt{n}$. It will turn out to be independent of a and σ.

It is obvious that for negative values of the argument, the distribution function $\Phi(y)$ is zero; for positive values, the function $\Phi(y)$ is equal to the probability that the point $(\xi_1, \xi_2, \ldots, \xi_n)$ falls inside the sphere

$$\sum_{k=1}^{n} (x_k - a)^2 = y^2 \cdot n \cdot \sigma^2.$$

Thus

$$\Phi(y) = \int \cdots \int_{\Sigma x_i^2 < y^2 n} \left(\frac{1}{\sqrt{2\pi}}\right)^n e^{-\sum_{i=1}^{n} \frac{x_i^2}{2}} dx_1 \, dx_2 \ldots dx_n.$$

To evaluate this integral, we pass to spherical coordinates, i.e., we make the substitution

$$x_1 = \varrho \cos\theta_1 \cos\theta_2 \ldots \cos\theta_{n-1},$$
$$x_2 = \varrho \cos\theta_1 \cos\theta_2 \ldots \sin\theta_{n-1},$$
$$\cdots\cdots\cdots\cdots\cdots\cdots\cdots\cdots$$
$$x_n = \varrho \sin\theta_1.$$

As a result of this substitution, we have:

$$\Phi(y) = \int_{-\frac{\pi}{2}}^{\frac{\pi}{2}} \cdots \int_{-\frac{\pi}{2}}^{\frac{\pi}{2}} \int_0^{y\sqrt{n}} \frac{1}{(\sqrt{2\pi})^n} e^{-\frac{\varrho^2}{2}} \varrho^{n-1} D(\theta_1 \ldots \theta_{n-1}) \, d\varrho \, d\theta_{n-1} \ldots d\theta_1$$

$$= C_n \int_0^{y\sqrt{n}} e^{-\frac{\varrho^2}{2}} \varrho^{n-1} d\varrho,$$

where the constant

§ 24. Functions of Random Variables

$$C_n = \frac{1}{(\sqrt{2\pi})^n} \int_{-\frac{\pi}{2}}^{\frac{\pi}{2}} \cdots \int_{-\frac{\pi}{2}}^{\frac{\pi}{2}} D(\theta_1 \ldots \theta_{n-1}) \, d\theta_{n-1} \ldots d\theta_1$$

depends only on n.

This constant can be readily evaluated by making use of the relation

$$\Phi(+\infty) = 1 = C_n \int_0^\infty e^{-\frac{\varrho^2}{2}} \varrho^{n-1} \, d\varrho = C_n \Gamma\left(\frac{n}{2}\right) \cdot 2^{\frac{n}{2}-1}$$

From this, we find:

$$\Phi(y) = \frac{1}{2^{\frac{n}{2}-1} \Gamma\left(\frac{n}{2}\right)} \int_0^{y\sqrt{n}} \varrho^{n-1} e^{-\frac{\varrho^2}{2}} \, d\varrho.$$

The density function of the random variable ζ for $y \geq 0$ is given by

$$\varphi(y) = \frac{\sqrt{2n}}{\Gamma\left(\frac{n}{2}\right)} \left(\frac{y\sqrt{n}}{\sqrt{2}}\right)^{n-1} e^{-\frac{ny^2}{2}}. \tag{6}$$

Hence, in the particular case $n = 1$, we naturally obtain a density function which is twice the density of the initially given normal law:

$$\varphi(y) = \sqrt{\frac{2}{\pi}} \, e^{-\frac{y^2}{2}} \quad (y \geq 0).$$

For $n = 3$, we obtain the familiar Maxwell distribution

$$\varphi(y) = \frac{3\sqrt{6}}{\sqrt{\pi}} y^2 e^{-\frac{3y^2}{2}}.$$

Either by computations analogous to those already performed or directly from formula (6), it is easy to derive the density function for the variable χ^2. For $x \geq 0$, this density is given by

$$p_n(x) = \frac{x^{\frac{n}{2}-1} e^{-\frac{x}{2}}}{2^{\frac{n}{2}} \Gamma\left(\frac{n}{2}\right)}. \tag{6'}$$

The distributions of variables related to χ^2 and frequently made use of in practice have been brought together in the following table:

Random Variables	Density for $x \geq 0$
$\chi^2 = \dfrac{1}{\sigma^2} \sum\limits_{k=1}^{n} (\xi_k - a)^2$	$\dfrac{x^{\frac{n}{2}-1} e^{-\frac{x}{2}}}{2^{\frac{n}{2}} \Gamma\left(\dfrac{n}{2}\right)}$
$\dfrac{1}{n}\chi^2 = \dfrac{1}{n\sigma^2} \sum\limits_{k=1}^{n} (\xi_k - a)^2$	$\dfrac{\left(\dfrac{n}{2}\right)^{\frac{n}{2}}}{\Gamma\left(\dfrac{n}{2}\right)} x^{\frac{n}{2}-1} e^{-\frac{nx}{2}}$
$\chi = \sqrt{\dfrac{1}{\sigma^2} \sum\limits_{k=1}^{n} (\xi_k - a)^2}$	$\dfrac{2}{2^{\frac{n}{2}} \Gamma\left(\dfrac{n}{2}\right)} x^{n-1} e^{-\frac{x^2}{2}}$
$\zeta = \dfrac{\chi}{\sqrt{n}} = \sqrt{\dfrac{1}{n\sigma^2} \sum\limits_{k=1}^{n} (\xi_k - a)^2}$	$\dfrac{\sqrt{2n}}{\Gamma\left(\dfrac{n}{2}\right)} \left(\dfrac{x\sqrt{n}}{\sqrt{2}}\right)^{n-1} e^{-\frac{nx^2}{2}}$

Example 4. The distribution function of the quotient. Let the probability density function of the variable (ξ, η) be $p(x, y)$. It is required to find the distribution function of the quotient $\zeta = \xi/\eta$.

By definition,

$$F_\zeta(x) = \mathsf{P}\left\{\dfrac{\xi}{\eta} < x\right\}.$$

If we regard ξ and η as the rectangular coordinates of a point in the plane, then $F_\zeta(x)$ is equal to the probability that the point (ξ, η) falls in the region corresponding to points whose coordinates satisfy the inequality $\xi/\eta < x$. This region is shaded in Fig. 16.

By the general formula, the required probability is

$$F_\zeta(x) = \int_0^\infty \int_{-\infty}^{zx} p(y, z)\, dy\, dz + \int_{-\infty}^0 \int_{zx}^\infty p(y, z)\, dy\, dz. \tag{7}$$

§ 24. Functions of Random Variables

It follows from this that if ξ and η are independent and $p_1(x)$ and $p_2(x)$ are their respective density functions, then

$$F_\zeta(x) = \int_0^\infty F_1(x\,z)\, p_2(z)\, dz + \int_{-\infty}^0 [1 - F_1(xz)]\, p_2(z)\, dz. \qquad (7')$$

By differentiation of (7), we find:

$$p_\zeta(x) = \int_0^\infty z\, p(z\,x,\, z)\, dz - \int_{-\infty}^0 z\, p\,(z\,x,\, z)\, dz. \qquad (8)$$

In particular, if ξ and η are independent, then

$$p_\zeta(x) = \int_0^\infty z\, p_1(z\,x)\, p_2(z)\, dz - \int_{-\infty}^0 z\, p_1(z\,x)\, p_2(z)\, dz. \qquad (8')$$

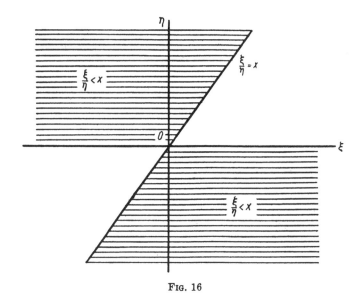

Fig. 16

Example 5. The random variable (ξ, η) is distributed according to the normal law

$$p(x, y) = \frac{1}{2\pi\sigma_1\sigma_2\sqrt{1-r^2}} \exp\left\{-\frac{1}{2(1-r^2)}\left[\frac{x^2}{\sigma_1^2} - 2r\frac{x\,y}{\sigma_1\sigma_2} + \frac{y^2}{\sigma_2^2}\right]\right\}$$

Find the distribution function for the quotient $\zeta = \xi/\eta$.

By equation (8), we have

$$p_\zeta(x) = \frac{1}{2\pi\sigma_1\sigma_2\sqrt{1-r^2}} \cdot \left[\int_0^\infty - \int_{-\infty}^0\right] z \exp\left\{-\frac{z^2}{2(1-r^2)}\left[\frac{\sigma_2^2 x^2 - 2r\sigma_1\sigma_2 x + \sigma_1^2}{\sigma_1^2 \sigma_2^2}\right]\right\} dz$$

$$= \frac{1}{\pi\sigma_1\sigma_2\sqrt{1-r^2}} \int_0^\infty z \exp\left\{-\frac{z^2}{2(1-r^2)} \cdot \frac{\sigma_2^2 x^2 - 2r\sigma_1\sigma_2 x + \sigma_1^2}{\sigma_1^2 \sigma_2^2}\right\} dz.$$

Let us make a substitution in the integral by setting

$$u = \frac{z^2}{2(1-r^2)} \cdot \frac{\sigma_2^2 x^2 - 2r\sigma_1\sigma_2 x + \sigma_1^2}{\sigma_1^2 \sigma_2^2}.$$

As a result, the expression for $p_\zeta(x)$ takes the form:

$$p_\zeta(x) = \frac{\sigma_1\sigma_2\sqrt{1-r^2}}{\pi(\sigma_2^2 x^2 - 2r\sigma_1\sigma_2 x + \sigma_1^2)} \int_0^\infty e^{-u} du = \frac{\sigma_1\sigma_2\sqrt{1-r^2}}{\pi(\sigma_2^2 x^2 - 2r\sigma_1\sigma_2 x + \sigma_1^2)};$$

if, in particular, the variables ξ and η are independent, then

$$p_\zeta(x) = \frac{\sigma_1\sigma_2}{\pi(\sigma_1^2 + \sigma_2^2 x^2)}.$$

The density function of the variable ζ is called the *Cauchy law*.

Example 6. "*Student's*" *distribution*. Find the distribution function of the quotient $\zeta = \xi/\eta$, where ξ and η are independent variables, ξ is distributed according to the normal law

$$p_\xi(x) = \sqrt{\frac{n}{2\pi}} e^{-\frac{nx^2}{2}},$$

and $\eta = \chi/\sqrt{n}$ (see Example 3), so that

$$p(y) = \frac{\sqrt{2n}}{\Gamma\left(\frac{n}{2}\right)} \left(\frac{y\sqrt{n}}{\sqrt{2}}\right)^{n-1} e^{-\frac{ny^2}{2}}.$$

By equation (8'), we have:

§ 24. Functions of Random Variables

$$p_\zeta(x) = \int_0^\infty z\sqrt{\frac{n}{2\pi}} e^{-\frac{nz^2x^2}{2}} \frac{\sqrt{2n}}{\Gamma\left(\frac{n}{2}\right)} \left(\frac{z\sqrt{n}}{\sqrt{2}}\right)^{n-1} e^{-\frac{nz^2}{2}} dz$$

$$= \frac{1}{\sqrt{\pi}\,\Gamma\left(\frac{n}{2}\right)} \int_0^\infty \left(\frac{z\sqrt{n}}{\sqrt{2}}\right)^{n-1} e^{-\frac{nz^2}{2}(x^2+1)} nz\, dz.$$

Making the substitution

$$u = \frac{nz^2}{2}(x^2+1),$$

we find:

$$p_\zeta(x) = \frac{(x^2+1)^{-\frac{n+1}{2}}}{\sqrt{\pi}\,\Gamma\left(\frac{n}{2}\right)} \int_0^\infty u^{\frac{n-1}{2}} e^{-u} du = \frac{\Gamma\left(\frac{n+1}{2}\right)}{\sqrt{\pi}\,\Gamma\left(\frac{n}{2}\right)} (x^2+1)^{-\frac{n+1}{2}}.$$

The probability density function

$$p_\zeta(x) = \frac{\Gamma\left(\frac{n+1}{2}\right)}{\sqrt{\pi}\,\Gamma\left(\frac{n}{2}\right)} (1+x^2)^{-\frac{n+1}{2}}$$

is referred to as *"Student's" law*.[4]

For $n = 1$, Student's law reduces to Cauchy's law.

Example 7. Rotation of the coordinate axes. Given the distribution function of the two-dimensional random variable (ξ, η), determine the distribution function for the quantities

$$\xi' = \xi \cos a + \eta \sin a, \qquad \eta' = -\xi \sin a + \eta \cos a. \tag{9}$$

Let $F(x, y)$ and $\Phi(x, y)$ denote the respective distribution functions of the variables (ξ, η) and (ξ', η'). If we regard (ξ, η) and (ξ', η') as the rectangular coordinates of a point in the plane, then it is readily apparent that the $\xi'\eta'$-coordinate system can be obtained from the $\xi\eta$-coordinate system by rotating the latter through the angle a. We shall restrict ourselves to the case $0 < a < \pi/2$, leaving to the reader the derivation of the analogous formulas for the remaining values of a.

---

[4] "Student" was the pseudonym of the English statistician W. L. Gosset, who by empirical means first discovered this law.

Let the coordinates of a point M in the $\xi\eta$ system be given by x and y and in the $\xi'\eta'$ system by x' and y'. Then the function $F(x, y)$ is equal to the probability that the point (ξ, η) will fall inside the 90° sector bounded by the half-rays AM and BM (Fig. 17), and the function $\Phi(x', y')$ is the

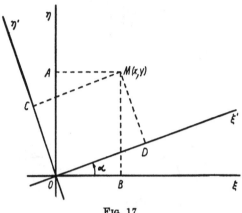

Fig. 17

probability that the point (ξ, η) will fall inside the 90° sector bounded by the half-rays CM and DM. The equations of the lines CM and DM in the $\xi\eta$-coordinate system are given as follows: the line CM by

$$\eta = (\xi - x) \tan \alpha + y$$

and the line DM by

$$\eta = -(\xi - x) \cot \alpha + y.$$

Since (x, y) and (x', y') are connected by the relations

$$x' = x \cos \alpha + y \sin \alpha, \qquad y' = -x \sin \alpha + y \cos x$$

these equations can be written in another form:

$$\eta = \xi \tan \alpha + \frac{y'}{\cos \alpha},$$
$$\eta = -\xi \cot \alpha + \frac{x'}{\sin \alpha}.$$

By virtue of our earlier discussion, we have

§ 24. Functions of Random Variables

$$\Phi(x', y') = \int\int p(\xi, \eta) \, d\eta \, d\xi,$$

the integral extending over the region interior to the angle CMD. It is easy to see that

$$\Phi(x', y') = \int_{-\infty}^{x} \int_{-\infty}^{\xi \tan \alpha + \frac{y'}{\cos \alpha}} p(\xi, \eta) \, d\eta \, d\xi + \int_{x}^{\infty} \int_{-\infty}^{-\xi \cot \alpha + \frac{x'}{\sin \alpha}} p(\xi, \eta) \, d\eta \, d\xi.$$

By differentiating this relation with respect to x' and y', we find:

$$\pi(x', y') = \frac{\partial^2 \Phi(x', y')}{\partial x' \partial y'} = p(x, y) =$$
$$= p(x' \cos \alpha - y' \sin \alpha, \ x' \sin \alpha + y' \cos \alpha). \qquad (10)$$

Example 8. A two-dimensional random variable (ξ, η) is distributed according to the normal law

$$p(x, y) = \frac{1}{2\pi\sigma_1\sigma_2\sqrt{1-r^2}} \exp\left\{-\frac{1}{2(1-r^2)}\left[\frac{x^2}{\sigma_1^2} - 2r\frac{xy}{\sigma_1\sigma_2} + \frac{y^2}{\sigma_2^2}\right]\right\}.$$

Determine the density function of the random variables

$$\xi' = \xi \cos \alpha + \eta \sin \alpha, \quad \eta' = -\xi \sin \alpha + \eta \cos \alpha.$$

By equation (10), we have

$$\pi(x', y') = p(x' \cos \alpha - y' \sin \alpha, \ x' \sin \alpha + y' \cos \alpha) =$$
$$= \frac{1}{2\pi\sigma_1\sigma_2\sqrt{1-r^2}} \exp\left\{-\frac{1}{2(1-r^2)}[A x'^2 - 2B x' y' + C y'^2]\right\},$$

where

$$A = \frac{\cos^2 \alpha}{\sigma_1^2} - 2r\frac{\cos \alpha \sin \alpha}{\sigma_1 \sigma_2} + \frac{\sin^2 \alpha}{\sigma_2^2},$$

$$B = \frac{\cos \alpha \sin \alpha}{\sigma_1^2} - r\frac{\sin^2 \alpha - \cos^2 \alpha}{\sigma_1 \sigma_2} - \frac{\cos \alpha \sin \alpha}{\sigma_2^2},$$

$$C = \frac{\sin^2 \alpha}{\sigma_1^2} + 2r\frac{\cos \alpha \sin \alpha}{\sigma_1 \sigma_2} + \frac{\cos^2 \alpha}{\sigma_2^2}.$$

We conclude from this that under rotation of axes a normal two-dimensional distribution goes over into a normal two-dimensional distribution.

We note that if the angle α is so selected that

$$\tan 2\alpha = \frac{2r\sigma_1\sigma_2}{\sigma_1^2 - \sigma_2^2},$$

then $B = 0$ and

$$\pi(x', y') = \frac{1}{2\pi\sigma_1\sigma_2\sqrt{1-r^2}} e^{-\frac{A x'^2}{2(1-r^2)} - \frac{C y'^2}{2(1-r^2)}}.$$

This relation implies that any normally distributed two-dimensional random variable can be reduced by means of a rotation of axes to a system of two normally distributed *independent* random variables. This result can be extended to n-dimensional random variables.

It is possible to prove a stronger theorem which exhaustively characterizes the normal probability distribution. Let there be a *non-degenerate* probability distribution in the plane (i.e., one not concentrated along a line). This distribution will be normal if and only if it is possible to choose two distinct planar coordinate systems $\xi_1\xi_2$ and $\eta_1\eta_2$ such that the coordinates ξ_1 and ξ_2, as well as η_1 and η_2, when considered as random variables with the given probability distribution, are independent.

§ 25. The Stieltjes Integral

We shall make considerable use in the sequel of the concept of the Stieltjes integral. To facilitate the investigations in the following sections, we give here the definition and fundamental properties of the Stieltjes integral, without stopping to give any proofs.

Suppose that on the interval (a, b) there are defined a function $f(x)$ and a non-decreasing function of bounded variation $F(x)$. In addition, let us suppose for definiteness, that the function $F(x)$ is continuous from the left. If a and b are finite, we subdivide the interval (a, b) by means of the points $a = x_0 < x_1 < x_2 < \ldots < x_n = b$ into a finite number of subintervals (x_i, x_{i+1}), and we form the sum

$$\sum_{i=1}^{n} f(\tilde{x}_i)\,[F(x_i) - F(x_{i-1})],$$

where $\tilde{x}_i$ is an arbitrary value chosen in the interval (x_{i-1}, x_i). We now

§ 25. The Stieltjes Integral

let the number of points of subdivision increase, simultaneously letting the length of the maximum subinterval approach zero. If the above sum then tends to a definite limit

$$J = \lim_{n \to \infty} \sum_{i=1}^{n} f(\tilde{x}_i) [F(x_i) - F(x_{i-1})], \tag{1}$$

then this limit is called the *Stieltjes integral* of the function $f(x)$ with respect to the function $F(x)$ and is denoted by the symbol

$$J = \int_a^b f(x) \, dF(x). \tag{2}$$

The improper Stieltjes integral—i.e., one whose interval of integration is infinite—is defined in the usual way: We consider the integral over some finite interval (a, b); the quantities a and b are then allowed to approach $-\infty$ and ∞ in an arbitrary way; if the resultant limit

$$\lim_{\substack{a \to -\infty \\ b \to \infty}} \int_a^b f(x) \, dF(x)$$

exists, then this limit is called the *Stieltjes integral* of the function $f(x)$ with respect to the function $F(x)$ in the interval $(-\infty, \infty)$ and is denoted by

$$\int f(x) \, dF(x).$$

It can be shown that if the function $f(x)$ is continuous and bounded, the limit of the sum (1) exists in the case of both finite and infinite limits of integration.

In certain cases, the Stieltjes integral also exists for unbounded functions $f(x)$. The consideration of such integrals is of considerable interest in the theory of probability (mathematical expectation, variance, moments, etc.).

Let us note that throughout the sequel *we consider the integral of the function $f(x)$ to exist if and only if the integral of $|f(x)|$ with respect to the same function $F(x)$ exists.*

For the purposes of probability theory, it is important to extend the definition of the Stieltjes integral to the case where the function $f(x)$ can have a finite or denumerable number of discontinuities. It can be shown[5] that any bounded function that has a finite or denumerable num-

---

[5] See, for example, *The Theory of Functions of a Real Variable*, by E. W. Hobson, Vol. I, 3rd ed., p. 542.

ber of discontinuities—and, in particular, any function of bounded variation—is integrable with respect to any function of bounded variation. For this, it is necessary to modify slightly the definition of a Stieltjes integral; namely, in forming the limit (1), one must consider only those sequences of subdivisions of the interval of integration for which every point of discontinuity of $f(x)$ is a point of division in all the subdivisions, with the possible exception of a finite number of them.

It should be noted that in fixing the limits of integration it is important to specify whether or not either of the endpoints of the interval of integration is included. Indeed, from the definition of the Stieltjes integral, we obtain the following relation (the symbol $a-0$ signifies that a is included in the interval of integration, and the symbol $a+0$ that a is excluded from it):

$$\int_{a-0}^{b} f(x)\, dF(x) = \lim_{n \to \infty} \sum_{i=1}^{n} f(\tilde{x}_i)\, [F(x_i) - F(x_{i-1})] =$$
$$= \lim_{n \to \infty} \sum_{i=2}^{n} f(\tilde{x}_i)\, [F(x_i) - F(x_{i-1})] + \lim_{x_1 \to x_0 = a} f(\tilde{x}_1)\, [F(x_1) - F(x_0)]$$
$$= \int_{a+0}^{b} f(x)\, dF(x) + f(a)\, [F(a+0) - F(a)].$$

Thus, if $f(a) \neq 0$ and the function $F(x)$ has a jump at $x = a$, then

$$\int_{a-0}^{b} f(x)\, dF(x) - \int_{a+0}^{b} f(x)\, dF(x) = f(a)\, [F(a+0) - F(a-0)].$$

This shows that a Stieltjes integral which is taken over an interval that reduces to a single point can yield a non-zero result. We agree that, in what follows, unless it is specifically stated to the contrary, the left-hand endpoint is to be included in the interval of integration and the right-hand endpoint excluded. This condition permits us to write the following equation:

$$\int_{a}^{b} dF(x) = F(b) - F(a).$$

In fact, we have, by definition,

$$\int_{a}^{b} dF(x) = \lim_{n \to \infty} \sum_{i=1}^{n} [F(x_i) - F(x_{i-1})] = \lim_{n \to \infty} [F(x_n) - F(x_0)] = F(b) - F(a)$$

§ 25. THE STIELTJES INTEGRAL

(recall that $F(x)$ is, by assumption, continuous from the left and therefore $F(b) = \lim_{\varepsilon \to 0} F(b-\varepsilon)$).

In particular, if $F(x)$ is the distribution function of a random variable ξ, then

$$\int_a^b dF(x) = F(b) - F(a) = \mathsf{P}\{a \leq \xi < b\},$$

$$\int_{-\infty}^b dF(x) = F(b) = \mathsf{P}\{\xi < b\}.$$

If $F(x)$ possesses a derivative of which it is an integral, then from the mean-value theorem

$$F(x_i) - F(x_{i-1}) = p(\tilde{x}_i)(x_i - x_{i-1}),$$

where $x_{i-1} < \tilde{x}_i < x_i$, the relation

$$\int_a^b f(x)\, dF(x) = \lim_{n \to \infty} \sum_{i=1}^n f(\tilde{x}_i)[F(x_i) - F(x_{i-1})]$$
$$= \lim_{n \to \infty} \sum_{i=1}^n f(\tilde{x}_i)\, p(\tilde{x}_i)(x_i - x_{i-1}) = \int_a^b f(x)\, p(x)\, dx$$

follows. We thus see that in this case a Stieltjes integral reduces to an ordinary integral.

If $F(x)$ has a jump at the point $x = c$, then by choosing the subdivisions so that for some value of the subscript $x_k < c < x_{k+1}$, we have:

$$\int_a^b f(x)\, dF(x) = \lim_{n \to \infty} \sum_{i=1}^k f(\tilde{x})[F(x_i) - F(x_{i-1})] +$$
$$+ \lim_{n \to \infty} f(c)[F(x_{k+1}) - F(x_k)] + \lim_{n \to \infty} \sum_{i=k+2}^n f(\tilde{x}_i)[F(x_i) - F(x_{i-1})]$$
$$= \int_a^c f(x)\, dF(x) + \int_{c+0}^b f(x)\, dF(x) + f(c)[F(c+0) - F(c-0)].$$

In particular, if the value of the function $F(x)$ changes only at the points $c_1, c_2, \ldots, c_n, \ldots$ then

$$\int_a^b f(x)\, dF(x) = \sum_{n=1}^\infty f(c_n)[F(c_n + 0) - F(c_n - 0)],$$

and the Stieltjes integral reduces to an infinite series.

Let us list the fundamental properties of the Stieltjes integral which we shall require in the sequel. The reader can easily supply the proofs of these properties by starting from the definition of a Stieltjes integral and using arguments similar to those employed in the theory of ordinary integrals.

1. For $a < c_1 < c_2 < \ldots < c_n < b$,

$$\int_a^b f(x)\, dF(x) = \sum_{i=0}^{n} \int_{c_i}^{c_{i+1}} f(x)\, dF(x) \qquad [a = c_0,\ b = c_{n+1}].$$

2. A constant factor may be removed from under the integral sign:

$$\int_a^b c\, f(x)\, dF(x) = c \int_a^b f(x)\, dF(x).$$

3. The integral of a sum of a finite number of functions is equal to the sum of their integrals:

$$\int_a^b \sum_{i=1}^{n} f_i(x)\, dF(x) = \sum_{i=1}^{n} \int_a^b f_i(x)\, dF(x).$$

4. If $f(x) \geqq 0$ and $b > a$, then

$$\int_a^b f(x)\, dF(x) \geqq 0.$$

5. If $F_1(x)$ and $F_2(x)$ are monotonic functions of bounded variation, and c_1 and c_2 are arbitrary constants, then

$$\int_a^b f(x)\, d[c_1 F_1(x) + c_2 F_2(x)] = c_1 \int_a^b f(x)\, dF_1(x) + c_2 \int_a^b f(x)\, dF_2(x).$$

6. If $F(x) = \int_c^x g(u)\, dG(u)$, where c is a constant, $g(u)$ a continuous function, and $G(u)$ a non-decreasing function of bounded variation, then

$$\int_a^b f(x)\, dF(x) = \int_a^b f(x)\, g(x)\, dG(x).$$

By use of the concept of Stieltjes integral, we can write a general formula for the distribution function of the sum of two independent random variables ξ_1 and ξ_2:

$$F(x) = \int F_1(x-z)\,dF_2(z) = \int F_2(x-z)\,dF_1(z)$$

as well as one for their quotient ξ_1/ξ_2:

$$F(x) = \int_0^\infty F_1(xz)\,dF_2(z) + \int_{-\infty}^0 [1 - F_1(xz)]\,dF_2(z),$$

under the assumption that $P\{\xi_2 = 0\} = 0$.

EXERCISES

1. Prove that if $F(x)$ is a distribution function, then for any $h \neq 0$ the functions

$$\Phi(x) = \frac{1}{h}\int_x^{x+h} F(z)\,dz \quad \text{and} \quad \Psi(x) = \frac{1}{2h}\int_{x-h}^{x+h} F(z)\,dz$$

are also distribution functions.

2. The random variable ξ has $F(x)$ as its distribution function ($p(x)$ is the density function). Find the distribution function (density function) for each of the random variables:

a) $\eta = a\xi + b$, where a and b are real numbers;
b) $\eta = \xi^{-1}$ ($P\{\xi = 0\} = 0$);
c) $\eta = \tan \xi$;
d) $\eta = \cos \xi$;
e) $\eta = f(\xi)$, where $f(x)$ is a continuous monotonic function not having any intervals of constancy.

3. A line is drawn from the point $(0, a)$ making an angle φ with the y-axis. Find the distribution function for the abscissa of the point of intersection of this line with the x-axis if

a) the angle φ is uniformly distributed in the interval $(0, \pi/2)$;
b) the angle φ is uniformly distributed in the interval $(-\pi/2, \pi/2)$.

4. A point is chosen at random on the circumference of a circle of radius R with center at the origin (in other words, the polar angle of the point chosen is uniformly distributed in the interval $(-\pi, \pi)$). Find the density function for

a) the abscissa of the point selected;
b) the length of the chord joining this point to the point $(-R, 0)$.

5. A point is chosen at random on the segment of the y-axis between the points $(0, 0)$ and $(0, R)$ (i.e., the ordinate of this point is uniformly distributed in the interval $(0, R)$). Through this point we draw the chord of the circle $x^2 + y^2 = R^2$ that is perpendicular to the y-axis. Determine the distribution of the length of this chord.

6. The diameter of a circle is measured approximately. Assuming that this quantity is uniformly distributed in the interval (a, b), find the distribution of the area of the circle.

7. The density function of a random variable ξ is given by the expression

$$p(x) = \frac{a}{e^{-x} + e^x}.$$

Find

a) the constant a;

b) the probability that in two independent observations ξ will take on values less than 1.

8. The distribution function of a random vector (ξ, η) is given by

a) $F(x, y) = F_1(x) F_2(y) + F_3(x)$;

b) $F(x, y) = F_1(x) F_2(y) + F_3(x) + F_4(y)$.

Can the functions $F_3(x)$ and $F_4(y)$ be arbitrary? Are the components ξ, η of the vector dependent or independent?

9. Two points are chosen at random in the interval $(0, a)$ (i.e., their abscissas are uniformly distributed in the interval $(0, a)$). Determine the distribution function for the distance between the two points.

10. n points are chosen in the interval $(0, a)$. Assuming that the points are chosen at random (i.e., each of them is distributed independently of the others and uniformly in $(0, a)$), find

a) the density function for the abscissa of the k-th point from the left;

b) the joint density function of the abscissas of the k-th and m-th points from the left $(k < m)$.

11. n independent trials are performed for the values of a random variable having a continuous distribution function, yielding the following values of the variable: $x_1, x_2, \ldots, x_n$. Determine the distribution function of each of the random variables:

a) $\eta_n = \max(x_1, x_2, \ldots, x_n)$;

b) $\zeta_n = \min(x_1, x_2, \ldots, x_n)$;

c) the k-th largest observed value;

d) the joint distribution of the k-th and m-th largest observed values.

12. The distribution function of the random vector $(\xi_1, \xi_2, \ldots, \xi_n)$ is $F(x_1, x_2, \ldots, x_n)$. A trial is made, the outcome of which is that the components of the vector assume the values $(z_1, z_2, \ldots, z_n)$. Find the distribution function of each of the random variables:

a) $\eta_n = \max(z_1, z_2, \ldots, z_n)$;

b) $\zeta_n = \min(z_1, z_2, \ldots, z_n)$.

13. The random variable ξ has a continuous distribution function $F(x)$. How is the random variable $\eta = F(\xi)$ distributed?

14. The random variables ξ and η are independent; their density functions are given by the relations:

$$p_\xi(x) = p_\eta(x) = 0 \qquad \text{for} \quad x \leq 0$$
$$p_\xi(x) = c_1 x^\alpha e^{-\beta x}, \qquad p_\eta(x) = c_2 x^\gamma e^{-\beta x} \qquad \text{for} \quad x > 0.$$

Find
a) the constants c_1 and c_2;
b) the density function of the sum $\xi + \eta$.

15. Find the distribution function for the sum of the independent variables ξ and η, the first of which is uniformly distributed in the interval $(-h, h)$, and the second of which has the distribution function $F(x)$.

16. The density function of the random vector (ξ, η, ζ) is given by

$$p(x, y, z) = \begin{cases} \dfrac{6}{(1+x+y+z)^4} & \text{for} \quad x > 0, \, y > 0, \, z > 0 \\ 0 & \text{otherwise.} \end{cases}$$

Find the distribution of the variable $\xi + \eta + \zeta$.

17. Determine the distribution function for the sum of the independent random variables ξ_1 and ξ_2 if their distributions are given by the conditions:
a) $F_1(x) = F_2(x) = 1/2 + (1/\pi) \arctan x$.
b) They are distributed uniformly in the intervals $(-5, 1)$ and $(1, 5)$, respectively;
c) $p_1(x) = p_2(x) = \dfrac{1}{2a} e^{-\dfrac{|x|}{a}}$.

18. The density functions of the independent random variables ξ and η are

a) $p_\xi(x) = p_\eta(x) = \begin{cases} 0 & \text{for} \quad x < 0 \\ a e^{-ax} & \text{for} \quad x > 0 \quad (a > 0) \end{cases}$

b) $p_\xi(x) = p_\eta(x) = \begin{cases} 0 & \text{for} \quad x < 0, \, x > a \\ \dfrac{1}{a} & \text{for} \quad 0 \leq x \leq a. \end{cases}$

Determine the density function of the variable $\zeta = \xi/\eta$.

19. Find the distribution function of the product of two independent random variables ξ and η in terms of their distribution functions $F_1(x)$ and $F_2(x)$.

20. The random variables ξ and η are independent and are distributed
a) uniformly in the interval $(-a, a)$;
b) normally, with parameters $a = 0$ and $\sigma = 1$. Determine the distribution function of their product.

21. The lengths ξ and η of two sides of a triangle are independent random variables. On the basis of their distribution functions $F_\xi(x)$, $F_\eta(x)$, find the distribution function of the third side if the angle between the two sides in question has the constant value α.

22. Prove that if the variables ξ and η are independent and their density functions are given by

$$p_\xi(x) = p_\eta(x) = \begin{cases} 0 & \text{for } x \leq 0 \\ e^{-x} & \text{for } x > 0 \end{cases},$$

then the variables $\xi + \eta$ and ξ/η are also independent.

23. Prove that if the variables ξ and η are independent and normally distributed with the parameters $a_1 = a_2 = 0$ and $\sigma_1 = \sigma_2 = \sigma$, then the variables

$$\zeta = \xi^2 + \eta^2 \quad \text{and} \quad \delta = \frac{\xi}{\eta}$$

are also independent.

24. Prove that if the variables ξ and η are independent and distributed according to the χ^2-law with parameters m and n, then the variables $\zeta = \xi + \eta$ and $\delta = \dfrac{\xi}{\eta}$ are independent.

25. The random variables $\xi_1, \xi_2, \ldots, \xi_n$ are independent and all have the same density function

$$p(x) = \frac{1}{\sigma\sqrt{2\pi}} e^{-\frac{(x-a)^2}{2\sigma^2}}.$$

Determine the two-dimensional density function of the variables

$$\eta = \sum_{k=1}^{n} \xi_k \quad \text{and} \quad \xi = \sum_{k=1}^{m} \xi_k \quad (m < n).$$

26. Prove that every distribution function has the following properties:

$$\lim_{x \to \infty} x \int_x^\infty \frac{1}{z} dF(z) = 0, \quad \lim_{x \to +0} x \int_x^\infty \frac{1}{z} dF(z) = 0,$$

$$\lim_{x \to -\infty} x \int_{-\infty}^x \frac{1}{z} dF(z) = 0, \quad \lim_{x \to -0} x \int_{-\infty}^x \frac{1}{z} dF(z) = 0.$$

27. Two series of independent trials are carried out for the values of a random variable ξ having a continuous distribution function $F(x)$; the random variable takes on the following values, arranged in increasing order of magnitude within each series:

$$x_1 < x_2 < \cdots < x_M, \qquad y_1 < y_2 < \cdots < y_N.$$

What is the probability of the inequality

$$y_\mu < x_{m+1} < y_{\mu+1},$$

where m and μ are given quantities $(0 < m < M, 0 < \mu < N)$?

28. The random variable ξ has a continuous distribution function $F(x)$. In n independent observations, ξ assumes the values $x_1 < x_2 < \ldots < x_n$, arranged in increasing order of magnitude. Find the density function of the variable

$$\eta = \frac{F(x_n) - F(x_2)}{F(x_n) - F(x_1)}.$$

CHAPTER V

NUMERICAL CHARACTERISTICS OF RANDOM VARIABLES

In the preceding chapter, we saw that the most complete characterization of a random variable is given by its distribution function. In fact, the distribution function simultaneously indicates what values a random variable may assume and with what probabilities. However, in a number of cases, one needs to know considerably less about the random variable, and a summary description will suffice. In probability theory and its applications, an important role is played by certain constants (numerical characteristics) obtained from the distribution functions of random variables in accordance with definite rules. Of these constants that serve to give a general quantitative characterization of random variables, the mathematical expectation, the variance, and the moments of various orders are particularly important.

§ 26. Mathematical Expectation

We begin our discussion by considering the following illustrative example. Suppose that when a certain gun is fired at a particular target one shot is needed to hit the target with probability p_1, two shots with probability p_2, 3 shots with probability p_3, and so on. Moreover, it is known that n shots definitely suffice to hit this target; thus, we know that

$$p_1 + p_2 + \ldots + p_n = 1.$$

We ask, How many shots, on the average, are needed to hit the target?

To answer this question, we argue as follows. Suppose that a very large number of shots are fired under the conditions specified above. Then, by Bernoulli's Theorem, we may assert that the relative frequency of the firings in which only a single shot is needed to hit the target is approximately p_1. Similarly, two shots are needed in approximately $100p_2\%$ of the firings, etc. Thus, "on the average" approximately

$$1 \cdot p_1 + 2 \cdot p_2 + \cdots + n \cdot p_n$$

shots are required to hit a single target.

Analogous questions involving the calculation of the average value of a random variable arise in the most varied of problems. This is why in the theory of probability we introduce the special numerical characteristic called the *mathematical expectation*. We shall give a definition of this concept first for discrete random variables, taking the example just considered as a model.

Let

$$x_1, x_2, \ldots, x_n, \ldots$$

denote the possible values of a discrete random variable ξ, and

$$p_1, p_2, \ldots, p_n, \ldots$$

their corresponding probabilities.

If the series $\sum_{n=1}^{\infty} x_n p_n$ converges *absolutely*, then its sum is called the *mathematical expectation*, or simply *expectation*, or *expected value* of the random variable ξ and is denoted by $\mathsf{E}\xi$.

For continuous random variables, the following definition is a natural one: If the random variable ξ is continuous and $p(x)$ is its density function, then the expectation of the variable ξ is given by the integral

$$\mathsf{E}\xi = \int x\, p(x)\, dx \tag{1}$$

in those cases in which the integral

$$\int |x|\, p(x)\, dx$$

exists.

The expectation of an arbitrary random variable ξ with a distribution function $F(x)$ is defined by the Stieltjes integral

$$\mathsf{E}\xi = \int x\, dF(x). \tag{2}$$

Using the definition of a Stieltjes integral, we can give a simple geometrical interpretation of the concept of expectation: the expectation is

§ 26. MATHEMATICAL EXPECTATION

the difference between the area bounded by the y-axis, the line $y = 1$, and the curve $y = F(x)$ in the interval $(0, \infty)$ and the area bounded by the x-axis, the y-axis, and the curve $y = F(x)$ in the interval $(-\infty, 0)$. In Fig. 18, the corresponding areas have been shaded and the algebraic sign that each takes when they are summed has been indicated. We note,

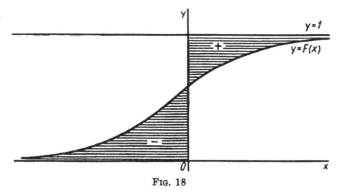

FIG. 18

incidentally, that the geometrical interpretation allows us to write the expectation in the following form:

$$\mathsf{E}\xi = -\int_{-\infty}^{0} F(x)\, dx + \int_{0}^{\infty} (1 - F(x))\, dx. \tag{3}$$

This observation enables us in many cases to find the expectation almost without computation. Thus, the expectation of the random variable distributed according to the law given at the end of § 22, is equal to one-half.

We note that, of the random variables considered earlier, the variable distributed according to the Cauchy law (Example 5, § 24) does not have an expectation.

We shall now discuss some examples.

Example 1. Determine the expectation of a random variable ξ distributed according to the normal law

$$p(x) = \frac{1}{\sigma\sqrt{2\pi}} \exp\left(-\frac{(x-a)^2}{2\sigma^2}\right).$$

By formula (2), we find:

$$\mathsf{E}\xi = \int x \frac{1}{\sigma\sqrt{2\pi}} \exp\left(-\frac{(x-a)^2}{2\sigma^2}\right) dx.$$

V. NUMERICAL CHARACTERISTICS OF RANDOM VARIABLES

The change of variables $z = (x-a)/\sigma$ reduces this integral to the form

$$\mathsf{E}\xi = \frac{1}{\sqrt{2\pi}} \int (\sigma z + a) e^{-\frac{z^2}{2}} dz = \frac{\sigma}{\sqrt{2\pi}} \int z e^{-\frac{z^2}{2}} dz + \frac{a}{\sqrt{2\pi}} \int e^{-\frac{z^2}{2}} dz.$$

Since

$$\int e^{-\frac{z^2}{2}} dz = \sqrt{2\pi} \quad \text{and} \quad \int z e^{-\frac{z^2}{2}} dz = 0,$$

it follows that

$$\mathsf{E}\xi = a.$$

We have thus obtained the following important result, which gives the probabilistic significance of one of the parameters defining the normal law: *The parameter a in the normal law of distribution of a random variable is equal to its expectation.*

Example 2. Find the expectation of a random variable ξ which is uniformly distributed in the interval (a, b).

We have:

$$\mathsf{E}\xi = \int_a^b x \frac{dx}{b-a} = \frac{b^2 - a^2}{2(b-a)} = \frac{a+b}{2}.$$

We see that the expectation of ξ coincides with the midpoint of the interval of possible values of the random variable.

Example 3. Find the expectation of the random variable ξ distributed according to the Poisson law

$$\mathsf{P}\{\xi = k\} = \frac{a^k e^{-a}}{k!} \quad (k = 0, 1, 2, \ldots).$$

We have:

$$\mathsf{E}\xi = \sum_{k=0}^{\infty} k \cdot \frac{a^k e^{-a}}{k!} = \sum_{k=1}^{\infty} k \cdot \frac{a^k e^{-a}}{k!} = a e^{-a} \sum_{k=1}^{\infty} \frac{a^{k-1}}{(k-1)!}$$

$$= a e^{-a} \sum_{k=0}^{\infty} \frac{a^k}{k!} = a.$$

If $F(x/B)$ is the conditional distribution function for a random variable ξ, then the integral

$$\mathsf{E}(\xi/B) = \int x \, dF(x/B) \tag{4}$$

is called the *conditional expectation of the random variable ξ with respect to the event B*.

§ 26. Mathematical Expectation

Let $B_1, B_2, \ldots, B_n$ be a complete group of mutually exclusive events and $F(x/B_1), F(x/B_2), \ldots, F(x/B_n)$ the conditional distribution functions of the variable ξ corresponding to these events. Let $F(x)$ denote the unconditional distribution function of the variable ξ; by the formula on total probability, we find:

$$F(x) = \sum_{k=1}^{n} \mathsf{P}(B_k) F(x/B_k).$$

This equation together with (4) enables us to obtain the following expression:

$$\mathsf{E}\xi = \sum_{k=1}^{n} \mathsf{P}(B_k)\mathsf{E}(\xi/B_k),$$

which obviously can also be written:

$$\mathsf{E}\xi = \mathsf{E}\{\mathsf{E}(\xi/B_k)\}. \tag{5}$$

In many cases, the formula just derived considerably simplifies the computation of the expectation.

Example 4. A workman operates n machines of the same kind set a feet apart in a row (Fig. 19). Assuming that in attending to the machines

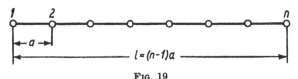

FIG. 19

the workman takes them in order of priority, find the average path taken between machines (the expectation of the path length).

Let us number the machines 1 to n, from left to right, and let B_k denote the event that the workman is at the machine numbered k. Since, by hypothesis, all the machines are of the same kind, the probability $p_i^{(k)}$ that the machine numbered i will be the next machine requiring the workman's attention is equal to $1/n$ ($1 \leq i \leq n$). The path length λ in this case is given by

$$\lambda_i^{(k)} = \begin{cases} (k-i)a & \text{for } k \geq i, \\ (i-k)a & \text{for } k < i. \end{cases}$$

By definition,

$$E(\lambda/B_k) = \frac{1}{n}\left(\sum_{i=1}^{k}(k-i)\,a + \sum_{i=k+1}^{n}(i-k)\,a\right)$$
$$= \frac{a}{n}\left(\frac{k(k-1)}{2} + \frac{(n-k)(n-k+1)}{2}\right)$$
$$= \frac{a}{2n}[2k^2 - 2(n+1)k + n(n+1)].$$

The probability that the workman is at the k-th machine is $1/n$, and therefore, by formula (5), we find:

$$E\lambda = \sum_{k=1}^{n}\frac{a}{2n^2}[2k^2 - 2(n+1)k + n(n+1)].$$

From the well-known formula

$$\sum_{k=1}^{\infty}k^2 = \frac{n(n+1)(2n+1)}{6},$$

we therefore obtain

$$E\lambda = \frac{a(n^2-1)}{3n} = \frac{l}{3}\left(1 + \frac{1}{n}\right),$$

where $l = (n-1)a$ signifies the distance between the end machines.

The mathematical expectation of an n-dimensional random variable $(\xi_1, \xi_2, \ldots, \xi_n)$ is defined as the set of n integrals

$$a_k = \iint \cdots \int x_k\, dF(x_1, \ldots, x_k, \ldots, x_n) = \int x\, dF_k(x) = E\,\xi_k,$$

where $F_k(x)$ is the distribution function of the variable ξ_k.[1]

Example 5. The density function of a two-dimensional random variable (ξ_1, ξ_2) is given by the expression (the two-dimensional normal distribution)

---

[1] We shall not give a formal definition of an n-dimensional Stieltjes integral, first, because we are actually going to consider only discrete variables and continuous variables and, second, because it is essentially the theory of the abstract Lebesgue integral rather than the general theory of the Stieltjes integral that is necessary in the theory of probability (for more details about this, see Chapter I of the monograph by Gnedenko and Kolmogorov, *Limit Distributions for Sums of Independent Random Variables* (Reading, Mass., 1954)).

$$p(x_1, x_2) = \frac{1}{2\pi\sigma_1\sigma_2\sqrt{1-r^2}} \exp\left\{-\frac{1}{2(1-r^2)}\left[\frac{(x_1-a)^2}{\sigma_1^2} - \frac{2r(x_1-a)(x_2-b)}{\sigma_1\sigma_2} + \frac{(x_2-b)^2}{\sigma_2^2}\right]\right\}.$$

Find its expectation.

By definition,

$$a_1 = \int\int x_1\, p(x_1, x_2)\, dx_1\, dx_2 = \int x_1\, p_1(x_1)\, dx_1$$

and

$$a_2 = \int\int x_2\, p(x_1, x_2)\, dx_1\, dx_2 = \int x_2\, p_2(x_2)\, dx_2.$$

In example 2 of § 23, we saw that

$$p_1(x_1) = \frac{1}{\sigma_1\sqrt{2\pi}} \exp\left\{-\frac{(x_1-a)^2}{2\sigma_1^2}\right\}, \quad p_2(x_2) = \frac{1}{\sigma_2\sqrt{2\pi}} \exp\left\{-\frac{(x_2-b)^2}{2\sigma_2^2}\right\},$$

and therefore, according to the results of Example 1 of the present section, we find:

$$a_1 = a, \quad a_2 = b.$$

Thus we have also succeeded in ascertaining the probabilistic meaning of the parameters a and b for the two-dimensional normal distribution.

§ 27. Variance

The *variance* of a random variable ξ is defined as the expectation of the square of the deviation of ξ from $\mathsf{E}\xi$.[2] We shall take the symbol $\mathsf{D}\xi$ as the notation for variance.[3] Thus, by definition:

$$\mathsf{D}\xi = \mathsf{E}(\xi - \mathsf{E}\xi)^2 = \int_{-\infty}^{\infty} x\, dF_\eta(x), \tag{1}$$

where $F_\eta(x)$ denotes the distribution function of the random variable $\eta = (\xi - \mathsf{E}\xi)^2$.

In practical problems, one makes use of another expression, namely:

$$\mathsf{D}\xi = \int (z - \mathsf{E}\xi)^2\, dF_\xi(z). \tag{2}$$

---

[2] The square root of the variance is called the standard deviation.
[3] The variance is also often denoted by $\mathrm{Var}(\xi)$.

The equivalence of formulas (1) and (2) follows immediately from the following general proposition.

THEOREM: *If $F_\xi(x)$ is the distribution function of a random variable ξ and $f(x)$ is a continuous function, then*

$$\mathsf{E}\, f(\xi) = \int f(x)\, dF_\xi(x).$$

We shall confine ourselves at this point to a proof of this theorem for the simplest special case

$$f(x) = (x-a)^k,$$

postponing the complete proof until § 29.

Let us introduce the notation

$$G(x) = \mathsf{P}\{(\xi-a)^k < x\};$$

then by definition

$$\mathsf{E}\,(\xi-a)^k = \int_{-\infty}^{\infty} x\, dG(x).$$

If k is an odd integer, then $(\xi-a)^k$ is a non-decreasing function of ξ, and therefore

$$G(x) = \mathsf{P}\{(\xi-a)^k < x\} = \mathsf{P}\{\xi - a < \sqrt[k]{x}\}$$
$$= \mathsf{P}\{\xi < a + \sqrt[k]{x}\} = F(a + \sqrt[k]{x}).$$

For odd k, we thus have:

$$\mathsf{E}\,(\xi-a)^k = \int x\, dF(a + \sqrt[k]{x}).$$

It is not difficult to show that the substitution $z = a + \sqrt[k]{x}$ reduces this integral to the form:

$$\mathsf{E}\,(\xi-a)^k = \int_{-\infty}^{\infty} (z-a)^k\, dF(z).$$

If, however, k is even, then $(\xi-a)^k$ is a non-negative quantity, and therefore $G(x) = 0$ for $x \leq 0$. For $x > 0$,

§ 27. Variance

$$G(x) = \mathsf{P}\{(\xi-a)^k < x\} = \mathsf{P}\{a - \sqrt[k]{x} < \xi < a + \sqrt[k]{x}\}$$
$$= F(a + \sqrt[k]{x}) - F(a - \sqrt[k]{x} + 0).$$

Thus, for even k, we have

$$\mathsf{E}\,(\xi-a)^k = \int_0^\infty x\,dF(a + \sqrt[k]{x}) - \int_0^\infty x\,dF(a - \sqrt[k]{x} + 0).$$

By making the substitutions $z = a + \sqrt[k]{x}$ in the first integral and $z = a - \sqrt[k]{x}$ in the second one, we can express $\mathsf{E}(\xi - a)^k$ in the form:

$$\mathsf{E}\,(\xi-a)^k = \int_{-\infty}^\infty (x-a)^k\,dF(x).$$

For practical purposes, it is useful to write expression (2) in an alternative form. Since

$$(z - \mathsf{E}\,\xi)^2 = z^2 - 2z\,\mathsf{E}\,\xi + (\mathsf{E}\,\xi)^2 \quad \text{and} \quad \mathsf{E}\,\xi = \int z\,dF_\xi(z),$$

formula (2) is also expressible as:

$$\mathsf{D}\,\xi = \int z^2\,dF_\xi(z) - (\int z\,dF_\xi(z))^2 = \mathsf{E}\,\xi^2 - (\mathsf{E}\,\xi)^2. \tag{3}$$

Inasmuch as the variance is a non-negative quantity, we conclude from the last relation that

$$\int z^2\,dF_\xi(z) \geqq (\int z\,dF_\xi(z))^2.$$

This inequality is a special case of the familiar Schwarz inequality.[4]

As is the case with mathematical expectation, the variance does not exist for all random variables. Thus, the Cauchy law, which we considered earlier (Example 5, § 24), does not have a finite variance.

---

[4] Also known as the Cauchy-Schwarz inequality and, in the Russian literature, as the Bunyakovsky inequality or the Bunyakovsky-Cauchy inequality. [*Trans.*]

V. Numerical Characteristics of Random Variables

Let us calculate the variance for a few distributions, by way of example.

Example 1. Find the variance of the random variable ξ which is distributed uniformly in the interval (a, b).

In our example,

$$\int x^2 \, dF_\xi(x) = \int_a^b \frac{x^2}{b-a} \, dx = \frac{b^3 - a^3}{3(b-a)} = \frac{b^2 + ab + a^2}{3}.$$

In the preceding section it was found that

$$\mathsf{E}\,\xi = \frac{a+b}{2}.$$

Hence,

$$\mathsf{D}\,\xi = \frac{a^2 + ab + b^2}{3} - \left(\frac{a+b}{2}\right)^2 = \frac{(b-a)^2}{12}.$$

We see that the variance depends only on the length of the interval (a, b) and is an increasing function of the length. The larger the interval of the values taken on by the random variable, i.e., the more spread out its values are, the larger will its variance be. Thus, the variance is important as a *measure of the dispersion* of the values of the random variable about its expectation.

Example 2. Find the variance of a random variable ξ distributed according to the normal law

$$p(x) = \frac{1}{\sigma \sqrt{2\pi}} \exp\left\{-\frac{(x-a)^2}{2\sigma^2}\right\}.$$

We know that $\mathsf{E}\,\xi = a$, and therefore

$$\mathsf{D}\,\xi = \int (x-a)^2 \, p(x) \, dx = \frac{1}{\sigma \sqrt{2\pi}} \int (x-a)^2 \, e^{-\frac{(x-a)^2}{2\sigma^2}} \, dx.$$

We now make a change of variable in the integral, setting

$$z = \frac{x-a}{\sigma};$$

as a result,

$$\mathsf{D}\,\xi = \frac{\sigma^2}{\sqrt{2\pi}} \int z^2 \, e^{-\frac{z^2}{2}} \, dz.$$

Integrating by parts, we find:

§ 27. Variance

$$\int z^2 e^{-\frac{z^2}{2}} dz = -ze^{-\frac{z^2}{2}}\Big|_{-\infty}^{\infty} + \int e^{-\frac{z^2}{2}} dz = \sqrt{2\pi}.$$

Thus, we finally obtain:

$$\mathsf{D}\,\xi = \sigma^2.$$

We have thus ascertained the probabilistic meaning of the second parameter in the normal law. We see that the *normal law is completely determined by the expectation and the variance*. This fact is widely used in theoretical investigations.

We note that in the case of a normally distributed random variable also, the variance allows us to judge as to how its values are dispersed. Although for any positive value of the variance a normally distributed random variable can assume *all* real values, the dispersion of its values will all the same be smaller, the smaller the variance; also, the closer these values are to the mathematical expectation, the larger are their probabilities. We noted this fact in the preceding chapter when we first encountered the normal law.

Example 3. Find the variance of the random variable λ considered in Example 4 of § 26.

Retaining the notation of Example 4, we find:

$$\mathsf{E}(\lambda^2/B_k) = \frac{1}{n}\left(\sum_{i=1}^{k}(k-i)^2 a^2 + \sum_{i=k+1}^{n}(i-k)^2 a^2\right)$$

$$= \frac{a^2}{6n}[(k-1)\cdot k(2k-1) + (n-k)(n-k+1)(2n-2k+1)]$$

$$= \frac{a^2}{6}[6k^2 - 6(n+1)k + (2n+1)(n+1)]$$

and, therefore,

$$\mathsf{E}(\lambda^2) = \frac{1}{n}\sum_{k=1}^{n}\mathsf{E}(\lambda^2/B_k) = \frac{a^2}{6n}[n(n+1)(2n+1) -$$

$$- 3(n+1)^2 n + n(n+1)(2n+1)] = \frac{a^2}{6}(n^2-1).$$

From this it follows that

$$\mathsf{D}(\lambda) = \mathsf{E}(\lambda^2) - (\mathsf{E}\,\lambda)^2 = \frac{a^2}{6}(n^2-1) - \frac{a^2(n^2-1)^2}{9n^2}$$

$$= \frac{a^2(n^2-1)(n^2+2)}{18n^2} = \frac{l^2}{18}\left(1 + \frac{2}{n} + \frac{4}{n^2} + \frac{6}{n^2(n-1)}\right).$$

V. NUMERICAL CHARACTERISTICS OF RANDOM VARIABLES

The *variance* (or *covariance matrix*) *of an n-dimensional random variable* $(\xi_1, \xi_2, \ldots, \xi_n)$ is defined as the collection of n^2 constants given by the formula:

$$b_{jk} = \int\int \cdots \int (x_j - \mathsf{E}\,\xi_j)(x_k - \mathsf{E}\,\xi_k)\, dF(x_1, x_2, \ldots, x_n) \qquad (4)$$

$$(1 \leq k \leq n, \quad 1 \leq j \leq n).$$

Since, for any real values of t_j $(1 \leq j \leq n)$,

$$\int \cdots \int \left\{\sum_{j=1}^{n} t_j (x_j - \mathsf{E}\,\xi_j)\right\}^2 dF(x_1, x_2, \ldots, x_n) = \sum_{j=1}^{n}\sum_{k=1}^{n} b_{jk}\, t_j\, t_k \geq 0,$$

it follows, as we know from the theory of quadratic forms, that the quantities b_{jk} satisfy the inequality

$$\begin{vmatrix} b_{11} & b_{12} & \cdots & b_{1k} \\ b_{21} & b_{22} & \cdots & b_{2k} \\ \cdots & \cdots & \cdots & \cdots \\ b_{k1} & b_{k2} & \cdots & b_{kk} \end{vmatrix} \geq 0 \quad \text{for} \quad k = 1, 2, \ldots, n.$$

It is obvious that

$$b_{kk} = \mathsf{D}\,\xi_k.$$

The quantities b_{jk} for $k \neq j$ are called *the mixed central moments of second order*, or *covariances*, of the variables ξ_j and ξ_k; obviously, $b_{jk} = b_{kj}$.

The following function of the moments of second order:

$$r_{ij} = \frac{b_{ij}}{\sqrt{b_{ii}\, b_{jj}}}$$

is called the *correlation coefficient* of the variables ξ_i and ξ_j. The correlation coefficient is a measure of the linear functional dependence of the variables ξ_i and ξ_j. The values of the correlation coefficient lie in the interval $(-1, 1)$.

The values ± 1 are assumed only if ξ_i is a linear function of ξ_j.

In fact, since

$$\mathsf{D}\left(\frac{\xi_i}{\sqrt{b_{ii}}} \pm \frac{\xi_j}{\sqrt{b_{jj}}}\right) = 2(1 \pm r_{ij}) \geq 0,$$

we have

$$-1 \leq r_{ij} \leq 1.$$

§ 27. Variance

The equality $r_{ij} = 1$ is possible if and only if

$$D\left(\frac{\xi_i}{\sqrt{b_{ii}}} - \frac{\xi_j}{\sqrt{b_{jj}}}\right) = 0.$$

However, the variance of a random variable can vanish only if the variable takes on a single value with probability one. Thus, if $r_{ij} = 1$, then

$$\frac{\xi_i}{\sqrt{b_{ii}}} - \frac{\xi_j}{\sqrt{b_{jj}}} = c$$

and, therefore,

$$\xi_i = \sqrt{\frac{b_{ii}}{b_{jj}}}\,\xi_j + \alpha \qquad (\alpha = c\sqrt{b_{ii}}).$$

Analogously, if $r_{ij} = -1$, then

$$\xi_i = -\sqrt{\frac{b_{ii}}{b_{jj}}}\,\xi_j + \alpha.$$

It is easy to show that *the correlation coefficient of two independent random variables ξ_i and ξ_j is equal to zero.* Similarly, a straightforward computation shows that *if random variables are linearly connected, then their correlation coefficient is equal to plus or minus one.*

Example 4. Find the variance of the two-dimensional random variable (ξ_1, ξ_2) distributed in accordance with the non-degenerate normal law

$$p(x, y) = \frac{1}{2\pi\sigma_1\sigma_2\sqrt{1-r^2}} \times$$

$$\times \exp\left\{-\frac{1}{2(1-r^2)}\left[\frac{(x-a)^2}{\sigma_1^2} - 2r\frac{(x-a)(y-b)}{\sigma_1\sigma_2} + \frac{(y-b)^2}{\sigma_2^2}\right]\right\}$$

By formula (4) and the results of Example 2 of the present section and Example 1 of § 26, we find:

$$D\,\xi_1 = \sigma_1^2, \qquad D\,\xi_2 = \sigma_2^2.$$

Further,

$$b_{12} = b_{21} = \iint (x-a)(y-b)\,p(x, y)\,dx\,dy =$$

$$= \frac{1}{2\pi\sigma_1\sigma_2\sqrt{1-r^2}} \int e^{-\frac{(y-b)^2}{2\sigma_2^2}}\,dy \times$$

$$\times \int (x-a)(y-b)\exp\left\{-\frac{1}{2(1-r^2)}\left(\frac{x-a}{\sigma_1} - r\frac{y-b}{\sigma_2}\right)^2\right\}dx.$$

By means of the substitutions

$$z = \frac{1}{\sqrt{1-r^2}}\left(\frac{x-a}{\sigma_1} - r\frac{y-b}{\sigma_2}\right), \quad t = \frac{y-b}{\sigma_2},$$

the expression for b_{12} can be reduced to the form

$$b_{12} = b_{21} = \frac{1}{2\pi}\int\int (\sigma_1\sigma_2\sqrt{1-r^2}\,tz + r\,\sigma_1\sigma_2\,t^2)\,e^{-\frac{t^2}{2}-\frac{z^2}{2}}\,dz\,dt =$$

$$= \frac{r\sigma_1\sigma_2}{2\pi}\int t^2 e^{-\frac{t^2}{2}}\,dt \int e^{-\frac{z^2}{2}}\,dz + \frac{\sigma_1\sigma_2\sqrt{1-r^2}}{2\pi}\int t\,e^{-\frac{t^2}{2}}\,dt \int z\,e^{-\frac{z^2}{2}}\,dz$$

$$= r\,\sigma_1\,\sigma_2\,.$$

From this, we find:

$$r = \frac{\int\int (x-a)(y-b)\,p(x,y)\,dx\,dy}{\sigma_1\sigma_2} = \frac{\mathsf{E}\{(\xi_1 - \mathsf{E}\,\xi_1)(\xi_2 - \mathsf{E}\,\xi_2)\}}{\sqrt{\mathsf{D}\,\xi_1\,\mathsf{D}\,\xi_2}}.$$

We thus see that *the two-dimensional normal law is completely determined*, just as is the one-dimensional case, *by the expectation and the variance*, i.e., it is determined by the five quantities $\mathsf{E}\xi_1$, $\mathsf{E}\xi_2$, $\mathsf{D}\xi_1$, $\mathsf{D}\xi_2$, and r.

§ 28. Theorems on Expectation and Variance

THEOREM 1: *The expectation of a constant is that constant.*

Proof. We may regard the constant C as a discrete random variable which can only assume the one value C with probability one; therefore,

$$\mathsf{E}C = C \cdot 1 = C.$$

THEOREM 2: *The expectation of the sum of two random variables is equal to the sum of their expectations:*

$$\mathsf{E}(\xi + \eta) = \mathsf{E}\xi + \mathsf{E}\eta.$$

Proof. Let us first consider the case of discrete random variables ξ and η. Let $a_1, a_2, \ldots, a_n, \ldots$ be the possible values of the variable ξ and $p_1, p_2, \ldots, p_n, \ldots$, their probabilities; and let $b_1, b_2, \ldots, b_k, \ldots$ the possible values of η and $q_1, q_2, \ldots, q_k, \ldots$, their probabilities. The possible values

§ 28. Theorems on Expectation and Variance

of the quantity $\xi + \eta$ are given by $a_n + b_k$ ($k, n = 1, 2, \ldots$). Let p_{nk} denote the probability that ξ assumes the value a_n and η the value b_k. By definition, the expectation

$$E(\xi + \eta) = \sum_{n,k=1}^{\infty} (a_n + b_k) p_{nk} = \sum_{n=1}^{\infty} \sum_{k=1}^{\infty} (a_n + b_k) p_{nk}$$

$$= \sum_{n=1}^{\infty} a_n \left(\sum_{k=1}^{\infty} p_{nk} \right) + \sum_{k=1}^{\infty} b_k \left(\sum_{n=1}^{\infty} p_{nk} \right).$$

Since, by the theorem on total probability,

$$\sum_{k=1}^{\infty} p_{nk} = p_n \quad \text{and} \quad \sum_{n=1}^{\infty} p_{nk} = q_k,$$

it follows that

$$\sum_{n=1}^{\infty} a_n \sum_{k=1}^{\infty} p_{nk} = \sum_{n=1}^{\infty} a_n p_n = E\xi$$

and

$$\sum_{k=1}^{\infty} b_k \sum_{n=1}^{\infty} p_{nk} = \sum_{k=1}^{\infty} b_k q_k = E\eta.$$

This completes the proof of the theorem for the case of discrete summands.

Similarly, if a two-dimensional density function $p(x, y)$ exists for the random variable (ξ, η) we find by formula (3) of § 24 that

$$E\zeta = E(\xi + \eta) = \int x \, dF_\zeta(x) = \int x \left(\int p(z, x - z) \, dz \right) dx$$
$$= \iint x \, p(z, x - z) \, dz \, dx = \iint (z + y) \, p(z, y) \, dz \, dy$$
$$= \iint z \, p(z, y) \, dz \, dy + \iint y \, p(z, y) \, dz \, dy$$
$$= \int z \, p_\xi(z) \, dz + \int y \, p_\eta(y) \, dy = E\xi + E\eta.$$

Theorem 2 will be proved for the general case in § 29.

COROLLARY 1: *The expectation of the sum of a finite number of random variables is equal to the sum of their expectations:*

$$E(\xi_1 + \xi_2 + \ldots + \xi_n) = E\xi_1 + E\xi_2 + \ldots + E\xi_n.$$

In fact, by virtue of the theorem just proved, we have

$$E(\xi_1 + \xi_2 + \ldots + \xi_n) = E\xi_1 + E(\xi_2 + \xi_3 + \ldots + \xi_n) =$$
$$= E\xi_1 + E\xi_2 + E(\xi_3 + \ldots + \xi_n) = \ldots = E\xi_1 + E\xi_2 + \ldots + E\xi_n.$$

COROLLARY 2: *Consider the sum*

$$\zeta_\mu = \xi_1 + \xi_2 + \cdots + \xi_\mu,$$

where μ is a random variable taking on only integral values, the random variables $\xi_1, \xi_2, \ldots$ are independent of μ, the expectation of μ is finite, and the series

$$\sum_{k=1}^{\infty} \mathsf{E}\, |\xi_k|\, \mathsf{P}\{\mu \geq k\}$$

converges. Then the expectation of the sum exists and is equal to

$$\mathsf{E}\,\zeta_\mu = \sum_{j=1}^{\infty} \mathsf{E}\,\xi_j\, \mathsf{P}\{\mu \geq j\}.$$

Proof. The conditional expectation of the sum, given that $\mu = k$, is in fact equal to

$$\mathsf{E}\,\{\zeta_\mu/\mu = k\} = \mathsf{E}\,\xi_1 + \mathsf{E}\,\xi_2 + \cdots + \mathsf{E}\,\xi_k.$$

The unconditional expectation is

$$\mathsf{E}\,\zeta_\mu = \sum_{k=1}^{\infty} \mathsf{E}\,\{\zeta_\mu/\mu = k\} \cdot \mathsf{P}\{\mu = k\} = \sum_{k=1}^{\infty} \mathsf{P}\{\mu = k\} \sum_{j=1}^{k} \mathsf{E}\,\xi_j$$

$$= \sum_{j=1}^{\infty} \mathsf{E}\,\xi_j \sum_{k=j}^{\infty} \mathsf{P}\{\mu = k\} = \sum_{j=1}^{\infty} \mathsf{E}\,\xi_j\, \mathsf{P}\{\mu \geq j\}.$$

If the terms $\xi_1, \xi_2, \xi_3, \ldots$ are identically distributed, i.e., if $\mathsf{P}\{\xi_1 < x\} = \mathsf{P}\{\xi_2 < x\} = \ldots = F(x)$, then

$$\mathsf{E}\,\zeta_\mu = \mathsf{E}\,\xi_1 \cdot \mathsf{E}\,\mu.$$

Indeed,

$$\mathsf{E}\,\zeta_\mu = \sum_{k=1}^{\infty} \mathsf{P}\{\mu = k\} \sum_{j=1}^{k} \mathsf{E}\,\xi_j = \mathsf{E}\,\xi_1 \sum_{k=1}^{\infty} k\, \mathsf{P}\{\mu = k\} = \mathsf{E}\,\xi_1 \cdot \mathsf{E}\,\mu.$$

Example 1. The number of cosmic particles that fall on a given portion of ground is a random variable μ which is subject to the Poisson law with parameter a; each particle has an energy ξ depending on chance. Find the mean energy $\mathscr{E}$ acquired by the ground in one unit of time.

According to Corollary 2, we have:

$$\mathsf{E}\,\mathscr{E} = \mathsf{E}\xi \cdot \mathsf{E}\mu = a\mathsf{E}\xi.$$

§ 28. THEOREMS ON EXPECTATION AND VARIANCE

Example 2. A particular target is fired at until it is hit for the n-th time. Assuming that the shots are fired independently of each other and that the probability of a hit in each round is equal to p, find the expectation of the number of shells expended.

Let ξ_k denote the number of shots fired from the $(k-1)$-st hit to the k-th hit (inclusive of the latter). It is obvious that the number of shells required for the n hits is

$$\xi = \xi_1 + \xi_2 + \ldots + \xi_n,$$

and therefore

$$\mathsf{E}\xi = \mathsf{E}\xi_1 + \mathsf{E}\xi_2 + \ldots + \mathsf{E}\xi_n$$

But

$$\mathsf{E}\xi_1 = \mathsf{E}\xi_2 = \ldots = \mathsf{E}\xi_n$$

and

$$\mathsf{E}\xi_1 = \sum_{k=1}^{\infty} k\, q^{k-1}\, p = \frac{p}{(1-q)^2} = \frac{1}{p},$$

and hence

$$\mathsf{E}\xi = \frac{n}{p}.$$

THEOREM 3: *The expectation of the product of two independent random variables ξ and η is equal to the product of their expectations.*

Proof: Suppose that the variables ξ and η are discrete. If $a_1, a_2, \ldots, a_k, \ldots$ are the possible values of ξ and $p_1, p_2, \ldots, p_k, \ldots$, their respective probabilities and if $b_1, b_2, \ldots, b_n, \ldots$ are the possible values of η and $q_1, q_2, \ldots, q_n, \ldots$, their respective probabilities, then the probability that ξ and η will take on the values a_k and b_n respectively is equal to $p_k q_n$. By the definition of expectation,

$$\mathsf{E}\,\xi\,\eta = \sum_{k,n} a_k\, b_n\, p_k\, q_n = \sum_{k=1}^{\infty}\sum_{n=1}^{\infty} a_k\, b_n\, p_k\, q_n$$

$$= \left(\sum_{k=1}^{\infty} a_k\, p_k\right)\left(\sum_{n=1}^{\infty} b_n\, q_n\right) = \mathsf{E}\,\xi\,\mathsf{E}\,\eta.$$

We leave to the reader the task, only slightly more complicated, of carrying out the proof of the theorem for the case of continuous variables.

We postpone until § 29 the proof of the theorem in the general case.

COROLLARY 1: *A constant factor may be removed from under the mathematical expectation sign*:

$$\mathsf{E}c\xi = c\mathsf{E}\xi.$$

This statement is obvious, since, whatever ξ may be, the constant c and the variable ξ may be considered to be independent variables.

THEOREM 4: *The variance of a constant is zero.*

Proof: According to Theorem 1,

$$\mathsf{D}c = \mathsf{E}(c - \mathsf{E}c)^2 = \mathsf{E}(c - c)^2 = \mathsf{E}0 = 0.$$

THEOREM 5: *If c is a constant, then*

$$\mathsf{D}\,c\,\xi = c^2\,\mathsf{D}\,\xi.$$

Proof: By virtue of the corollary to Theorem 3, we have

$$\mathsf{D}\,c\,\xi = \mathsf{E}\,[c\,\xi - \mathsf{E}\,c\,\xi]^2 = \mathsf{E}\,[c\,\xi - c\,\mathsf{E}\,\xi]^2 =$$
$$= \mathsf{E}\,c^2\,[\xi - \mathsf{E}\,\xi]^2 = c^2\,\mathsf{E}\,[\xi - \mathsf{E}\,\xi]^2 = c^2\,\mathsf{D}\,\xi.$$

THEOREM 6: *The variance of the sum of two independent random variables ξ and η is equal to the sum of their variances*:

$$\mathsf{D}\,(\xi + \eta) = \mathsf{D}\,\xi + \mathsf{D}\,\eta.$$

Proof: In fact,

$$\mathsf{D}\,(\xi + \eta) = \mathsf{E}\,[\xi + \eta - \mathsf{E}\,(\xi + \eta)]^2 = \mathsf{E}\,[(\xi - \mathsf{E}\,\xi) + (\eta - \mathsf{E}\,\eta)]^2$$
$$= \mathsf{D}\,\xi + \mathsf{D}\,\eta + 2\,\mathsf{E}\,(\xi - \mathsf{E}\,\xi)\,(\eta - \mathsf{E}\,\eta).$$

Since the variables ξ and η are independent, the variables $\xi - \mathsf{E}\xi$ and $\eta - \mathsf{E}\eta$ are also independent; hence

$$\mathsf{E}\,(\xi - \mathsf{E}\,\xi)\,(\eta - \mathsf{E}\,\eta) = \mathsf{E}\,(\xi - \mathsf{E}\,\xi) \cdot \mathsf{E}\,(\eta - \mathsf{E}\,\eta) = 0.$$

COROLLARY 1: *If $\xi_1, \xi_2, \ldots, \xi_n$ are random variables such that each of them is independent of the sum of the preceding ones, then*

$$\mathsf{D}\,(\xi_1 + \xi_2 + \cdots + \xi_n) = \mathsf{D}\,\xi_1 + \mathsf{D}\,\xi_2 + \cdots + \mathsf{D}\,\xi_n.$$

COROLLARY 2: *The variance of the sum of a finite number of pairwise independent random variables $\xi_1, \xi_2, \ldots, \xi_n$ is equal to the sum of their variances.*

§ 28. Theorems on Expectation and Variance

Proof: We have
$$D(\xi_1 + \xi_2 + \cdots + \xi_n) = E\left(\sum_{k=1}^{n}(\xi_k - E\,\xi_k)\right)^2 =$$
$$= E\sum_{j=1}^{n}\sum_{k=1}^{n}(\xi_k - E\,\xi_k)(\xi_j - E\,\xi_j) = \sum_{k=1}^{n}\sum_{j=1}^{n} E\,(\xi_k - E\,\xi_k)(\xi_j - E\,\xi_j)$$
$$= \sum_{k=1}^{n} D\,\xi_k + \sum_{k \neq j} E\,(\xi_k - E\,\xi_k)(\xi_j - E\,\xi_j).$$

Because of the independence of any pair of quantities ξ_j and ξ_k ($j \neq k$), it follows that
$$E(\xi_k - E\,\xi_k)(\xi_j - E\,\xi_j) = 0$$
for $j \neq k$. With this, obviously, the proof is complete.

Example 3. The ratio
$$\frac{\xi - E\,\xi}{\sqrt{D\,\xi}}$$
is called the *normalized deviation* of the random variable ξ, or simply, the normalized variable corresponding to ξ. Prove that
$$D\left(\frac{\xi - E\,\xi}{\sqrt{D\,\xi}}\right) = 1.$$

Since ξ and $E\xi$, regarded as random variables, are independent, by virtue of Theorems 5 and 6, we therefore have
$$D\left(\frac{\xi - E\,\xi}{\sqrt{D\,\xi}}\right) = \frac{D\,\xi + D(-E\,\xi)}{D\,\xi} = \frac{D\,\xi}{D\,\xi} = 1.$$

Example 4. If ξ and η are independent random variables, then
$$D(\xi - \eta) = D\,\xi + D\,\eta.$$

By virtue of Theorems 5 and 6, $D(-\eta) = (-1)^2 D\eta = D\eta$, and so $D(\xi - \eta) = D\xi + D\eta$.

Example 5. Theorems 2 and 6 enable us to compute in a very simple way the expectation and the variance of μ, the number of occurrences of an event A in n independent trials.

Let p_k be the probability of occurrence of the event A in the k-th trial. Let μ_k denote the number of times the event A occurs in the k-th trial. Obviously, μ_k is a random variable assuming the values 0 or 1 with the probabilities $q_k = 1 - p_k$ and p_k, respectively.

The variable μ is thus representable in the form

$$\mu = \mu_1 + \mu_2 + \cdots + \mu_n.$$

Since

$$\mathsf{E}\,\mu_k = 0 \cdot q_k + 1 \cdot p_k = p_k,$$

$$\mathsf{D}\,\mu_k = \mathsf{E}\,\mu_k^2 - (\mathsf{E}\,\mu_k)^2 = 0 \cdot q_k + 1 \cdot p_k - p_k^2 = p_k(1-p_k) = p_k\,q_k.$$

we can conclude from the theorems proved above that

$$\mathsf{E}\,\mu = p_1 + p_2 + \cdots + p_n \quad \text{and} \quad \mathsf{D}\,\mu = p_1\,q_1 + \cdots + p_n\,q_n$$

For the case of Bernoulli trials, $p_k = p$ and therefore

$$\mathsf{E}\mu = np \quad \text{and} \quad \mathsf{D}\mu = npq.$$

We note that it follows from this that

$$\mathsf{E}\frac{\mu}{n} = p, \quad \mathsf{D}\frac{\mu}{n} = \frac{pq}{n}.$$

Example 6. Let us find the expectation and variance of the number of times an event E occurs in n trials connected in a homogeneous Markov chain.

As in the preceding example, let μ_k be the number of times the event E occurs in the k-th trial. The number of occurrences of the event in n trials is

$$\mu = \mu_1 + \mu_2 + \ldots + \mu_n.$$

But

$$\mathsf{E}\mu = \sum_{k=1}^{n} \mathsf{E}\mu_k = \sum_{k=1}^{n} p_k.$$

By formula (1') of § 20,

$$p_k = p + (p_1 - p)\delta^{k-1}$$

Thus,

$$\mathsf{E}\mu = np + \sum_{k=1}^{n}(p_1 - p)\delta^{k-1} =$$
$$= np + (p_1 - p)\frac{1-\delta^n}{1-\delta}.$$

§ 28. Theorems on Expectation and Variance

By definition,

$$D\mu = E(\mu - E\mu)^2 = E\left(\sum_{k=1}^{n}(\mu_k - p_k)\right)^2 =$$
$$= \sum_{k=1}^{n} E(\mu_k - p_k)^2 + 2 \sum_{j>i} E(\mu_i - p_i)(\mu_j - p_j).$$

But

$$D\mu_k = p_k q_k = pq + (q-p)(p_1-p)\delta^{k-1} - (p_1-p)^2\delta^{2k-2},$$

where

$$q_k = 1 - p_k = q - (p_1 - p)\frac{1-\delta^n}{1-\delta}.$$

Furthermore,

$$E(\mu_i - p_i)(\mu_j - p_j) = E\mu_i\mu_j - p_i p_j.$$

Since the probability of the equality $\mu_i\mu_j = 1$ is clearly $p_i p_j^{(i)}$, it follows that

$$E(\mu_i - p_i)(\mu_j - p_j) = p_i(p_j^{(i)} - p_j).$$

Using formulas (1') and (2') of § 20, we find that

$$E(\mu_i - p_i)(\mu_j - p_j) = pq\delta^{j-i} + (p_1-p)(q-p)\delta^{j-1} - (p_1-p)^2\delta^{i+j-2}.$$

Now

$$\sum_{k=1}^{n} D\mu_k = npq + (q-p)(p_1-p)\frac{1-\delta^n}{1-\delta} - (p_1-p)^2\frac{1-\delta^{2n}}{1-\delta^2}$$

and

$$\sum_{j>i} E(\mu_i - p_i)(\mu_j - p_j) = npq\frac{\delta}{1-\delta} - pq\frac{\delta}{1-\delta}\left(1 + \frac{\delta - \delta^n}{1-\delta}\right) +$$
$$+ \frac{(p_1-p)(q-p)}{1-\delta}\left(\frac{\delta - \delta^{n+1}}{1-\delta} - n\delta^n\right) - \frac{(p_1-p)^2}{1-\delta} \cdot \frac{\delta(1-\delta^{n-1})(1-\delta^n)}{1-\delta^2}.$$

Thus,

$$D\mu = npq\frac{1+\delta}{1-\delta} + a_n,$$

where a_n is a certain quantity which remains bounded as n increases.

§ 29. The Definition of Mathematical Expectation in Kolmogorov's Axiomatic Treatment

The present section should be omitted at a first reading, since it requires a deeper knowledge of the theory of integration. The general interpretation which is presented here is the natural development of the concepts of random event, probability, and random variable as formulated by Kolmogorov (see §§ 9 and 21). In this interpretation, the concept of expectation leads in a natural way to the abstract Lebesgue integral.

By definition, the mathematical expectation of a random variable $\xi = f(e)$ is the integral

$$\mathsf{E}\,\xi = \int_{\mathscr{E}} f(e)\,\mathsf{P}\{de\}.$$

Its conditional expectation, under the hypothesis B, is equal to

$$\mathsf{E}\{\xi|B\} = \int_{\mathscr{E}} f(e)\,\mathsf{P}\{de|B\}.$$

It is easily shown that this definition is equivalent to the following one:

$$\mathsf{E}\{\xi|B\} = \int_{B} f(e)\,\mathsf{P}\{de\} \cdot \frac{1}{\mathsf{P}\{B\}},$$

which is often more suitable in practical applications.

We note that if the event B is representable as the sum of a finite or denumerable number of disjoint (mutually exclusive) events B_k:

$$B = B_1 + B_2 + \cdots,$$

then

$$\int_{B} f(e)\,\mathsf{P}\{de\} = \sum_{k} \int_{B_k} f(e)\,\mathsf{P}\{de\}.$$

It is also worth noting that, whereas we previously had to carry out a rather lengthy argument in proving the theorem on the expectation of a sum, this theorem is now a consequence of the relation

$$\int (f+g)\mathsf{P}\{de\} = \int f\mathsf{P}\{de\} + \int g\mathsf{P}\{de\}.$$

§ 29. Definition of Mathematical Expectation

We previously derived the formula

$$E(\xi \cdot \eta) = E\xi \cdot E\eta \qquad (1)$$

for two independent random variables ξ and η only for the case of discrete variables and the case of continuous variables.

To treat the general case, we now define discrete random variables ξ_n and η_n by means of the relations

$$\xi_n = \frac{m}{n} \quad \text{for} \quad \frac{m}{n} \leq \xi < \frac{m+1}{n},$$

$$\eta_n = \frac{k}{n} \quad \text{for} \quad \frac{k}{n} \leq \eta < \frac{k+1}{n}.$$

Then

$$E(\xi_n \cdot \eta_n) = E\xi_n \cdot E\eta_n.$$

From the well-known theorems on Lebesgue integrals concerning passing to the limit under the integral sign, we easily conclude that

$$\lim E\xi_n = E\xi, \quad \lim E\eta_n = E\eta,$$
$$\lim E(\xi_n \cdot \eta_n) = E(\xi \cdot \eta).$$

Thus, formula (1) is proved for the general case.

We now apply the results obtained to derive a formula which generalizes Corollary 2 of Theorem 2, § 28. This expression will be obtained from the following theorem, which is due to Kolmogorov and Prokhorov.

Let there be given a sequence of random variables

$$\xi_1, \xi_2, \ldots, \xi_n, \ldots$$

and let

$$\zeta_\nu = \xi_1 + \xi_2 + \cdots + \xi_\nu,$$

denote the sum of the first ν variables, the number of terms ν itself being a random variable.

Let S_m denote the event that $\nu = m$, and set

$$p_m = P\{S_m\}, \qquad P_n = P\{\nu \geq n\} = \sum_{m=n}^{\infty} p_m.$$

THEOREM: *If for $n > m$ the random variable ξ_n and the event S_m are independent, the expectations*
$$a_n = \mathsf{E}\xi_n$$
exist (which implies that the quantities $c_n = \mathsf{E}|\xi_n|$ are finite), and the series
$$\sum_{n=1}^{\infty} c_n P_n$$
converges, then the expectation of the variable ζ_ν exists and is given by
$$\mathsf{E}\,\zeta_\nu = \sum_{n=1}^{\infty} p_n A_n,$$
where
$$A_n = \mathsf{E}\,\zeta_n = a_1 + a_2 + \cdots + a_n.$$

Proof: By virtue of the given assumptions,
$$\sum_{n=1}^{\infty} p_n A_n = \sum_{n=1}^{\infty} P_n a_n.$$

Since ξ_n is independent of the event $\{\nu < n\}$, it is also independent of the complementary event $\{\nu \geq n\}$, and therefore
$$a_n = \mathsf{E}\xi_n = \mathsf{E}\{\xi_n/\nu \geq n\}.$$

Taking into account the last two equations and also the properties of conditional expectations that were cited before, we can write the following sequence of equations:
$$\sum_{n=1}^{\infty} p_n A_n = \sum_{n=1}^{\infty} \mathsf{P}\{\nu \geq n\}\, \mathsf{E}\,\{\xi_n/\nu \geq n\} = \sum_{n=1}^{\infty} \int_{\{\nu \geq n\}} \xi_n\, \mathsf{P}\{de\}$$
$$= \sum_{n=1}^{\infty} \sum_{m=n}^{\infty} \int_{\{\nu = m\}} \xi_n\, \mathsf{P}\{de\}.$$

But since the variable $|\xi_n|$ and the event $\{\nu \geq n\}$ are also independent, it follows that
$$\sum_{n=1}^{\infty} \sum_{m=n}^{\infty} \left| \int_{\{\nu = m\}} \xi_n\, \mathsf{P}\{de\} \right| \leq \sum_{n=1}^{\infty} \sum_{m=n}^{\infty} \int_{S_m} |\xi_n|\, \mathsf{P}\{de\} =$$
$$= \sum_{n=1}^{\infty} \int_{\{\nu \geq n\}} |\xi_n|\, \mathsf{P}\{de\} = \sum_{n=1}^{\infty} \mathsf{P}\{\nu \geq n\}\, \mathsf{E}\,\{|\xi_n|/\nu \geq n\} =$$
$$= \sum_{n=1}^{\infty} \mathsf{P}\{\nu \geq n\}\, \mathsf{E}\,|\xi_n| = \sum_{n=1}^{\infty} P_n c_n < +\infty.$$

The estimate just obtained allows us to write the equation

$$\sum_{n=1}^{\infty}\sum_{m=n}^{\infty}\int_{S_m}\xi_n\,P(de) = \sum_{m=1}^{\infty}\sum_{n=1}^{m}\int_{S_m}\xi_n\,P(de) = \sum_{m=1}^{\infty}\int_{S_m}\zeta_m\,P(de).$$

Since

$$\mathsf{E}\,\zeta_\nu = \int_U \zeta_\nu\,P(de) = \sum_{m=1}^{\infty}\int_{S_m}\zeta_m\,P(de),$$

the preceding equation proves the theorem.

COROLLARY: *Under the assumptions of the preceding theorem, but with* $a = a_1 = a_2 = \ldots$, *we have*

$$\mathsf{E}\,\zeta_\nu = a\,\mathsf{E}\,\nu = a\sum_{n=1}^{\infty} n\,p_n.$$

§ 30. Moments

The expectation of the variable $(\xi - a)^k$ is called the *moment of k-th order*, or *k-th moment*, of the random variable ξ:

$$\nu_k(a) = \mathsf{E}(\xi - a)^k. \tag{1}$$

If $a = 0$, then the moment is called the *k-th moment about the origin*. It is clear that the first moment about the origin is the expectation of the variable ξ.

If $a = \mathsf{E}\xi$, then the moment is called the *central moment of k-th order*. It is easy to see that the central moment of first order vanishes and that the central moment of second order is nothing other than the variance.

We shall denote moments about the origin by ν_k, and central moments by μ_k; in both cases, the subscript indicates the order of the moment.

There exists a simple relation between central moments and moments about the origin. In fact,

$$\mu_n = \mathsf{E}\,(\xi - \mathsf{E}\,\xi)^n = \sum_{k=0}^{n}\binom{n}{k}(-\mathsf{E}\,\xi)^{n-k}\mathsf{E}\,\xi^k = \sum_{k=0}^{n}\binom{n}{k}(-\mathsf{E}\,\xi)^{n-k}\nu_k. \tag{2}$$

Since $\nu_1 = \mathsf{E}\xi$, we have:

$$\mu_n = \sum_{k=2}^{n}(-1)^{n-k}\binom{n}{k}\nu_k\,\nu_1^{n-k} + (-1)^{n-1}(n-1)(\nu_1)^n. \tag{3}$$

Let us write out explicitly the relation between the moments for the first four values of n:

$$\left.\begin{aligned} \mu_0 &= 1, \\ \mu_1 &= 0, \\ \mu_2 &= \nu_2 - \nu_1^2, \\ \mu_3 &= \nu_3 - 3\nu_2\nu_1 + 2\nu_1^3, \\ \mu_4 &= \nu_4 - 4\nu_3\nu_1 + 6\nu_2\nu_1^2 - 3\nu_1^4. \end{aligned}\right\} \quad (3')$$

These first few moments play an especially important part in statistics.

The quantity

$$m_k = \mathsf{E}\,|\,\xi - a\,|^k \tag{4}$$

is referred to as the *absolute moment* of k-th order.

According to Theorem 1 of § 27,

$$\nu_k(a) = \int (x-a)^k\, dF(x). \tag{5}$$

Since we have agreed to say that a random variable ξ has an expectation only if the integral defining the expectation converges absolutely, it is clear that the k-th moment about the origin of the variable ξ exists if and only if the integral

$$\int |x|^k\, dF_\xi(x)$$

converges. From this remark, it follows that if the random variable ξ has a moment of k-th order, then it also has moments of all positive orders less than k. In fact, since for $r < k$, $|x|^r < |x|^k$ when $|x| > 1$, we have:

$$\int |x|^r\, dF_\xi(x) = \int_{|x|\leq 1} |x|^r\, dF_\xi(x) + \int_{|x|>1} |x|^r\, dF_\xi(x) \leq$$

$$\leq \int_{|x|\leq 1} |x|^r\, dF_\xi(x) + \int_{|x|>1} |x|^k\, dF_\xi(x).$$

The first integral on the right-hand side of this inequality is finite, because the limits of integration are finite and the integrand is bounded; the second integral converges by assumption.

Example. Determine the central and the absolute central moments of a random variable distributed according to the normal law:

$$p(x) = \frac{1}{\sigma\sqrt{2\pi}} \exp\left\{-\frac{(x-a)^2}{2\sigma^2}\right\}.$$

§ 30. MOMENTS

We have:

$$\mu_k = \frac{1}{\sigma\sqrt{2\pi}} \int (x-a)^k \exp\left\{-\frac{(x-a)^2}{2\sigma^2}\right\} dx = \frac{\sigma^k}{\sqrt{2\pi}} \int x^k e^{-\frac{x^2}{2}} dx.$$

For odd k,

$$\mu_k = 0,$$

because the integrand is an odd function.

For even k,

$$\mu_k = m_k = \sqrt{\frac{2}{\pi}} \sigma^k \int_0^\infty x^k e^{-\frac{x^2}{2}} dx.$$

The substitution $x^2 = 2z$, reduces this integral to the form

$$\mu_k = m_k = \sqrt{\frac{2}{\pi}} \sigma^k 2^{\frac{k-1}{2}} \int_0^\infty z^{\frac{k-1}{2}} e^{-z} dz = \sqrt{\frac{2}{\pi}} \sigma^k 2^{\frac{k-1}{2}} \Gamma\left(\frac{k+1}{2}\right) =$$

$$= \sigma^k (k-1)(k-3)\ldots 1 = \sigma^k \frac{k!}{2^{k/2}\left(\frac{k}{2}\right)!}.$$

For odd values of k, the absolute moment is

$$m_k = \sqrt{\frac{2}{\pi}} \sigma^k \int_0^\infty x^k e^{-\frac{x^2}{2}} dx = \sqrt{\frac{2}{\pi}} \sigma^k 2^{\frac{k-1}{2}} \Gamma\left(\frac{k+1}{2}\right) =$$

$$= \sqrt{\frac{2}{\pi}} 2^{\frac{k-1}{2}} \left(\frac{k-1}{2}\right)! \sigma^k.$$

The moments of a distribution cannot be arbitrary quantities. As a matter of fact, the quadratic form

$$J_n = \int \left(\sum_{k=0}^n t_k(x-a)^k\right)^2 dF(x) = \sum_{j=0}^n \sum_{k=0}^n \nu_{k+j}(a) t_k t_j \geq 0$$

is non-negative regardless of the values of the constants $t_0, t_1, \ldots, t_n$; therefore, the first $2n$ moments $\nu_j(a)$ must satisfy the determinantal inequalities:

$$\begin{vmatrix} \nu_0(a) & \nu_1(a) & \ldots & \nu_k(a) \\ \nu_1(a) & \nu_2(a) & \ldots & \nu_{k+1}(a) \\ \ldots & \ldots & \ldots & \ldots \\ \nu_k(a) & \nu_{k+1}(a) & \ldots & \nu_{2k}(a) \end{vmatrix} \geqq 0 \qquad (k=0,1,2,\ldots,n).$$

The absolute moments also have to satisfy analogous inequalities.

We prove one more theorem concerning absolute moments, as follows.

THEOREM: *If the random variable ξ has an absolute moment of k-th order, then for arbitrary t and τ $(0 < t < \tau < k)$,*

$$\sqrt[t]{m_t} \leqq \sqrt[\tau]{m_\tau} \leqq \sqrt[k]{m_k},$$

where

$$m_t = \mathsf{E}\,|\xi - a|^t$$

and a is any real number.

Proof: Let us first prove the theorem for the case where t, τ, and k are rational numbers. Specifically, let

$$t = \frac{p}{q}, \qquad \tau = \frac{s}{q}, \qquad k = \frac{w}{q},$$

where, by hypothesis,

$$p < s < w.$$

Now, let r be some positive integer less than w. Let us consider the positive-semidefinite quadratic form

$$m_{\frac{r-1}{q}} u^2 + 2 m_{\frac{r}{q}} uv + m_{\frac{r+1}{q}} v^2 = \int \left[u\,|x-a|^{\frac{r-1}{2q}} + v\,|x-a|^{\frac{r+1}{2q}} \right]^2 dF(x).$$

The condition that it be positive-semidefinite is, as is well known, that the inequality

$$m_{\frac{r}{q}}^2 \leqq m_{\frac{r-1}{q}} \cdot m_{\frac{r+1}{q}}$$

hold. This inequality can obviously be written in the form:

§ 30. Moments

$$m_{\frac{r}{q}}^{2r} \leq m_{\frac{r-1}{q}}^{r} \cdot m_{\frac{r+1}{q}}^{r}.$$

If we assign to r the set of values from 1 to r, we obtain the sequence of inequalities

$$m_{\frac{1}{q}}^{2} \leq m_0\, m_{\frac{2}{q}},$$

$$m_{\frac{2}{q}}^{2\cdot 2} \leq m_{\frac{1}{q}}^{2}\, m_{\frac{3}{q}}^{2},$$

$$\cdots\cdots\cdots\cdots\cdots$$

$$m_{\frac{r}{q}}^{2r} \leq m_{\frac{r-1}{q}}^{r}\, m_{\frac{r+1}{q}}^{r}.$$

If we multiply these inequalities together and observe that m_0 is always equal to one, we obtain, after cancelling out a suitable common factor from both sides of the resulting inequality,

$$m_{\frac{r}{q}}^{r+1} \leq m_{\frac{r+1}{q}}^{r}.$$

Hence

$$m_{\frac{r}{q}}^{\frac{1}{r}} \leq m_{\frac{r+1}{q}}^{\frac{1}{r+1}},$$

or

$$m_{\frac{r}{q}}^{\frac{q}{r}} \leq m_{\frac{r+1}{q}}^{\frac{q}{r+1}}.$$

This inequality, obviously, proves the theorem for the case of rational t, τ, and k.

Since the function m_t is continuous with respect to the argument t in the interval $0 \leq t \leq k$, we can satisfy ourselves by means of a limit argument that the theorem remains valid for any values of t, τ, and k.

We note that the theorem just proved implies the following important property of moments:

$$m_1 \leq m_2^{\frac{1}{2}} \leq m_3^{\frac{1}{3}} \leq \cdots \leq m_k^{\frac{1}{k}} \leq m_{k+1}^{\frac{1}{k+1}} \leq \cdots.$$

In the examples of the preceding sections, where the form of the distribution function was known in advance, the first two moments of the random variable completely determined the function (this was true for the normal, Poisson, uniform distributions, etc.). In mathematical statistics, an important role is played by distribution laws that depend on more than two parameters. If it is known in advance that a random variable is subject to a law of a well-defined form and only the values of the parameters are unknown, then in the most important cases these unknown parameters can be determined in terms of the first few moments. However, if we do not know the form of the distribution function then, generally speaking, a knowledge of the first few moments or even of all the moments of integral order does not enable us to determine the desired distribution function. It turns out to be possible to construct examples in which distinct distributions have identical moments of all integral orders. In this connection, there arises the question (*The Problem of Moments*): Given the sequence of constants

$$c_0 = 1, \ c_1, \ c_2, \ c_3, \ \ldots \ ;$$

1) Under what conditions does there exist a distribution function $F(x)$ for which all n of the equalities

$$c_n = \int x^n \, dF(x)$$

hold, and

2) When is this function unique?

There exists at present a complete solution to this problem, but we shall not dwell further on this issue, since it is outside the scope of our book.

Among the other numerical characteristics, a most important role is played by the so-called *semi-invariants*, or *cumulants*; we shall postpone their definition until Chapter VII, merely observing the following at this point. The moment of a sum of independent random variables is not, in general, equal to the sum of the moments of the summands. We have the relation

$$\mathsf{E}(\xi + \eta)^n = \sum_{k=0}^{n} \binom{n}{k} \mathsf{E}\,\xi^k \, \mathsf{E}\,\eta^{n-k}$$

for the moment of the sum of two independent variables ξ and η.

The semi-invariants of different orders possess the property that the semi-invariant of a sum of independent variables is equal to the sum of the semi-invariants of the same order of the summands. It turns out that the semi-invariant of any order k is a rational function of the moments of orders less than or equal to k.

EXERCISES

1. A random variable ξ assumes only non-negative integral values with the probabilities

a) $P\{\xi = k\} = \dfrac{a^k}{(1+a)^{k+1}}$, where a is a positive constant (this distribution is called the *Pascal distribution*).

b) $p_k = P\{\xi = k\} = \left(\dfrac{\alpha}{1+\alpha\beta}\right)^k \dfrac{1(1+\beta)\dots(1+(k-1)\beta)}{k!} p_0$ for all $k > 0$.

where $\alpha > 0$, $\beta > 0$, and

$$p_0 = P\{\xi = 0\} = (1+\alpha\beta)^{-\frac{1}{\beta}}.$$

This distribution is called the *Polya distribution*.

Determine $E\xi$ and $D\xi$.

2. Let μ be the number of occurrences of an event A in n independent trials in each of which $P(A) = p$. Find

a) $E\mu^3$, b) $E\mu^4$, and c) $E|\mu - np|$.

3. The probability of occurrence of an event A in the i-th trial is equal to p_i. Let μ be the number of times the event A occurs in the first n independent trials. Find

a) $E\mu$, b) $D\mu$, c) $E\left(\mu - \sum_{i=1}^{n} p_i\right)^3$, d) $E\left(\mu - \sum_{i=1}^{n} p_i\right)^4$.

4. Prove that under the conditions of the preceding problem the maximum value of $D\mu$ for a given value of $a = \dfrac{1}{n}\sum_{i=1}^{n} p_i$ is attained when $p_1 = p_2 = \dots = p_n = a$.

5. Let μ be the number of times an event A occurs in n independent trials in each of which $P(A) = p$. Further, let the variable η be equal to 0 or 1 depending on whether μ proves to be even or odd. Determine $E\eta$.

6. The density function of the random variable ξ is

$$p(x) = \frac{1}{2\alpha} e^{-\frac{|x-a|}{\alpha}}$$

(the *Laplace distribution*). Find $E\xi$ and $D\xi$.

7. The density function of the magnitude of the velocity of a molecule is given by the *Maxwell distribution*

$$p(x) = \frac{4x^2}{\alpha^3 \sqrt{\pi}} e^{-\frac{x^2}{\alpha^2}} \quad \text{for } x > 0$$

and $p(x) = 0$ for $x \leq 0$ (α is a positive constant). Find the average speed and the average kinetic energy of a molecule (the mass of a molecule is m), and the variances of the speed and kinetic energy.

8. The probability density of the distance x from a reflecting wall at which a molecule in Brownian motion will be found at time t if it was at a distance of x_0 from the wall at time t_0, is given by the expression

$$p(x) = \frac{1}{2\sqrt{\pi D t}} \left[e^{-\frac{(x+x_0)^2}{4Dt}} + e^{-\frac{(x-x_0)^2}{4Dt}} \right] \quad \text{for } x \geq 0,$$

$$p(x) = 0 \quad \text{for } x < 0.$$

Find the expectation and variance of the magnitude of the displacement of the molecule during the time from $t = t_0$ to t.

9. Prove that an arbitrary random variable ξ whose possible values lie in the interval (a, b) satisfies the following inequalities:

$$a \leq \mathsf{E}\xi \leq b, \; \mathsf{D}\xi \leq (b-a)^2/4.$$

10. Let $x_1, x_2, \ldots, x_k$ be the possible values of a random variable ξ. Prove that

a) $\dfrac{\mathsf{E}\,\xi^{n+1}}{\mathsf{E}\,\xi^n} \to \max\limits_{1 \leq i \leq k} x_j$, b) $\sqrt[n]{\mathsf{E}\,\xi^n} \to \max\limits_{1 \leq i \leq k} x_j$

for $n \to \infty$

11. Let $F(x)$ be the distribution function of ξ. Prove that if $\mathsf{E}\xi$ exists then

$$\mathsf{E}\,\xi = \int_0^\infty [1 - F(x) - F(-x)]\,dx$$

and that

$$\lim_{x \to -\infty} x F(x) = \lim_{x \to +\infty} x[1 - F(x)] = 0$$

is a necessary condition for the existence of $\mathsf{E}\xi$.

12. Two points are picked at random in the interval $(0, l)$. Determine the expectation and variance of the distance between them and the expectation of the n-th power of the distance.

13. A random variable ξ is distributed according to the logarithmic normal law, i.e., for $x > 0$ the density function of ξ is

$$p(x) = \frac{1}{\beta x \sqrt{2\pi}} e^{-\frac{1}{2\beta^2}(\log x - a)^2} \quad \text{for } x > 0$$

($p(x) = 0$ for $x \leq 0$). Find $\mathsf{E}\xi$ and $\mathsf{D}\xi$.

(Kolmogorov has shown that the distribution of the size of the particles that result from pulverization is logarithmically normal.) [5]

14. A random variable ξ is uniformly distributed in the interval $(0, b)$. Find $\mathsf{E}|\xi - a|^k$, where $a = \mathsf{E}\xi$.

---

[5] For details of the lognormal distribution, see J. Aitchison and J. A. Brown, *The Lognormal Distribution* (Cambridge, 1957).

EXERCISES

15. There are 2^n tickets contained in a box; the number i $(i = 0, 1, \ldots, n)$ is marked on $\binom{n}{i}$ of them. m tickets are withdrawn at random and s is the sum of the numbers marked on them; find $\mathsf{E}s$ and $\mathsf{D}s$.

16. The random variables $\xi_1, \xi_2, \ldots, \xi_{n+m}$ $(n > m)$ are independent, identically distributed, and have finite variances. Determine the correlation coefficient of the two sums $s = \xi_1 + \xi_2 + \ldots + \xi_n$ and $\sigma = \xi_{m+1} + \xi_{m+2} + \ldots + \xi_{m+n}$.

17. The random variables ξ and η are independent and normally distributed with the same parameters a and σ. Determine the correlation coefficient of the variables $\alpha\xi + \beta\eta$ and $\alpha\xi - \beta\eta$, and also their joint distribution.

18. The random vector (ξ, η) is normally distributed; $\mathsf{E}\xi = a$, $\mathsf{E}\eta = b$, $\mathsf{D}\xi = \sigma_1^2$, $\mathsf{D}\eta = \sigma_2^2$, and R is the correlation coefficient of ξ and η. Prove that $R = \cos q\pi$, where $q = \mathsf{P}\{(\xi - a)(\eta - b) < 0\}$.

19. Let x_1 and x_2 be the results of two independent observations of a normally distributed variable ξ. Prove that $\mathsf{E}\max(x_1, x_2) = a + \sigma/\sqrt{\pi}$, where $a = \mathsf{E}\xi$ and $\sigma^2 = \mathsf{D}\xi$.

20. A random vector (ξ, η) is normally distributed, with $\mathsf{E}\xi = \mathsf{E}\eta = 0$, $\mathsf{D}\xi = \mathsf{D}\eta = 1$, and $\mathsf{E}\xi\eta = R$.

Prove that

$$\mathsf{E} \max(\xi, \eta) = \sqrt{\frac{1-R}{\pi}}$$

21. The unevenness in the length of cotton fibre is defined as the quantity

$$\lambda = \frac{a'' - a'}{a},$$

where a is the expectation of the fibre length, a'' is the expectation of the length of those fibres whose length is larger than a, and a' is the expectation of those fibres whose length is less than a. Determine the relation between the quantities

a) λ, a, and $\mathsf{E}|\xi - a|$;
b) λ, a, and σ, if ξ is distributed normally, where ξ is the fibre length.

CHAPTER VI

THE LAW OF LARGE NUMBERS

§ 31. Mass Phenomena and the Law of Large Numbers

The vast experience accumulated by mankind teaches us that a phenomenon with probability very close to one is almost certain to take place. Similarly, an event whose probability of occurrence is very small (in other words, very close to zero) happens very rarely. This fact plays a basic role in all inferences of a practical nature in probability theory. For, in practical work, this *empirical fact* gives one the right to consider events with small probabilities as being *practically impossible* and those with probabilities very close to one as being *practically certain*. However, the perfectly natural question, What should the probability of an event be if it is to be regarded as being practically impossible (or: practically certain), does not admit of a unique answer. This is quite understandable, since in any practical application it is necessary to take into account the importance of the kind of event with which we are dealing.

Thus, for example, if the distance between two villages were measured and found to be 15,340 ft. and if the error in this measurement were greater than or equal to 50 ft., with a probability of 0.02, then we could ignore the possibility of such an error and consider the distance as actually being equal to 15,340 ft. In this example, we thus consider an event with probability 0.02 as having no practical importance, and we do not take it into account in our practical work. At the same time, there are other cases where one cannot disregard an event having a probability of 0.02, or even a still smaller probability. Thus, if in the construction of a large hydro-electric plant requiring an enormous expenditure of material and manpower it were ascertained that the probability of a catastrophic flood under given conditions were equal to 0.02, then this probability would be considered large, and would have to be taken into account in the planning of the project rather than be disregarded as in the preceding example.

§ 31. Mass Phenomena and Law of Large Numbers

Thus, only practical considerations can suggest the criteria according to which various events are to be considered as being practically impossible or practically certain.

At the same time, it is necessary to observe that any event that has a positive probability, no matter how small, can indeed occur; and if its probability of occurrence is the same in each trial and the number of trials is very large, then the probability of the event occurring at least once can come as close to one as desired. This fact should always be borne in mind. However, if the probability of some event is very small, it is very hard to believe that it will occur in some particular trial *specified beforehand*. Thus, if a dealer says that four players are each going to be dealt cards, all of one suit on the very first deal, then it is natural to suspect that the dealer was influenced by certain definite information— for example, that the cards were arranged in some specific order known to him. Our confidence in this is based on the fact that the probability of such a distribution when the cards are properly shuffled is equal to $(13!)^4 4!/52! < 4.8 \cdot 10^{-30}$, i.e., it is extremely small. Nevertheless, the fact that such a distribution of cards has occurred is a matter of record. This example illustrates rather well the distinction between the concept of practical impossibility and that of, so to speak, categorical impossibility.

From this discussion, it is obvious that events with probabilities close to zero, or to one, are of great importance in practical applications as well as in general theoretical problems. Hence one of the fundamental problems of probability theory is clearly the establishment of laws which hold with probability close to one; an important part must be played here by those laws which arise as the result of imposing a large number of independent or weakly dependent chance factors. The law of large numbers is one such proposition in probability theory and, moreover, one of the most important.

It would now be natural that we take the law of large numbers to mean the aggregate of theorems which state that some event that depends on a set of random events whose number increases without limit and each of which has only a negligible effect upon it, will occur with a probability as close to one as desired.

This general description of theorems of the type of the law of large numbers can be formulated in a somewhat more definite way. Let there be given a sequence of random variables

$$\xi_1, \xi_2, \ldots, \xi_n, \ldots . \qquad (1)$$

Consider the random variables ζ_n, which are certain given symmetric functions of the first n variables in the sequence (1):

$$\zeta_n = f_n(\xi_1, \xi_2, \ldots, \xi_n).$$

If there exists a sequence of constants $a_1, a_2, \ldots, a_n$ such that for any $\varepsilon > 0$

$$\lim_{n \to \infty} P\{|\zeta_n - a_n| < \varepsilon\} = 1, \qquad (2)$$

then the sequence (1) obeys the law of large numbers relative to the given functions f_n.

However, it is customary to invest the concept of the law of large numbers with a much more specific content. Namely, we confine ourselves to the case where f_n is the *arithmetic mean* of the variables $\xi_1, \xi_2, \ldots, \xi_n$.

If each of the quantities a_n in relation (2) has the same value a, then we say that the sequence of random variables ζ_n *converges in probability* to a. In these terms, (2) means that $\zeta_n - a_n$ converges in probability to zero.

In observing particular phenomena, we observe them with all their individual peculiarities, which thus obscure the manifestation of the laws that hold when like phenomena are observed a large number of times. It was already observed a long time ago that factors not concerned with the overall nature of a process and which only show up in its individual realizations cancel each other out when the average of a large number of observations is considered.

Subsequently, this empirical result was noted more and more frequently and this, moreover, without any attempt, as a rule, to find a theoretical explanation for it. However, many authors did not demand such an explanation because in their day the existence of laws governing both natural and social phenomena was regarded as nothing other than the manifestation of the laws of divine order.

Certain authors have hitherto understated the content of the law of large numbers and have even distorted its methodological significance by reducing it simply to an experimentally observable law. Actually, the enduring scientific value of the investigations of Tchebychev, Markov, and others who have done research on the law of large numbers has not been the fact that they have observed the empirical stability of the mean but that they have determined general sufficient conditions for the statistical stability of the mean.

To show how the law of large numbers operates, we give the following illustrative example. From the standpoint of modern physics, a gas is composed of an enormous number of molecules in ceaseless chaotic motion. One cannot predict how fast each individual molecule will be moving, nor where it will be at any given moment of time. However, under given conditions as to the state of the gas, we can compute the proportion of molecules that have a given velocity or the proportion that are to be found in a given volume. But this, as a matter of fact, is exactly what the physicist needs to know, since the basic characteristics of a gas: its pressure, temperature, viscosity, etc. are determined not by the complex behavior of a single molecule but by the collective action of all the molecules. Thus, the pressure of a gas measures the overall effect of the molecules impinging on a unit area in a unit period of time. The number of impinging molecules and their velocities vary in a way depending on chance; however, by virtue of the law of large numbers (in Tchebychev's form), the pressure is almost a constant. This "equalizing" effect of the law of large numbers is observed in physical phenomena with exceptional exactness. It suffices to recall that, say, under ordinary conditions, even very precise measurements barely allow one to observe deviations from Pascal's law of hydrostatic pressure. This extremely good agreement between the theoretical results and experiment has even furnished the opponents of the molecular theory of matter with a unique argument: if matter did have a molecular structure, then deviations from Pascal's law should also be observed. These deviations, the so-called fluctuations of pressure, were indeed successfully observed after it had been learned how to isolate a comparatively small number of molecules, as a result of which the influence of the individual molecule is still strong enough and not completely neutralized.

§ 32. Tchebychev's Form of the Law of Large Numbers

We shall now formulate and prove the theorems of Tchebychev, Markov, and others; the method employed here is due to Tchebychev.

TCHEBYCHEV'S INEQUALITY: *For every random variable ξ having a finite variance and for every $\varepsilon > 0$, the inequality*

$$\mathsf{P}\{|\xi - \mathsf{E}\xi| \geqq \varepsilon\} \leqq \mathsf{D}\xi/\varepsilon^2 \qquad (1)$$

holds.

Proof: If $F(x)$ denotes the distribution function of the random variable ξ, then

$$\mathsf{P}\{|\xi - \mathsf{E}\,\xi| \geq \varepsilon\} = \int\limits_{|x - \mathsf{E}\,\xi| \geq \varepsilon} dF(x).$$

Since in the region of integration,

$$\frac{|x - \mathsf{E}\,\xi|}{\varepsilon} \geq 1,$$

it follows that

$$\int\limits_{|x - \mathsf{E}\,\xi| \geq \varepsilon} dF(x) \leq \frac{1}{\varepsilon^2} \int\limits_{|x - \mathsf{E}\,\xi| \geq \varepsilon} (x - \mathsf{E}\,\xi)^2\, dF(x).$$

By extending the integration to all values of x, we merely strengthen the inequality:

$$\int\limits_{|x - \mathsf{E}\,\xi| \geq \varepsilon} dF(x) \leq \frac{1}{\varepsilon^2} \int (x - \mathsf{E}\,\xi)^2\, dF(x) = \frac{\mathsf{D}\,\xi}{\varepsilon^2}.$$

This proves Tchebychev's inequality.

TCHEBYCHEV'S THEOREM: *If $\xi_1, \xi_2, \ldots, \xi_n, \ldots$ is a sequence of pairwise independent random variables possessing finite variances which are uniformly bounded:*

$$\mathsf{D}\,\xi_1 \leq C, \quad \mathsf{D}\,\xi_2 \leq C, \ldots, \mathsf{D}\,\xi_n \leq C, \ldots,$$

then for any positive constant ε,

$$\lim_{n \to \infty} \mathsf{P}\left\{\left|\frac{1}{n}\sum_{k=1}^{n} \xi_k - \frac{1}{n}\sum_{k=1}^{n} \mathsf{E}\,\xi_k\right| < \varepsilon\right\} = 1. \tag{2}$$

Proof: By the hypotheses of the theorem, we know that

$$\mathsf{D}\left(\frac{1}{n}\sum_{k=1}^{n} \xi_k\right) = \frac{1}{n^2}\sum_{k=1}^{n} \mathsf{D}\,\xi_k$$

and therefore

$$\mathsf{D}\left(\frac{1}{n}\sum_{k=1}^{n} \xi_k\right) \leq \frac{C}{n}.$$

According to Tchebychev's inequality, we have

§ 32. Tchebyshev's Form of Law of Large Numbers

$$P\left\{\left|\frac{1}{n}\sum_{k=1}^{n}\xi_k - \frac{1}{n}\sum_{k=1}^{n}E\,\xi_k\right| < \varepsilon\right\} \geq 1 - \frac{D\left(\frac{1}{n}\sum_{k=1}^{n}\xi_k\right)}{\varepsilon^2} \geq 1 - \frac{C}{n\,\varepsilon^2}.$$

Letting $n \to \infty$, we obtain in the limit:

$$\lim_{n\to\infty} P\left\{\left|\frac{1}{n}\sum_{k=1}^{n}\xi_k - \frac{1}{n}\sum_{k=1}^{n}E\,\xi_k\right| < \varepsilon\right\} \geq 1.$$

But since a probability can not exceed one, the theorem follows from this.

We note some important special cases of Tchebychev's Theorem.

1. BERNOULLI'S THEOREM: *Let μ be the number of occurrences of an event A in n independent trials and p the probability of occurrence of the event A in each trial. Then for any $\varepsilon > 0$,*

$$\lim_{n\to\infty} P\left\{\left|\frac{\mu}{n} - p\right| < \varepsilon\right\} = 1. \tag{3}$$

Proof: For, by introducing the random variables μ_k that give the number of occurrences of the event A in the k-th trial, we have:

$$\mu = \mu_1 + \mu_2 + \cdots \mu_n.$$

And since

$$E\,\mu_k = p, \quad D\,\mu_k = pq \leq \frac{1}{4},$$

Bernoulli's Theorem is the simplest special case of Tchebychev's Theorem.

Since in practice unknown probabilities frequently have to be approximated empirically, a large number of experiments have been conducted in order to verify the agreement between Bernoulli's Theorem and experiment. To this end, events were considered whose probabilities could for one reason or another be regarded as known and for which one could easily conduct trials and ensure their independence and the constancy of the probabilities in each trial. All such experiments yielded excellent agreement with the theory. We shall cite the outcomes of a few of these easily reproducible experiments.

A deck of 52 playing cards was divided at random into two equal parts one hundred times. Table 11 gives the outcomes of this experiment:

TABLE 11

Trial Number	Number of Red Cards	Number of Favorable Cases	Relative Frequency	Trial Number	Number of Red Cards	Number of Favorable Cases	Relative Frequency
1	17	0	0.00	26	16	6	0.23
2	13	1	0.50	27	15	6	0.22
3	13	2	0.66	28	12	6	0.21
4	13	3	0.75	29	15	6	0.21
5	14	3	0.60	30	12	6	0.20
6	13	4	0.66	31	14	6	0.19
7	11	4	0.57	32	13	7	0.22
8	14	4	0.50	33	11	7	0.21
9	12	4	0.44	34	14	7	0.21
10	15	4	0.40	35	14	7	0.20
11	13	5	0.45	36	13	8	0.22
12	11	5	0.42	37	12	8	0.22
13	11	5	0.38	38	13	9	0.24
14	14	5	0.36	39	11	9	0.23
15	15	5	0.33	40	16	9	0.22
16	14	5	0.31	41	13	10	0.24
17	14	5	0.29	42	10	10	0.24
18	15	5	0.28	43	13	11	0.26
19	13	6	0.32	44	11	11	0.25
20	18	6	0.30	45	12	11	0.24
21	14	6	0.29	46	14	11	0.24
22	16	6	0.27	47	14	11	0.23
23	12	6	0.26	48	12	11	0.23
24	11	6	0.25	49	12	11	0.22
25	11	6	0.24	50	11	11	0.22

TABLE 11 (Cont'd)

Trial Number	Number of Red Cards	Number of Favorable Cases	Relative Frequency	Trial Number	Number of Red Cards	Number of Favorable Cases	Relative Frequency
51	13	12	0.24	76	11	19	0.25
52	13	13	0.25	77	14	19	0.25
53	13	14	0.27	78	13	20	0.26
54	17	14	0.26	79	14	20	0.26
55	14	14	0.26	80	15	20	0.25
56	12	14	0.25	81	15	20	0.25
57	13	15	0.26	82	17	20	0.24
58	12	15	0.26	83	13	21	0.25
59	13	16	0.27	84	13	22	0.26
60	13	17	0.28	85	13	23	0.27
61	14	17	0.28	86	11	23	0.27
62	16	17	0.27	87	12	23	0.27
63	14	17	0.27	88	11	23	0.26
64	13	18	0.28	89	15	23	0.26
65	10	18	0.28	90	11	23	0.26
66	15	18	0.27	91	16	23	0.25
67	12	18	0.27	92	13	24	0.26
68	15	18	0.27	93	16	24	0.26
69	12	18	0.26	94	15	24	0.26
70	17	18	0.26	95	11	24	0.25
71	15	18	0.25	96	12	24	0.25
72	12	18	0.25	97	17	24	0.25
73	15	18	0.25	98	10	24	0.25
74	11	18	0.24	99	15	24	0.24
75	13	19	0.25	100	9	24	0.24

the first column indicates the number of the trial; the second, the number of red cards appearing in one of the half-decks; the third, the cumulative number of times this half-deck contained an equal number of red and black cards; and finally, the fourth column gives the value of the relative frequency of this event.

In Example 3 of § 5, it was shown that when a deck is divided in half the probability of obtaining an equal number of red and black cards in each half-deck has the value

$$p = \frac{(26!)^4}{52!\,(13!)^4} \approx 0.22.$$

The curve in Fig. 20 gives a graphical representation of how the relative frequency μ/n varies as a function of the number of the trial. At

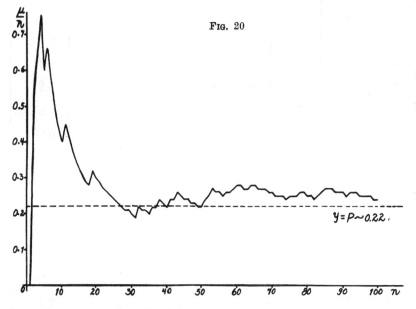

Fig. 20

first, when the number of trials is small, the polygonal line from time to time deviates considerably from the line $y = p \approx 0.22$. Later on, as the number of trials increases, the polygonal line on the whole approaches this line more and more closely.

In this case, the final deviation (for $n = 100$) of the relative frequency from the probability is fairly significant (approximately 0.02). By Laplace's Theorem, the probability of getting a deviation greater than or equal to this is given by

§ 32. Tchebyshev's Form of Law of Large Numbers

$$P\left\{\left|\frac{\mu}{n}-p\right| \geq 0.02\right\} = P\left\{\left|\frac{\mu-np}{\sqrt{npq}}\right| \geq 0.02\sqrt{\frac{n}{pq}}\right\} \approx 1 - 2\,\Phi\left(0.02\sqrt{\frac{n}{pq}}\right)$$

$$= 1 - 2\,\Phi\left(0.02\sqrt{\frac{100}{0.22\cdot 0.78}}\right) = 1 - 2\,\Phi(0.483) \approx 0.63.$$

Thus, if the experiment in question is repeated a large number of times, an error not less than the one produced in our experiment will be obtained in approximately two-thirds of the cases.

The 18th-century French naturalist, Buffon, tossed a coin 4,040 times, obtaining heads 2,048 times. The relative frequency of occurrence of heads in Buffon's experiment was approximately 0.507.

The English statistician, Karl Pearson, tossed a coin 12,000 times and in doing so obtained heads 6,019 times. The relative frequency of heads in Pearson's experiment was 0.5016.

On another occasion, he tossed a coin 24,000 times, with heads turning up 12,012 times; here, the relative frequency of heads turned out to be 0.5005. In all of the experiments cited, the relative frequency deviated by very little from the probability 0.5.

2. POISSON'S THEOREM: *If in a sequence of independent trials the probability of occurrence of an event A in the k-th trial is p_k, then*

$$\lim_{n\to\infty} P\left\{\left|\frac{\mu}{n} - \frac{p_1+p_2+\cdots+p_n}{n}\right| < \varepsilon\right\} = 1,$$

where, as usual, μ denotes the number of times the event A occurs in the first n trials.

By introducing the random variables μ_k that give the number of times the event A occurs in the k-th trial and by noting that

$$\mathsf{E}\,\mu_k = p_k, \qquad \mathsf{D}\,\mu_k = p_k\,q_k \leq \frac{1}{4},$$

we can satisfy ourselves that Poisson's Theorem is a special case of Tchebychev's Theorem.

3. *If the sequence of pairwise independent random variables $\xi_1, \xi_2, \ldots, \xi_n, \ldots$ is such that*

$$\mathsf{E}\xi_1 = \mathsf{E}\xi_2 = \ldots = \mathsf{E}\xi_n = \ldots = a$$

and

$$\mathsf{D}\xi_i \leq c \qquad (i = 1, 2, \ldots, n, \ldots),$$

then for any constant $\varepsilon > 0$

$$\lim_{n \to \infty} \mathsf{P}\left\{\left|\frac{1}{n}\sum_{k=1}^{n}\xi_k - a\right| < \varepsilon\right\} = 1.$$

This special case of Tchebychev's Theorem serves as a basis for the rule of the arithmetic mean, which is in constant use in the theory of measurement. Suppose that some physical quantity a is being measured. Repeating the measurement n times under identical conditions, the observer obtains the results $x_1, x_2, \ldots, x_n$ which do not completely coincide. As an approximate value for a, it is natural to take the arithmetic mean of the results observed:

$$a \approx \frac{x_1 + x_2 + \cdots + x_n}{n}.$$

If the measurements are free of any systematic error, i.e., if

$$\mathsf{E}x_1 = \mathsf{E}x_2 = \ldots = \mathsf{E}x_n = a,$$

and if there is no uncertainty about the observed values themselves, then according to the law of large numbers we can obtain a value as close to the required quantity a as we wish and with a probability as close to one as we wish by making the number of measurements n sufficiently large.

Tchebychev's inequality enables us to obtain a stronger result in the case of identically distributed independent summands.

KHINTCHINE'S THEOREM: *If $\xi_1, \xi_2, \ldots$ are identically distributed independent random variables which have finite expectations ($\mathsf{E}\xi_n = a$), then as $n \to \infty$,*

$$\mathsf{P}\left\{\left|\frac{1}{n}\sum_{k=1}^{n}\xi_k - a\right| < \varepsilon\right\} \to 1.$$

Proof: We make use of a method first employed by Markov in 1907, which is called *the method of truncation*. This method is frequently made use of in modern probability theory.

Let us define new random variables according to the following rule: Let δ be a fixed positive quantity and, for $k = 1, 2, \ldots, n$, let

$$\eta_k = \xi_k, \quad \zeta_k = 0, \quad \text{if} \quad |\xi_k| < \delta n;$$

$$\eta_k = 0, \quad \zeta_k = \xi_k, \quad \text{if} \quad |\xi_k| \geq \delta n.$$

§ 32. Tchebyshev's Form of Law of Large Numbers

It is clear that for any k $(1 \leq k \leq n)$

$$\xi_k = \eta_k + \zeta_k.$$

The expectation and variance

$$a_n = \mathsf{E}\,\eta_k = \int_{-\delta n}^{\delta n} x\,dF(x),$$

$$\mathsf{D}\,\eta_k = \int_{-\delta n}^{\delta n} x^2\,dF(x) - a_n^2 \leq \delta n \int_{-\delta n}^{\delta n} |x|\,dF(x) \leq \delta b n,$$

where $b = \int_{-\infty}^{+\infty} |x|\,dF(x)$, both exist for the variables η_k. Since

$$a_n \to a$$

as $n \to \infty$, it follows that for any $\varepsilon > 0$

$$|a_n - a| < \varepsilon \tag{4}$$

for sufficiently large values of n. By virtue of Tchebychev's inequality, we have:

$$\mathsf{P}\left\{\left|\frac{1}{n}\sum_{k=1}^{n}\eta_k - a_n\right| \geq \varepsilon\right\} \leq \frac{b\delta}{\varepsilon^2}.$$

And applying inequality (4), we find:

$$\mathsf{P}\left\{\left|\frac{1}{n}\sum_{k=1}^{n}\eta_k - a\right| \geq 2\varepsilon\right\} \leq \frac{b\delta}{\varepsilon^2}.$$

Now

$$\mathsf{P}\{\zeta_k \neq 0\} = \int_{|x| \geq \delta n} dF(x) \leq \frac{1}{\delta n}\int_{|x| \geq \delta n} |x|\,dF(x).$$

Since the expectation exists, the right-hand side can be made smaller than δ/n for n sufficiently large. Thus,

$$\mathsf{P}\left\{\sum_{k=1}^{n}\zeta_k \neq 0\right\} \leq \sum_{k=1}^{n}\mathsf{P}\{\zeta_k \neq 0\} \leq \delta.$$

Now

$$\frac{1}{n}\sum_{k=1}^{n}\xi_k - a = \frac{1}{n}\sum_{k=1}^{n}\eta_k - a + \frac{1}{n}\sum_{k=1}^{n}\zeta_k.$$

and

$$P\left\{\left|\sum_{k=1}^{n}\zeta_k\right|>0\right\}=P\left\{\sum_{k=1}^{n}\zeta_k\neq 0\right\},$$

and therefore

$$P\left\{\left|\frac{1}{n}\sum_{k=1}^{n}\xi_k-a\right|\geq 2\varepsilon\right\}\leq$$

$$\leq P\left\{\left|\frac{1}{n}\sum_{k=1}^{n}\eta_k-a\right|\geq 2\varepsilon\right\}+P\left\{\sum_{k=1}^{n}\zeta_k\neq 0\right\}\leq\frac{b\delta}{\varepsilon^2}+\delta.$$

Since ε and δ are arbitrary, the right-hand side can be made arbitrarily small; the theorem is thus proved.

We next give the statement of a theorem of Markov; its proof is an obvious consequence of Tchebychev's inequality.

MARKOV'S THEOREM: *If the sequence of random variables*

$$\xi_1, \xi_2, \ldots, \xi_n, \ldots,$$

is such that

$$\frac{1}{n^2}D\left(\sum_{k=1}^{n}\xi_k\right)\to 0 \tag{5}$$

for $n\to\infty$, *then for any positive constant* ε

$$\lim_{n\to\infty}P\left\{\left|\frac{1}{n}\sum_{k=1}^{n}\xi_k-\frac{1}{n}\sum_{k=1}^{n}E\,\xi_k\right|<\varepsilon\right\}=1.$$

If the random variables $\xi_1, \xi_2, \ldots, \xi_n$ are pairwise independent, then Markov's condition takes on the following form: As $n\to\infty$,

$$\frac{1}{n^2}\sum_{k=1}^{n}D\,\xi_k\to 0.$$

From this, it is clear that Tchebychev's Theorem is a special case of Markov's Theorem.

As an immediate consequence of Markov's Theorem, we obtain the following theorem, also due to Markov.

THEOREM: *Let μ be the number of times the event E occurs in n trials connected in a homogeneous Markov chain and let $p_1, p_2, \ldots$ be the probability of occurrence of the event E in the first, second, etc. trial, respectively. Then for any $\varepsilon > 0$*

$$\lim_{n\to\infty}P\left\{\left|\frac{\mu}{n}-\frac{1}{n}\sum_{k=1}^{n}p_k\right|<\varepsilon\right\}=1. \tag{6}$$

§ 33. Necessary and Sufficient Condition

The proof of this theorem is obvious in virtue of the results of Example 6, § 28.

Since
$$\frac{1}{n}\sum_{k=1}^{n} p_k = p + o(1),$$

by the results of that example, it follows that (6) is equivalent to the equation

$$\lim_{n \to \infty} \mathsf{P}\left\{\left|\frac{\mu}{n} - p\right| < \varepsilon\right\} = 1.$$

In this form, the theorem in question is completely analogous to Bernoulli's Theorem.

§ 33. A Necessary and Sufficient Condition for the Law of Large Numbers

We have already indicated that the law of large numbers is one of the fundamental propositions of the theory of probability. It is therefore easily understandable why so much effort has gone into establishing the most general conditions upon the variables $\xi_1, \xi_2, \ldots, \xi_n, \ldots$ under which the law of large numbers holds.

The history of the problem is as follows. At the end of the 17th and beginning of the 18th centuries, James Bernoulli proved the theorem now bearing his name. The theorem of Bernoulli was first published posthumously in 1715 in the treatise *Ars conjectandi*. Later on, at the beginning of the 19th century, Poisson proved an analogous theorem under more general conditions. Until the middle of the 19th century, no further progress was made. In 1866, the great Russian mathematician Tchebychev discovered the method which was presented in the preceding section. Later, Markov observed that the reasoning of Tchebychev allowed a more general result to be obtained (see § 32).

For a long time, further efforts did not yield any fundamental advances, and only in 1926 did Kolmogorov derive conditions that were necessary and sufficient for a sequence of mutually independent random variables $\xi_1, \xi_2, \ldots, \xi_n \ldots$ to obey the law of large numbers. In 1928, Khintchine showed that if the random variables ξ_n were not only independent but also identically distributed, then the existence of the expectation $\mathsf{E}\xi_n$ was a sufficient condition for the law of large numbers to apply.

234   VI. THE LAW OF LARGE NUMBERS

In recent years, much work has been devoted to the determination of those conditions which it is necessary to impose on *dependent* variables in order that they satisfy the law of large numbers. Markov's Theorem belongs to this class of propositions.

By making use of Tchebychev's method, one can easily obtain a condition analogous to Markov's which is not only sufficient but also necessary for the law of large numbers to hold for a sequence of arbitrary random variables.

THEOREM: *In order for the sequence* $\xi_1, \xi_2, \xi_3, \ldots$ *of random variables* (not necessarily independent) *to satisfy the relation*

$$\lim_{n \to \infty} \mathsf{P}\left\{ \left| \frac{1}{n} \sum_{k=1}^{n} \xi_k - \frac{1}{n} \sum_{k=1}^{n} \mathsf{E}\,\xi_k \right| < \varepsilon \right\} = 1 \tag{1}$$

for any positive quantity ε, *it is necessary and sufficient that*

$$\mathsf{E}\left\{ \frac{\left(\sum_{k=1}^{n}(\xi_k - \mathsf{E}\,\xi_k)\right)^2}{n^2 + \left(\sum_{k=1}^{n}(\xi_k - \mathsf{E}\,\xi_k)\right)^2} \right\} \to 0 \tag{2}$$

as $n \to \infty$.

Proof: Suppose first that (2) is satisfied; we shall then show that (1) is also satisfied. Let $\Phi_n(x)$ denote the distribution function of the variable

$$\eta_n = \frac{1}{n} \sum_{k=1}^{n} (\xi_k - \mathsf{E}\,\xi_k).$$

It is easy to verify the following chain of relations:[1]

$$\mathsf{P}\left\{ \left| \frac{1}{n}\sum_{k=1}^{n}(\xi_k - \mathsf{E}\,\xi_k)\right| \geq \varepsilon \right\} = \mathsf{P}\{|\eta_n| \geq \varepsilon\} =$$

$$= \int_{|x| \geq \varepsilon} d\Phi_n(x) \leq \frac{1+\varepsilon^2}{\varepsilon^2} \int_{|x| \geq \varepsilon} \frac{x^2}{1+x^2} d\Phi_n(x) \leq$$

$$\leq \frac{1+\varepsilon^2}{\varepsilon^2} \int \frac{x^2}{1+x^2} d\Phi_n(x) = \frac{1+\varepsilon^2}{\varepsilon^2} \mathsf{E}\,\frac{\eta_n^2}{1+\eta_n^2}.$$

This inequality shows that the conditions of the theorem are sufficient.

---
[1] We can write the last equality on the basis of the formula

$$\mathsf{E}\,f(\xi) = \int f(x)\,dF_\xi(x)$$

(see Theorem 1, § 27).

§ 33. Necessary and Sufficient Condition

Let us now show that condition (2) is sufficient. It is easy to see that

$$P\{|\eta_n| \geq \varepsilon\} = \int_{|x| \geq \varepsilon} d\Phi_n(x) \geq \int_{|x| \geq \varepsilon} \frac{x^2}{1+x^2} d\Phi_n(x) =$$

$$= \int \frac{x^2}{1+x^2} d\Phi_n(x) - \int_{|x| < \varepsilon} \frac{x^2}{1+x^2} d\Phi_n(x) \geq$$

$$\geq \int \frac{x^2}{1+x^2} d\Phi_n(x) - \varepsilon^2 = \mathsf{E}\frac{\eta_n^2}{1+\eta_n^2} - \varepsilon^2. \qquad (3)$$

Thus

$$0 \leq \mathsf{E}\frac{\eta_n^2}{1+\eta_n^2} \leq \varepsilon^2 + P\{|\eta_n| \geq \varepsilon\}.$$

By first choosing ε sufficiently small and then n sufficiently large, we can make the right-hand side of the last inequality arbitrarily small.

We note that all of the theorems that were proved in the preceding section follow in a simple way from the general theorem just proved. In fact, since the inequality

$$\frac{\eta_n^2}{1+\eta_n^2} \leq \eta_n^2 = \left[\frac{1}{n}\sum_{k=1}^{n}(\xi_k - \mathsf{E}\,\xi_k)\right]^2$$

holds for any n and any ξ_k, we have the inequality

$$\mathsf{E}\frac{\eta_n^2}{1+\eta_n^2} \leq \frac{1}{n^2}\mathsf{D}\sum_{k=1}^{n}\xi_k,$$

provided the variance exists.

Thus, if Markov's condition is satisfied, then so is condition (2), and therefore the sequence $\xi_1, \xi_2, \ldots, \xi_n, \ldots$ is subject to the law of large numbers.

We should nevertheless note that in more complicated situations, such as when the variables ξ_k are not assumed to have finite variances, the theorem just proved is of very little use in actually verifying the applicability of the law of large numbers, because condition (2) applies not to the individual variables but to their sum. Apparently, however, we have no right to expect that, without making any assumptions about the variables ξ_k or about the existence of some relation between them, we can nonetheless succeed in finding necessary and sufficient conditions, and conditions, moreover, which are convenient in application.

One fundamental difficulty is encountered in the practical application of the theorems just proved: can we consider a manufacturing process or a phenomenon we are investigating to be taking place under the influence of independent causes? Does not the very concept of independence contradict our basic idea that all phenomena of the external world are interrelated? Whenever we undertake a mathematical study of some natural phenomenon, technical process, or social phenomenon, we must first obtain our basic premises by making a profound study of the essence of the phenomenon, of its qualitative characteristics. We must take into account any change in the external conditions under which our phenomenon occurs, and we must change the mathematical apparatus and the premises underlying its application as soon as it is discovered that the conditions under which the phenomenon is realized have changed.

While acknowledging the interrelation of all natural processes, dialectical materialism indicates, at the same time, that these relationships can either be *essential* or *unessential* to a phenomenon. An examination of all the existent relations, including those that are unessential, simply leads to an obscurement of the phenomenon and a delay in mastering it in view of the sheer complexity of the situation. Therefore, we should not take any unessential relations into consideration and our entire attention should be concentrated on those which are essential to some given phenomenon.

Discarding the unessential relations between the causes of a phenomenon makes it possible to use independent random variables as a working tool. How successful has been our schematization of a phenomenon and how successful our selection of the mathematical tools for studying it, can only be judged by the agreement that exists between the theory that we set up and practice. If our theoretical results differ essentially from those of experiment, then we have to re-examine our premises. In particular, if the question concerns the applicability of the law of large numbers, then perhaps we might have to reject the assumption that our effective causes are independent and go over to the assumption that they are at least weakly dependent.

But we have already stated that our accumulative experience with using the laws of large numbers has shown that the condition of independence is a satisfactory one in many important problems in natural science and engineering.

§ 34. The Strong Law of Large Numbers

Frequently, the quite unjustified inference is made from Bernoulli's Theorem that the relative frequency of an event A tends to the probability of the event A when the number of trials is increased indefinitely. The fact is that Bernoulli's Theorem merely establishes this: that when the number of trials n is made sufficiently large, the probability of the one single inequality

$$\left|\frac{\mu}{n} - p\right| < \varepsilon$$

becomes larger than $1 - \eta$ for arbitrary positive η. In 1909, the French mathematician Émile Borel proved a deeper theorem, now known as *the strong law of large numbers*. In order to formulate and prove Borel's Theorem as well as the more general theorems of Kolmogorov, we have to introduce the important notion of *convergence of a sequence of random variables*.

Let there be a sequence of random variables defined over the same set of elementary events U:

$$\xi_n = f_n(e) \qquad (e \in U). \tag{1}$$

Consider the set A of all elementary events e for which the sequence $f_n(e)$ converges. Let $f(e)$ denote the limit of $f_n(e)$ at the point e. If A_{nk}^r denotes the set of those e for which the inequality

$$|f_{n+k}(e) - f(e)| < \frac{1}{r} \tag{2}$$

is satisfied, then we have

$$A = \prod_{r=1}^{\infty} \sum_{n=1}^{\infty} \prod_{k=1}^{\infty} A_{nk}^r. \tag{3}$$

Actually, if the sequence of functions $f_n(e)$ converges at the point e, then: 1) inequality (2) has to be satisfied for every k if n is sufficiently large; 2) it has to be satisfied beginning with a certain value of n; 3) it has to be satisfied for sufficiently large n for any value of r. Equation (3) is precisely the analytic representation of all of these three requirements. By virtue of equation (3) and the definition of a random event, the subset A belongs to the field of random events. We now define a random variable ξ as follows: If $e \in A$, then $\xi = f(e)$; if, however, $e \in \bar{A}$, then $\xi = 0$.

VI. THE LAW OF LARGE NUMBERS

If the probability of the random event A is one, then we say that *the sequence of random variables ξ_n converges to the random variable ξ almost certainly* (or that *it converges with probability one*).[2]

If a sequence ξ_n converges to ξ almost certainly, this fact will be written in the form

$$\mathsf{P}\{\xi_n \to \xi\} = 1. \tag{4}$$

Clearly, this equation may also be written in the alternative form:

$$\mathsf{P}\{\xi_n \nrightarrow \xi\} = 0. \tag{4'}$$

The last expression means that the probability of finding a number r such that the inequality

$$|\xi_{n+k} - \xi| \geq \frac{1}{r}$$

holds for all values of n and at least one value of $k = k(n)$ is zero.

We shall now give a sufficient condition for a sequence of random variables to converge with probability one.

LEMMA: *If*

$$\sum_{n=1}^{\infty} \mathsf{P}\left\{|\xi_n - \xi| \geq \frac{1}{r}\right\} \tag{5}$$

converges for any positive integer r, then (4) holds—or, what is the same, (4').

---

[2] The concept of almost certain convergence corresponds precisely to the concept of convergence almost everywhere in the theory of functions of a real variable.

An important part is also played in probability theory by the so-called *convergence in probability*. A sequence of random variables ξ_n converges in probability to the random variable ξ if for any $\epsilon > 0$, the probability of the inequality

$$|\xi - \xi_n| \leq \epsilon$$

approaches one as $n \to \infty$.

Convergence in probability is the analog of the convergence in measure of a sequence of functions in the theory of functions of a real variable.

Clearly, the law of large numbers states that under certain conditions the sum $\frac{1}{n}\sum_{k=1}^{n}(\xi_k - \mathsf{E}\,\xi_k)$ converges in probability to zero.

Proof: Let E_n^r denote the event that the inequality

$$|\xi_n - \xi| \geq \frac{1}{r}$$

is satisfied, and set

$$S_n^r = \sum_{k=1}^{\infty} E_{n+k}^r.$$

From

$$\mathsf{P}\{S_n^r\} \leq \sum_{k=1}^{\infty} \mathsf{P}\{E_{n+k}^r\} = \sum_{l=n+1}^{\infty} \mathsf{P}\left\{|\xi_l - \xi| \geq \frac{1}{r}\right\},$$

we conclude by (5) that

$$\lim_{n \to \infty} \mathsf{P}\{S_n^r\} = 0. \tag{6}$$

Now let

$$S^r = S_1^r S_2^r S_3^r \ldots.$$

Since the event S^r implies every one of the events S_n^r, we obtain by virtue of (6):

$$\mathsf{P}(S^r) = 0. \tag{7}$$

Finally, let us set

$$S = S^1 + S^2 + S^3 + \ldots.$$

As we can readily prove, this event implies that a value of r can be found such that the inequality

$$|\xi_{n+k} - \xi| \geq \frac{1}{r}$$

will be satisfied for every value of n ($n = 1, 2, 3, \ldots$) and for at least one value of k ($k = k(n)$). Since

$$\mathsf{P}(S) \leq \sum_{r=1}^{\infty} \mathsf{P}\{S^r\},$$

in view of (7), we have:

$$\mathsf{P}\{S\} = 0,$$

Q.E.D.

By repeating verbatim the argument used in the theorem just proved, we can obtain the slightly stronger result:

VI. THE LAW OF LARGE NUMBERS

If a sequence of integers $1 = n_1 < n_2 < n_3 < \ldots$ *exists such that the series*

$$\sum_{k=1}^{\infty} \mathsf{P}\left\{\max_{n_k \leq n < n_{k+1}} |\xi_n - \xi| \geq \frac{1}{r}\right\}$$

converges for any positive integer r, *then the sequence of random variables* $\xi_1, \xi_2, \ldots$ *converges to* ξ *almost certainly.*

We now proceed to apply the above concept and the lemma just proved.

BOREL'S THEOREM: *Let* μ *be the number of times an event* A *occurs in* n *independent trials in each of which it may occur with probability* p. *Then for* $n \to \infty$,

$$\mathsf{P}\left\{\frac{\mu}{n} \to p\right\} = 1.$$

Proof: According to Lemma 1, to prove the theorem it suffices to show that the series

$$\sum_{n=1}^{\infty} \mathsf{P}\left\{\left|\frac{\mu}{n} - p\right| \geq \frac{1}{r}\right\} \tag{8}$$

converges for any natural number r. With this in mind, we observe that it is possible to establish the following inequality in a way analogous to the one used to prove Tchebychev's inequality (§ 32): For any random variable for which the expectation $\mathsf{E}(\xi - \mathsf{E}\xi)^4$ exists,

$$\mathsf{P}\{|\xi - \mathsf{E}\xi| \geq \varepsilon\} \leq \frac{1}{\varepsilon^4} \mathsf{E}(\xi - \mathsf{E}\xi)^4.$$

Thus,

$$\mathsf{P}\left\{\left|\frac{\mu}{n} - p\right| \geq \frac{1}{r}\right\} \leq r^4 \, \mathsf{E}\left(\frac{\mu}{n} - p\right)^4.$$

Let us introduce, as we have done many times before, the auxiliary random variables μ_i corresponding to the number of occurrences of the event A in the i-th trial. Since

$$\frac{\mu}{n} - p = \frac{1}{n} \sum_{i=1}^{n} (\mu_j - p),$$

it follows that

$$\mathsf{E}\left(\frac{\mu}{n} - p\right)^4 = \frac{1}{n^4} \sum_{i=1}^{n} \sum_{j=1}^{n} \sum_{k=1}^{n} \sum_{l=1}^{n} \mathsf{E}(\mu_i - p)(\mu_j - p)(\mu_k - p)(\mu_l - p). \tag{9}$$

§ 34. STRONG LAW OF LARGE NUMBERS

Since $E(\mu_i - p) = 0$, all the terms in this sum that contain at least one of the factors $\mu_i - p$ to the first power vanish. Thus, only the terms of the form $E(\mu_i - p)^4$ and $E(\mu_i - p)^2(\mu_s - p)^2$ do not vanish. Clearly

$$E(\mu_i - p)^4 = pq(p^3 + q^3)$$

and

$$E(\mu_i - p)^2 (\mu_s - p)^2 = p^2 q^2 \quad (i \neq s).$$

There are n summands of the first form and $3n(n-1)$ summands of the second form. In fact, as to the latter form, i can coincide with j, with k, or with l, and in doing so can take on any value from 1 up to n; s can then only assume one of $n-1$ values, since $s \neq i$.

Thus, we find

$$E\left(\frac{\mu}{n} - p\right)^4 = \frac{pq}{n^4}[n(p^3 + q^3) + 3pq(n^2 - n)] < \frac{1}{4n^2},$$

and this implies the convergence of the series (8). The theorem is thus proved.

Borel's Theorem has been the starting point for an important series of investigations devoted to finding the conditions under which the so-called strong law of large numbers holds.

We shall say that *the sequence of random variables* $\xi_1, \xi_2, \xi_3, \ldots$ *obeys the strong law of large numbers if*

$$\frac{1}{n}\sum_{k=1}^{n}\xi_k - \frac{1}{n}\sum_{k=1}^{n}E\xi_k \to 0$$

with probability one as $n \to \infty$.

A broad but, at the same time, simple sufficient condition for the strong law of large numbers to hold is given by the following theorem.

KOLMOGOROV'S THEOREM: *If the sequence of mutually independent random variables* $\xi_1, \xi_2, \ldots$ *satisfies the condition*

$$\sum_{n=1}^{\infty}\frac{D\xi_n}{n^2} < +\infty,$$

then it obeys the strong law of large numbers.

We shall give a simple proof this theorem, which was discovered recently by Hájek and Rényi. It makes use of an interesting generalization of Kolmogorov's Inequality.

INEQUALITY OF HÁJEK-RÉNYI: *If the mutually independent random variables $\eta_1, \eta_2, \ldots$ have finite variances, if $\mathsf{E}\eta_k = 0$, and if c_k ($k = 1, 2, \ldots$) is a non-increasing sequence of positive constants, then for any positive integers m and n, with $m < n$ and arbitrary $\varepsilon > 0$,*

$$\mathsf{P}\{\max_{m \leq k \leq n} c_k \, |\, \eta_1 + \eta_2 + \ldots + \eta_k\,| \geq \varepsilon\} \leq \frac{1}{\varepsilon^2} \left(c_m^2 \sum_{k=1}^{m} \mathsf{D}\eta_k + \sum_{k=m+1}^{n} c_k^2 \mathsf{D}\eta_k\right).$$

Proof: Consider the quantity

$$\zeta = \sum_{k=m}^{n-1} S_k^2 (c_k^2 - c_{k+1}^2) + c_n^2 S_n^2,$$

where $S_k = \sum_{j=1}^{k} \eta_j$. By a simple computation, we may satisfy ourselves that

$$\mathsf{E}\zeta = c_m^2 \sum_{k=1}^{m} \mathsf{D}\eta_k + \sum_{k=m+1}^{n} c_k^2 \mathsf{D}\eta_k.$$

Let us consider the event E_i ($i = m, m+1, \ldots, n$) consisting in the fact that

$$c_j \,|\, S_j\,| < \varepsilon \quad \text{for} \quad m \leq j < i$$

and ·

$$c_i \,|\, S_i\,| \geq \varepsilon.$$

Since the event

$$\max_{m \leq k \leq n} c_k \,|\, S_k\,| \geq \varepsilon$$

implies that

$$c_k \,|\, S_k\,| \geq \varepsilon$$

for at least one value of k, it is equivalent to the event $\sum_{k=m}^{n} E_k$. However, by the mutual exclusiveness of the events E_k we obtain the equation

$$\mathsf{P}\{\max_{m \leq k \leq n} c_k \,|\, S_k\,| \geq \varepsilon\} = \sum_{k=m}^{n} \mathsf{P}\{E_k\}.$$

Now let E_0 denote the event that

$$c_j \,|\, S_j\,| < \varepsilon \quad \text{for} \quad m \leq j \leq n.$$

By equation (5) of § 26,

$$E\zeta = \sum_{i=m}^{n} E(\zeta/E_i) \cdot P(E_i) + E(\zeta/E_0)P(E_0) \geqq$$

$$\geqq \sum_{i=m}^{n} E(\zeta/E_i) \cdot P(E_i).$$

It is further evident that for $k \geqq i$,

$$E(S_k^2/E_i) = E\{S_i^2 + 2\sum_{j>i} S_i \eta_j + \sum_{j>i} \eta_j^2 + 2\sum_{j>h>i} \eta_j \eta_h / E_i\}$$

$$\geqq E\{S_i^2 + 2\sum_{j>i} S_i \eta_j + 2\sum_{j>h>i} \eta_j \eta_h / E_i\}.$$

But since the occurrence of the event E_i only imposes a restriction on the first i of the variables η_j, and the following ones, under this condition, remain independent of one another and of S_i, it follows that for $j > i$,

$$E(S_i\eta_j/E_i) = E(\eta_j/E_i) \cdot E(S_i/E_i) = 0,$$

and for $j > h > i$,

$$E(\eta_i\eta_h/E_i) = 0.$$

Thus

$$E(S_k{}^2/E_i) \geqq E(S_i{}^2/E_i).$$

Since the event E_i has been realized,

$$E(S_i{}^2/E_i) \geqq \frac{\varepsilon^2}{c_i{}^2}.$$

But

$$E(\zeta/E_i) = \sum_{k=m}^{n-1} E(S_k^2/E_i)(c_k{}^2 - c_{k+1}^2) + c_n{}^2 E(S_n^2/E_i) \geqq$$

$$\geqq \sum_{k=i}^{n-1} E(S_k^2/E_i)(c_k{}^2 - c_{k+1}^2) + c_n{}^2 E(S_n^2/E_i),$$

and therefore

$$E(\zeta/E_i) \geqq \frac{\varepsilon^2}{c_i{}^2}\left[\sum_{k=i}^{n-1}(c_k{}^2 - c_{k+1}^2) + c_n{}^2\right] = \varepsilon^2.$$

The inequality (10) now gives

$$\mathsf{E}\zeta \geq \varepsilon^2 \sum_{i=m}^{n} \mathsf{P}(E_i) .$$

With this, the inequality of Hájek-Rényi is proved.

In the special case $m = 1$, $c_1 = c_2 = \ldots = c_n = 1$, we obtain Kolmogorov's Inequality by setting $\eta_k = \xi_k - \mathsf{E}\xi_k$.

KOLMOGOROV'S INEQUALITY: *If the mutually independent random variables* $\xi_1, \xi_2, \ldots, \xi_n$ *have finite variances, then the probability of the simultaneous realization of the inequalities*

$$\left|\sum_{s=1}^{k} (\xi_s - \mathsf{E}\,\xi_s)\right| < \varepsilon \qquad (k = 1, 2, \ldots, n)$$

is not less than

$$1 - \frac{1}{\varepsilon^2} \sum_{k=1}^{n} \mathsf{D}\,\xi_k .$$

As a second special case of the Hájek-Rényi inequality, we take $c_k = 1/k$. Here

$$\mathsf{P}\left\{\max_{m \leq k \leq n} \frac{1}{k}\left|\sum_{s=1}^{k} \eta_s\right| \geq \varepsilon\right\} \leq \frac{1}{\varepsilon^2}\left(\frac{1}{m^2}\sum_{k=1}^{m} \mathsf{D}\eta_k + \sum_{k=m+1}^{n} \frac{\mathsf{D}\eta_k}{k^2}\right) .$$

By letting $n \to \infty$, we find from this that

$$\mathsf{P}\left\{\max_{m \leq k} \frac{1}{k}\left|\sum_{s=1}^{k} \eta_s\right| \geq \varepsilon\right\} \leq \frac{1}{\varepsilon^2}\left(\frac{1}{m^2}\sum_{k=1}^{m} \mathsf{D}\eta_k + \sum_{k=m+1}^{\infty} \frac{\mathsf{D}\eta_k}{k^2}\right) .$$

If we assume that the series $\sum_{k=1}^{\infty} \mathsf{D}\eta_k/k^2$ converges, then from the last inequality it follows that

$$\lim_{m \to \infty} \mathsf{P}\left\{\max_{m \leq k} \frac{1}{k}\left|\sum_{s=1}^{k} \eta_s\right| \geq \varepsilon\right\} = 0 .$$

But this means that

$$\mathsf{P}\left\{\lim_{m \to \infty} \frac{1}{k}\sum_{s=1}^{k} \eta_s = 0\right\} = 1 .$$

The strong law of large numbers in Kolmogorov's form is thus proved.

This theorem clearly implies the following result:

§ 34. Strong Law of Large Numbers

COROLLARY: *If the variances of the random variables ξ_k are bounded by the same constant C, then the sequence of mutually independent random variables $\xi_1, \xi_2, \xi_3, \ldots$ obeys the strong law of large numbers.*

We thus see that the strong law of large numbers holds not only for a Bernoulli scheme in which the probability of occurrence of the event A in each trial is the same (Borel's Theorem), but also in the case of a Poisson scheme (the probability of the event A depends on the number of the trial).

On the basis of Kolmogorov's Theorem, we obtain one final result, also due to Kolmogorov.

THEOREM: *The existence of the expectation is a necessary and sufficient condition for the strong law of large numbers to hold for a sequence of identically distributed and mutually independent random variables.*

Proof: The existence of the expectation entails the finiteness of the integral $\int |x|\, dF(x)$, where $F(x)$ is the common distribution function of the random variables ξ_n. (We shall denote the expectation of ξ_n simply by $\mathsf{E}\xi$). Therefore

$$\sum_{n=1}^{\infty} \mathsf{P}\{|\xi| \geq n\} = \sum_{n=1}^{\infty} \sum_{k \geq n} \mathsf{P}\{k \leq |\xi| < k+1\} = \qquad (11)$$

$$= \sum_{k=1}^{\infty} k \mathsf{P}\{k \leq |\xi| < k+1\} \leq \sum_{k=0}^{\infty} \int_{k \leq |x| < k+1} |x|\, dF(x) = \int |x|\, dF(x) < \infty.$$

Let us introduce the truncated variables

$$\xi_n^* = \begin{cases} \xi_n \text{ for } |\xi_n| < n, \\ 0 \text{ for } |\xi_n| \geq n. \end{cases}$$

We then obtain

$$\mathsf{D}\xi_n^* \leq \mathsf{E}\xi_n^{*2} = \int_{-n}^{+n} x^2\, dF(x) \leq \sum_{k=0}^{n-1} (k+1)^2\, \mathsf{P}\{k \leq |\xi| < k+1\}$$

and

$$\sum_{n=1}^{\infty} \frac{\mathsf{D}\xi_n^*}{n^2} \leq \sum_{n=1}^{\infty} \sum_{k=0}^{n-1} \frac{(k+1)^2}{n^2} \mathsf{P}\{k \leq |\xi| < k+1\}$$

$$\leq \sum_{k=1}^{\infty} \mathsf{P}\{k-1 \leq |\xi| < k\}\, k^2 \sum_{n=k}^{\infty} \frac{1}{n^2}.$$

VI. The Law of Large Numbers

Since
$$\sum_{n=k}^{\infty} \frac{1}{n^2} < \frac{1}{k^2} + \frac{1}{k} < \frac{2}{k},$$

it easily follows from (11) that
$$\sum_{n=1}^{\infty} \frac{\mathsf{D}\xi_n^*}{n^2} < \infty,$$

i.e., the sequence ξ_n^* satisfies the strong law of large numbers.

Next, since
$$\mathsf{E}\xi_k^* = \int_{|x|<k} x \, dF(x),$$

we see that $\mathsf{E}\xi_k^* \to \mathsf{E}\xi$ as $k \to \infty$. Now consider the identity
$$\frac{1}{n}\sum_{k=1}^{n} \mathsf{E}\xi_k^* = \mathsf{E}\xi + \frac{1}{n}\sum_{k=1}^{n} (\mathsf{E}\xi_k^* - \mathsf{E}\xi).$$

Let $\varepsilon > 0$ and $N(\varepsilon)$ be sufficiently large that $|\mathsf{E}\xi_k^* - \mathsf{E}\xi| < \varepsilon$ when $k \geq N(\varepsilon)$. Then
$$\left|\frac{1}{n}\sum_{k=1}^{n} \mathsf{E}\xi_k^* - \mathsf{E}\xi\right| \leq \frac{1}{n}\sum_{k < N(\varepsilon)} |\mathsf{E}\xi_k^* - \mathsf{E}\xi| + \varepsilon.$$

Letting first $n \to \infty$ and then $\varepsilon \to 0$, we have
$$\frac{1}{n}\sum_{k=1}^{n} \mathsf{E}\xi_k^* \to \mathsf{E}\xi.$$

It remains to show that this yields our result. To do this, it clearly suffices to prove that the probability that one or more of the inequalities
$$\xi_n \neq \xi_n^*$$

hold, with $n \geq N$, tends to zero as $N \to \infty$. Indeed,
$$\mathsf{P}\{\xi_n \neq \xi_n^* \text{ for any } n \geq N\} \leq \sum_{n \geq N} \mathsf{P}\{\xi_n \neq \xi_n^*\} =$$

$$= \sum_{n=N}^{\infty} \mathsf{P}\{|\xi_n| \geq n\} \leq \sum_{n=N}^{\infty} (n - N + 1) \mathsf{P}\{n \leq |\xi_n| < n+1\} \leq$$

$$\leq \sum_{n=N}^{\infty} n \int_{n \leq |x| < n+1} dF(x) \leq \sum_{n=N}^{\infty} \int_{n \leq |x| < n+1} |x| \, dF(x) = \int_{|x| \geq N} |x| \, dF(x).$$

§ 34. Strong Law of Large Numbers

By the hypothesis of the theorem, the right-hand side of this inequality may be made arbitrarily small by choosing N sufficiently large.

For the necessity part of the proof we need the following basic proposition:

BOREL-CANTELLI LEMMA: *Let $E_1, E_2, \ldots$ be a sequence of events, and put*
$$E = \prod_{n=1}^{\infty} \sum_{k=n}^{\infty} E_k.$$

(i) *If $\sum_{k=1}^{\infty} \mathsf{P}\{E_k\} < \infty$, then $\mathsf{P}\{E\} = 0$.*

(ii) *If the events in each finite subsequence of $E_1, E_2, \ldots$ are mutually independent, and $\sum_{k=1}^{\infty} \mathsf{P}\{E_k\} = \infty$, then $\mathsf{P}\{E\} = 1$.*

Remark: Clearly, E is the set of all elementary events that belong to infinitely many of the sets $E_1, E_2, \ldots$. Hence the occurrence of E is equivalent to the occurrence of infinitely many of the E_n.

To prove (i), we observe that $E \subset \sum_{k=n}^{\infty} E_k$ for $n = 1, 2, \ldots$. Hence, if $\sum_{k=1}^{\infty} \mathsf{P}\{E_k\} < \infty$,
$$0 \leq \mathsf{P}\{E\} \leq \mathsf{P}\{\sum_{k=n}^{\infty} E_k\} \leq \sum_{k=n}^{\infty} \mathsf{P}\{E_k\} \to 0 \text{ as } n \to \infty,$$
which implies that $\mathsf{P}\{E\} = 0$.

Suppose now that $\sum_{k=1}^{\infty} \mathsf{P}\{E_k\} = \infty$. Put $A_n = \prod_{k=n}^{\infty} \overline{E}_k$. Clearly $A_1 \subset A_2 \subset \ldots$, and $\overline{E} = \sum_{n=1}^{\infty} A_n$. By the Corollary to Lemma 1, § 62,
$$\mathsf{P}\{\overline{E}\} = \lim_{n \to \infty} \mathsf{P}\{A_n\}.$$

Now by the assumption of independence,
$$0 \leq \mathsf{P}\{A_n\} = \mathsf{P}\{\prod_{k=n}^{\infty} \overline{E}_k\} \leq \mathsf{P}\{\prod_{k=n}^{N} \overline{E}_k\} = \prod_{k=n}^{N} \mathsf{P}\{\overline{E}_k\} =$$
$$= \prod_{k=n}^{N} (1 - \mathsf{P}\{E_k\}) \leq \exp\left(-\sum_{k=n}^{N} \mathsf{P}\{E_k\}\right).$$

But the quantity on the right approaches zero as $N \to \infty$, since $\sum_{k=1}^{\infty} \mathsf{P}\{E_k\} = \infty$. Hence $\mathsf{P}\{\overline{E}\} = 0$, which is equivalent to $\mathsf{P}\{E\} = 1$.

To complete the proof of the theorem, let ξ be any random variable whose distribution function is $F(x)$. We obtain upper and lower bounds for $|\xi|$ by setting $\xi_* = n$ and $\xi^* = n+1$ if $n \leq |\xi| < n+1$ ($n = 0, 1, 2, \ldots$). Then,

$$E\xi_* = \sum_{j=1}^{\infty} jP\{\xi_* = j\} = \sum_{n=1}^{\infty}\sum_{j=n}^{\infty} P\{\xi_* = j\} =$$

$$= \sum_{n=1}^{\infty}\sum_{j=n}^{\infty} P\{j \leq |\xi| < j+1\} = \sum_{n=1}^{\infty} P\{|\xi| \geq n\} = \sum_{n=1}^{\infty} P\{|\xi_n| \geq n\};$$

and by a similar argument,

$$E\xi^* = \sum_{n=1}^{\infty} P\{|\xi| \geq n-1\} = 1 + \sum_{n=1}^{\infty} P\{|\xi| \geq n\} = 1 + \sum_{n=1}^{\infty} P\{|\xi_n| \geq n\}.$$

Since $\xi_* \leq |\xi| \leq \xi^*$, it follows that

$$\sum_{n=1}^{\infty} P\{|\xi_n| \geq n\} \leq E|\xi| \leq 1 + \sum_{n=1}^{\infty} P\{|\xi_n| \geq n\}. \tag{12}$$

Let E_n denote the event that $|\xi_n| \geq n$. If $\frac{1}{n}\sum_{j=1}^{n} \xi_j$ converges with probability 1 to some constant c as $n \to \infty$, then

$$\frac{\xi_n}{n} = \frac{1}{n}\sum_{j=1}^{n} \xi_j - \frac{n-1}{n} \cdot \frac{1}{n-1}\sum_{j=1}^{n-1} \xi_j \to 0$$

with probability 1, which means that ξ_n/n is arbitrarily small if n is sufficiently large. Hence the probability that infinitely many of the events E_n occur is 0. Since the independence of the random variables ξ_n implies the independence of the events E_n, we conclude from the Borel-Cantelli Lemma that

$$\sum_{n=1}^{\infty} P\{E_n\} = \sum_{n=1}^{\infty} P\{|\xi_n| \geq n\} < \infty.$$

Hence, by (12), $E|\xi|$ is finite, and from the sufficiency argument it follows that $c = E\xi$.

The strong law of large numbers plays a very fundamental role in the theory of probability and its applications. Let us suppose for the moment that in the case of, say, identically distributed summands having finite expectations, the strong law did not hold. Then, with probability as close to one as we please, we could assert that instances will recur in which the arithmetic mean of the observed values will be far removed from the expectation. And this could happen even if the observations were made without systematical error and with complete definiteness. Under such circumstances, could the arithmetic mean of the observed values be considered as approaching the quantity being measured? And could we assume, under these circumstances, that the arithmetic mean can be taken as an approximation for the measured quantity? This seems quite doubtful.

EXERCISES

1. Prove that if the random variable ξ is such that $Ee^{a\xi}$ exists (a is a positive constant), then

$$P\{\xi \geq \varepsilon\} \leq \frac{E\,e^{a\xi}}{e^{a\varepsilon}}.$$

2. Let $f(x)$ be a positive non-decreasing function. Prove that if $E(f(|\xi - E\xi|))$ exists, then

$$P\{|\xi - E\xi| \geq \varepsilon\} \leq \frac{Ef(|\xi - E\xi|)}{f(\varepsilon)}.$$

3. A sequence of independent and identically distributed random variables $\{\xi_k\}$ is defined by the relations

a) $P\{\xi_k = 2^{n - 2\log n - 2\log\log n}\} = \dfrac{1}{2^n}$ $(n = 1, 2, 3, \ldots)$,

b) $P\{\xi_k = n\} = \dfrac{c}{n^2 \log^2 n}$, $\left(n \geq 2;\ c^{-1} = \sum_{n=2}^{\infty} \dfrac{1}{n^2 \log^2 n}\right)$.

Prove that the law of large numbers is applicable to each of the indicated sequences.

4. Prove that the law of large numbers is applicable to the sequence of independent random variables $\{\xi_n\}$ for which

$$P\{\xi_n = n^\alpha\} = P\{\xi_n = -n^\alpha\} = \frac{1}{2}$$

if and only if $\alpha < 1/2$.

5. Prove that if the independent random variables $\xi_1, \xi_2, \ldots, \xi_n, \ldots$ are such that

$$\max_{1 \leq k \leq n} \int_{|x| \geq A} |x|\, dF_k(x) \to 0 \text{ for } A \to \infty,$$

then the law of large numbers is applicable to the sequence $\{\xi_n\}$.

Hint: Use the method employed in the proof of Khintchine's Theorem.

6. By making use of the result of the preceding example, prove that if for a sequence of independent random variables $\{\xi_n\}$ there exist numbers $\alpha > 1$ and β such that $E|\xi_n|^\alpha \leq \beta$ then the law of large numbers holds for the sequence $\{\xi_n\}$ (Markov's Theorem).

7. Given the sequence of random variables $\xi_1, \xi_2, \ldots$ for which $D\xi_n \leq C$ and $R_{ij} \to 0$ as $|i - j| \to \infty$ (R_{ij} is the correlation coefficient of ξ_i and ξ_j). Prove that the law of large numbers holds for the given sequence (Bernstein's Theorem).

CHAPTER VII

CHARACTERISTIC FUNCTIONS

In the preceding chapters, we saw that extensive use could be made in the theory of probability of the methods and analytical tools of various branches of mathematical analysis. A simple solution to very many problems in probability theory, particularly those connected with sums of independent random variables, is obtainable by means of *characteristic functions*, the theory of which is developed in analysis, where it goes by the more familiar name of *Fourier transformations*. The present chapter is devoted to a presentation of the fundamental properties of characteristic functions.

§ 35. The Definition and Simplest Properties of Characteristic Functions

The expectation of the random variable $e^{it\xi}$ is called the *characteristic function* of the random variable ξ.[1] If $F(x)$ is the distribution function of the variable ξ, then by the theorem of § 27, its characteristic function is

$$f(t) = \int e^{itx} dF(x). \tag{1}$$

In the sequel, we agree to denote a distribution function and the characteristic function that corresponds to it, by a capital letter and the corresponding lower case letter, respectively.

From the fact that $|e^{itx}| = 1$ for all real values of t, it follows that the integral (1) exists for all distribution functions; hence, a characteristic function can be defined for every random variable.

---

[1] t is a real parameter. The expectation of a complex random variable $\xi + i\eta$ is defined as $E\xi + iE\eta$. It is easily verified that Theorems 1, 2, and 3 of § 28 are also valid in this case.

§ 35. Definition and Simplest Properties

Theorem 1: *A characteristic function is uniformly continuous over the whole real line and satisfies the following relations:*

$$f(0) = 1, \quad |f(t)| \leq 1 \quad (-\infty < t < \infty). \tag{2}$$

Proof: The relations (2) follow immediately from the definition of a characteristic function. For by (1),

$$f(0) = \int 1 \cdot dF(x) = 1$$

and

$$|f(t)| = \left|\int e^{itx} dF(x)\right| \leq \int |e^{itx}| dF(x) = \int dF(x) = 1.$$

It remains to prove the uniform continuity of the function $f(t)$. For this purpose, consider the difference

$$f(t+h) - f(t) = \int e^{itx} (e^{ixh} - 1) dF(x)$$

and let us estimate its absolute value. We have:

$$|f(t+h) - f(t)| \leq \int |e^{ixh} - 1| dF(x).$$

Let ε be an arbitrary positive quantity; choose A sufficiently large that

$$\int_{|x|>A} dF(x) < \frac{\varepsilon}{4}$$

and a corresponding h sufficiently small that for $|x| < A$,

$$|e^{ixh} - 1| < \frac{\varepsilon}{2}.$$

Then

$$|f(t+h) - f(t)| \leq \int_{-A}^{A} |e^{ixh} - 1| dF(x) + 2 \int_{|x| \geq A} dF(x) \leq \varepsilon.$$

This inequality proves the theorem.

Theorem 2: *If $\eta = a\xi + b$, where a and b are constants, then*

$$f_\eta(t) = f_\xi(at) e^{ibt},$$

where $f_\xi(t)$ and $f_\eta(t)$ are the characteristic functions of the variables ξ and η.

Proof: In fact,

$$f_\eta(t) = \mathsf{E}\, e^{it\eta} = \mathsf{E}\, e^{it(a\xi+b)} = e^{itb}\, \mathsf{E}\, e^{ita\xi} = e^{itb} f_\xi(at).$$

THEOREM 3: *The characteristic function of the sum of two independent random variables is equal to the product of their characteristic functions.*

Proof: Let ξ and η be independent random variables and let $\zeta = \xi + \eta$. Then clearly not only ξ and η, but also $e^{it\xi}$ and $e^{it\eta}$ are independent random variables. From this it follows that

$$\mathsf{E}\, e^{it\zeta} = \mathsf{E}\, e^{it(\xi+\eta)} = \mathsf{E}\, e^{it\xi} e^{it\eta} = \mathsf{E}\, e^{it\xi}\, \mathsf{E}\, e^{it\eta}.$$

This proves the theorem.

COROLLARY: *If*

$$\xi = \xi_1 + \xi_2 + \ldots + \xi_n,$$

where each term is independent of the sum of the preceding ones, then the characteristic function of the variable ξ is equal to the product of the characteristic functions of the summands.

The application of characteristic functions rests to a considerable extent on the property formulated in Theorem 3. As we have seen in §§ 24 and 25, the addition of independent random variables leads to a very complex operation, to the convolution of the distribution functions of the summands. In terms of characteristic functions, this complex operation is replaced by a very simple one, namely, the multiplication of characteristic functions.

THEOREM 4: *If a random variable ξ has an absolute moment of n-th order, then its characteristic function has an n-th derivative, and for $k \leqq n$,*

$$f^{(k)}(0) = i^k\, \mathsf{E}\, \xi^k. \tag{3}$$

Proof: Formal differentiation of the characteristic function k times ($k \leqq n$), in fact, results in the expression

$$f^{(k)}(t) = i^k \int x^k\, e^{itx}\, dF(x). \tag{4}$$

But

$$\left| \int x^k\, e^{itx}\, dF(x) \right| \leqq \int |x|^k\, dF(x),$$

and, by the hypothesis of the theorem, it is therefore bounded. From this it follows that the integral (4) exists and hence that the differentiation is valid. By setting $t = 0$ in (4), we find:

$$f^{(k)}(0) = i^k \int x^k\, dF(x).$$

§ 35. Definition and Simplest Properties

The expectation and variance can be very simply expressed in terms of the derivatives of the logarithm of the characteristic function. Thus, let us set

$$\psi(t) = \log f(t).$$

Then

$$\psi'(t) = \frac{f'(t)}{f(t)}$$

and

$$\psi''(t) = \frac{f''(t) \cdot f(t) - [f'(t)]^2}{f^2(t)}.$$

Taking into account equation (3) and the fact that $f(0) = 1$, we find:

$$\psi'(0) = f'(0) = i \, \mathsf{E}\, \xi$$

and

$$\psi''(0) = f''(0) - [f'(0)]^2 = i^2 \, \mathsf{E}\, \xi^2 - [i \, \mathsf{E}\, \xi]^2 = -\mathsf{D}\, \xi.$$

Hence,

$$\left. \begin{array}{l} \mathsf{E}\, \xi = \dfrac{1}{i} \psi'(0), \\ \mathsf{D}\, \xi = -\psi''(0). \end{array} \right\} \quad (5)$$

The k-th derivative of the logarithm of the characteristic function at the point 0, multiplied by i^{-k}, is called the *semi-invariant*, or *cumulant*, *of k-th order of the random variable*.

As follows immediately from Theorem 3, the semi-invariant of a sum of independent random variables is equal to the sum of the semi-invariants of the individual terms.

We have just seen that the first two semi-invariants are the expectation and the variance, i.e., the first-order moment and a certain rational function of the moments of first and second order. By carrying out the computation, one can easily satisfy oneself that a semi-invariant of any order k is an entire rational function of the first k moments. For illustrative purposes, we give the explicit expressions for the semi-invariants of the third and fourth orders:

$$i^{-3} \psi'''(0) = \mathsf{E}\, \xi^3 - 3\, \mathsf{E}\, \xi^2 \cdot \mathsf{E}\, \xi + 2[\mathsf{E}\, \xi]^3,$$

$$i^{-4} \psi^{IV}(0) = \mathsf{E}\, \xi^4 - 4\, \mathsf{E}\, \xi^3\, \mathsf{E}\, \xi - 3[\mathsf{E}\, \xi^2]^2 + 12\, \mathsf{E}\, \xi^2 [\mathsf{E}\, \xi]^2 - 6[\mathsf{E}\, \xi]^4.$$

Let us now discuss a few examples of characteristic functions.

VII. Characteristic Functions

Example 1. A random variable ξ is normally distributed with an expectation a and variance σ^2. The characteristic function of the variable ξ is

$$\varphi(t) = \frac{1}{\sigma \sqrt{2\pi}} \int e^{itx - \frac{(x-a)^2}{2\sigma^2}} dx.$$

As a result of the substitution

$$z = \frac{x-a}{\sigma} - it\sigma,$$

$\varphi(t)$ reduces to the form

$$\varphi(t) = e^{iat - \frac{\sigma^2 t^2}{2}} \frac{1}{\sqrt{2\pi}} \int_{-\infty - it\sigma}^{\infty - it\sigma} e^{-\frac{z^2}{2}} dz.$$

Using the well-known fact that for any real value of α

$$\int_{-\infty - i\alpha}^{\infty - i\alpha} e^{-\frac{z^2}{2}} dz = \sqrt{2\pi},$$

we thus find:

$$\varphi(t) = e^{iat - \frac{\sigma^2 t^2}{2}}.$$

By use of Theorem 4, we can easily compute the central moments for a normal distribution and in this alternative way obtain the result of the example discussed in § 30.

Example 2. Determine the characteristic function of a random variable ξ distributed according to the Poisson law.

By hypothesis, the variable ξ assumes only integral values, where

$$\mathsf{P}\{\xi = k\} = \frac{\lambda^k e^{-\lambda}}{k!} \quad (k = 0, 1, 2, \ldots),$$

λ being some positive constant.

The characteristic function of the variable ξ is

$$f(t) = \mathsf{E}\, e^{it\xi} = \sum_{k=0}^{\infty} e^{ikt} \mathsf{P}\{\xi = k\} = \sum_{k=0}^{\infty} e^{ikt} \frac{\lambda^k}{k!} e^{-\lambda}$$

$$= e^{-\lambda} \sum_{k=0}^{\infty} \frac{(\lambda e^{it})^k}{k!} = e^{-\lambda + \lambda e^{it}} = e^{\lambda (e^{it} - 1)}.$$

§ 35. DEFINITION AND SIMPLEST PROPERTIES

Hence, according to (5), we find:

$$\mathsf{E}\,\xi = \frac{1}{i}\psi'(0) = \lambda, \qquad \mathsf{D}\,\xi = -\psi''(0) = \lambda.$$

The first of these relations we previously obtained directly (§ 26, Example 3).

Example 3. A random variable ξ is uniformly distributed in the interval $(-a, a)$. Its characteristic function is

$$f(t) = \int_{-a}^{a} e^{itx}\frac{dx}{2a} = \frac{\sin at}{at}.$$

Example 4. Find the characteristic function of the variable μ representing the number of times an event A occurs in n independent trials in each of which the probability of occurrence of the event A is p.

The quantity μ is representable as the sum

$$\mu = \mu_1 + \mu_2 + \cdots + \mu_n$$

of n independent variables each of which assumes only the two values 0 and 1, with probability $q = 1 - p$ and p, respectively. The variable μ_k assumes the value 1 if the event A occurs in the k-th trial and the value 0 if A does not occur in the k-th trial.

The characteristic function of the variable μ_k is given by

$$f_k(t) = \mathsf{E}\,e^{it\mu_k} = e^{it\cdot 0}\,q + e^{it\cdot 1}\,p = q + p\,e^{it}.$$

According to Theorem 3, the characteristic function of the variable μ is given by

$$f(t) = \prod_{k=1}^{n} f_k(t) = (q + p\,e^{it})^n.$$

Let us now determine the characteristic function of the normalized variable $\eta = (\mu - np)/\sqrt{npq}$. By Theorem 2, we have

$$f_\eta(t) = e^{-it\sqrt{\frac{np}{q}}} f\!\left(\frac{t}{\sqrt{npq}}\right) = e^{-it\sqrt{\frac{np}{q}}}\!\left(q + p\,e^{i\frac{t}{\sqrt{npq}}}\right)^n$$

$$= \left(q\,e^{-it\sqrt{\frac{p}{nq}}} + p\,e^{it\sqrt{\frac{q}{np}}}\right)^n.$$

Example 5. A characteristic function satisfies the condition

$$f(-t) = \overline{f(t)}.$$

Indeed,

$$f(-t) = \int e^{-itx}\,dF(x) = \overline{\int e^{itx}\,dF(x)} = \overline{f(t)}.$$

§ 36. The Inversion Formula and The Uniqueness Theorem

We have seen that the characteristic function of a random variable ξ can always be found from its distribution function; it is important for us that the converse theorem also hold: a distribution function is uniquely determined by its characteristic function.

THEOREM 1: *Let $f(t)$ and $F(x)$ be the characteristic function and distribution function respectively of a random variable ξ. If x_1 and x_2 are points of continuity of the function $F(x)$, then*

$$F(x_2) - F(x_1) = \frac{1}{2\pi} \lim_{c\to\infty} \int_{-c}^{c} \frac{e^{-itx_1} - e^{-itx_2}}{it} f(t)\,dt. \tag{1}$$

Proof: From the definition of a characteristic function, it follows that the integral

$$J_c = \frac{1}{2\pi} \int_{-c}^{c} \frac{e^{-itx_1} - e^{-itx_2}}{it} f(t)\,dt$$

is equal to

$$J_c = \frac{1}{2\pi} \int_{-c}^{c}\!\!\int \frac{1}{it}[e^{it(z-x_1)} - e^{it(z-x_2)}]\,dF(z)\,dt.$$

The order of integration in the last integral may be changed, since the integral with respect to z converges absolutely and the integral with respect to t has finite limits of integration. Thus

§ 36. Inversion Formula and Uniqueness Theorem

$$J_c = \frac{1}{2\pi} \int \left[\int_{-c}^{c} \frac{e^{it(z-x_1)} - e^{it(z-x_2)}}{it} dt \right] dF(z)$$

$$= \frac{1}{2\pi} \int \left[\int_{0}^{c} \frac{e^{it(z-x_1)} - e^{-it(z-x_1)} - e^{it(z-x_2)} + e^{-it(z-x_2)}}{it} dt \right] dF(z)$$

$$= \frac{1}{\pi} \int_{-\infty}^{\infty} \int_{0}^{c} \left[\frac{\sin t(z-x_1)}{t} - \frac{\sin t(z-x_2)}{t} \right] dt \, dF(z) .$$

From analysis, we know that as $c \to \infty$

$$\frac{1}{\pi} \int_{0}^{c} \frac{\sin \alpha t}{t} dt \to \begin{cases} \frac{1}{2} & \text{if } \alpha > 0, \\ -\frac{1}{2} & \text{if } \alpha < 0, \end{cases} \qquad (2)$$

that the convergence is uniform with respect to α in every interval $\alpha > \delta > 0$ (or $\alpha < -\delta$), and that for $|\alpha| \leq \delta$ and all values of c,

$$\left| \frac{1}{\pi} \int_{0}^{c} \frac{\sin \alpha t}{t} dt \right| < 1 . \qquad (3)$$

For definiteness, let us suppose that $x_1 < x_2$ and let us represent the integral J_c in the form of a sum, as follows:

$$J_c = \int_{-\infty}^{x_1-\delta} + \int_{x_1-\delta}^{x_1+\delta} + \int_{x_1+\delta}^{x_2-\delta} + \int_{x_2-\delta}^{x_2+\delta} + \int_{x_2+\delta}^{\infty} \psi(c, z; x_1, x_2) \, dF(z),$$

where for brevity we have introduced the notation

$$\psi(c, z; x_1, x_2) = \frac{1}{\pi} \int_{0}^{c} \left\{ \frac{\sin t(z-x_1)}{t} - \frac{\sin t(z-x_2)}{t} \right\} dt$$

and $\delta > 0$ is so chosen that $x_1 + \delta < x_2 - \delta$.

In the interval $-\infty < z < x_1 - \delta$ the inequalities $z - x_1 < -\delta$ and $z - x_2 < -\delta$ hold. Therefore, on the basis of (2), we conclude that

$$\int_{-\infty}^{x_1-\delta} \psi(c, z; x_1, x_2) \, dF(z) \to 0$$

as $c \to \infty$. Analogously, for $x_2 + \delta < z < \infty$ and $c \to \infty$,

$$\int_{x_2+\delta}^{\infty} \psi(c, z; x_1, x_2)\, dF(z) \to 0.$$

Further, since the inequalities $z - x_1 > \delta$ and $z - x_2 < -\delta$ hold in the interval $x_1 + \delta < z < x_2 - \delta$, it follows from (2) that as $c \to \infty$

$$\int_{x_1+\delta}^{x_2-\delta} \psi(c, z; x_1, x_2)\, dF(z) \to \int_{x_1+\delta}^{x_2-\delta} dF(z) = F(x_2 - \delta) - F(x_1 + \delta).$$

Finally, in view of (3), we can apply the estimates

$$\left| \int_{x_1-\delta}^{x_1+\delta} \psi(c, z; x_1, x_2)\, dF(z) \right| < 2 \int_{x_1-\delta}^{x_1+\delta} dF(z) = 2\,[F(x_1 + \delta) - F(x_1 - \delta)]$$

and

$$\left| \int_{x_2-\delta}^{x_2+\delta} \psi(c, z; x_1, x_2)\, dF(z) \right| < 2 \int_{x_1-\delta}^{x_1+\delta} dF(z) = 2\,[F(x_2 + \delta) - F(x_2 - \delta)].$$

Thus we find that for any $\delta > 0$

$$\overline{\lim_{c \to \infty}}\, J_c = F(x_2 - \delta) - F(x_1 + \delta) + R_1(\delta, x_1, x_2)$$

and

$$\underline{\lim_{c \to \infty}}\, J_c = F(x_2 - \delta) - F(x_1 + \delta) + R_2(\delta, x_1, x_2),$$

where

$$|R_i(\delta, x_1, x_2)| < 2\,\{F(x_1 + \delta) - F(x_1 - \delta) + F(x_2 + \delta) - F(x_2 - \delta)\}$$
$$(i = 1, 2).$$

Now let $\delta \to 0$. From this and from the fact that the function $F(x)$ is continuous at the points x_1 and x_2 follow the equalities

$$\lim_{\delta \to 0} F(x_1 + \delta) = \lim_{\delta \to 0} F(x_1 - \delta) = F(x_1)$$

and

$$\lim_{\delta \to 0} F(x_2 + \delta) = \lim_{\delta \to 0} F(x_2 - \delta) = F(x_2).$$

And since J_c does not depend on δ, we have:

$$\lim_{c \to \infty} J_c = F(x_2) - F(x_1).$$

The equation (1) is referred to as the *inversion formula*. We now use this formula to prove the following important proposition.

§ 36. Inversion Formula and Uniqueness Theorem

Theorem 2 (The Uniqueness Theorem): *A distribution function is uniquely determined by its characteristic function.*

Proof: From Theorem 1, it immediately follows that the formula

$$F(x) = \frac{1}{2\pi} \lim_{y \to -\infty} \lim_{c \to \infty} \int_{-c}^{+c} \frac{e^{-ity} - e^{-itx}}{it} f(t)\, dt$$

is applicable at each point of continuity of $F(x)$, where the limit in y is evaluated with respect to any set of points which are points of continuity of the function $F(x)$.

As an application of the last theorem, we shall prove the following statement.

Example 1. If the independent random variables ξ_1 and ξ_2 are normally distributed, then their sum $\xi = \xi_1 + \xi_2$ is also normally distributed. If

$$\mathsf{E}\,\xi_1 = a_1,\; \mathsf{D}\,\xi_1 = \sigma_1^2;\; \mathsf{E}\,\xi_2 = a_2,\; \mathsf{D}\,\xi_2 = \sigma_2^2,$$

then the characteristic functions of the variables ξ_1 and ξ_2 are given by

$$f_1(t) = e^{ia_1 t - \frac{1}{2}\sigma_1^2 t^2}, \quad f_2(t) = e^{ia_2 t - \frac{1}{2}\sigma_2^2 t^2}.$$

By Theorem 3 of § 35, the characteristic function $f(t)$ of their sum is

$$f(t) = f_1(t) \cdot f_2(t) = e^{it(a_1 + a_2) - \frac{1}{2}(\sigma_1^2 + \sigma_2^2) t^2}.$$

This is just the characteristic function for a normal law with an expectation $a = a_1 + a_2$ and variance $\sigma^2 = \sigma_1^2 + \sigma_2^2$. On the basis of the Uniqueness Theorem we can conclude that the distribution function of the variable ξ is normal.

The converse proposition, due to Cramér and stated without proof in § 24 of this book, can be formulated in terms of characteristic functions as follows: *If $f_1(t)$ and $f_2(t)$ are characteristic functions and if*

$$f_1(t) \cdot f_2(t) = e^{-\frac{t^2}{2}},$$

then

$$f_1(t) = e^{iat - \sigma^2 \frac{t^2}{2}}, \quad f_2(t) = e^{-iat - \frac{(1-\sigma^2) t^2}{2}} \quad (0 \leq \sigma \leq 1).$$

Example 2. The independent random variables ξ_1 and ξ_2 are distributed according to the Poisson law, where

$$P\{\xi_1 = k\} = \frac{\lambda_1^k e^{-\lambda_1}}{k!}, \quad P\{\xi_2 = k\} = \frac{\lambda_2^k e^{-\lambda_2}}{k!}.$$

We shall prove that the random variable $\xi = \xi_1 + \xi_2$ is distributed according to the Poisson law with the parameter $\lambda = \lambda_1 + \lambda_2$.

In Example 2 of the preceding section, indeed, we found that the characteristic functions of the random variables ξ_1 and ξ_2 were

$$f_1(t) = e^{\lambda_1(e^{it}-1)}, \quad f_2(t) = e^{\lambda_2(e^{it}-1)}$$

By virtue of Theorem 3 of the preceding section, the characteristic function of their sum $\xi = \xi_1 + \xi_2$ is

$$f(t) = f_1(t) \cdot f_2(t) = e^{(\lambda_1 + \lambda_2)(e^{it}-1)};$$

i.e., it is the characteristic function for some Poisson law. By the uniqueness theorem, the unique distribution which has $f(t)$ as its characteristic function is the Poisson law for which

$$P\{\xi = k\} = \frac{(\lambda_1 + \lambda_2)^k e^{-(\lambda_1+\lambda_2)}}{k!} \quad (k \geqq 0).$$

Raikov has proved the deeper converse theorem: if the sum of two independent random variables is distributed according to the Poisson law, then each summand is also distributed according to the Poisson law.

Example 3. A characteristic function is real if and only if its corresponding distribution function is symmetric, i.e., the distribution function satisfies the condition

$$F(x) = 1 - F(-x + 0)$$

for all values of x.

If a distribution function is symmetric, then its characteristic function is real. This is proved by the simple computation:

§ 36. Inversion Formula and Uniqueness Theorem

$$f(t) = \int e^{itx} \, dF(x) =$$
$$= -\int_0^\infty e^{-itx} \, dF(-x+0) + \int_0^\infty e^{itx} \, dF(x) + F(+0) - F(-0)$$
$$= \int_0^\infty (e^{-itx} + e^{itx}) \, dF(x) + F(+0) - F(-0)$$
$$= 2\int_0^\infty \cos tx \, dF(x) + F(+0) - F(-0) = \int \cos tx \, dF(x).$$

(We recall here that we have agreed to include the lower limit and to exclude the upper limit from the interval of integration.)

To prove the converse, we consider the random variable $\eta = -\xi$. The distribution function for the variable η is

$$G(x) = \mathsf{P}\{\eta < x\} = \mathsf{P}\{\xi > -x\} = 1 - F(-x+0).$$

The characteristic functions of the variables ξ and η are connected by the relation

$$g(t) = \mathsf{E}\, e^{it\eta} = \mathsf{E}\, e^{-it\xi} = \overline{\mathsf{E}\, e^{it\xi}} = \overline{f(t)}.$$

Since by hypothesis $f(t)$ is real, $\overline{f(t)} = f(t)$, and this implies that

$$g(t) = f(t).$$

From the uniqueness theorem, we now conclude that the distribution functions of the variables ξ and η coincide, i.e., that

$$F(x) = 1 - F(-x+0),$$

Q.E.D.

We prove the following fact, which we shall need later.

Theorem 3: *If the characteristic function $f(t)$ is summable (Lebesgue integrable) over the entire line, then the distribution function $F(x)$ that corresponds to it is continuous, its derivative $p(x)$ is continuous, and*

$$p(x) = F'(x) = \frac{1}{2\pi} \int e^{-itx} f(t) \, dt.$$

Proof: If the function $f(t)$ is summable over the entire line, the function
$$\frac{e^{-itx_1} - e^{-itx_2}}{it} f(t)$$
also has this property, and therefore the inversion formula is expressible in the form
$$F(x_2) - F(x_1) = \frac{1}{2\pi} \int \frac{e^{-itx_1} - e^{-itx_2}}{it} f(t) dt.$$

Now let h be such that $x_1 = x - h$ and $x_2 = x + h$ are points of continuity of $F(x)$. After a simple formal transformation, we arrive at the equation
$$F(x+h) - F(x-h) = \frac{2h}{2\pi} \int \frac{\sin th}{th} e^{-itx} f(t) dt. \qquad (4)$$

Since $|(\sin th)/th| \leq 1$, it follows that
$$F(x+h) - F(x-h) < \frac{h}{\pi} \int |f(t)| dt.$$

Now let $h \to 0$. The last inequality then goes over into the equation
$$F(x+0) - F(x-0) = 0,$$
which proves that $F(x)$ is continuous.

Now (4) can be written in the following form:
$$\frac{F(x+h) - F(x-h)}{2h} = \frac{1}{2\pi} \int \frac{\sin th}{th} e^{-itx} f(t) dt. \qquad (5)$$

Since the integrand tends to $e^{-itx} f(t)$ as $h \to 0$, it follows from the familiar Lebesgue Theorem for passing to the limit under the integral sign that
$$\lim_{h \to 0} \frac{1}{2\pi} \int \frac{\sin th}{th} e^{-itx} f(t) dt = \frac{1}{2\pi} \int e^{-itx} f(t) dt.$$

Since the limit of the right-hand side of equation (5) exists, the limit of the left-hand side also exists. Thus, for every value of x we have

$$p(x) = \lim_{h \to 0} \frac{F(x+h) - F(x-h)}{2h} = \frac{1}{2\pi} \int e^{-itx} f(t) dt.$$

By a straightforward calculation, it follows that

$$|p(x+h) - p(x)| \leq \frac{1}{\pi} \int |\sin(th/2)| \cdot |f(t)| dt.$$

Let ε be a given positive quantity. We choose A sufficiently large that

$$\frac{1}{\pi} \int_{|t|>A} |f(t)| dt < \frac{\varepsilon}{2}.$$

The integral

$$\frac{1}{\pi} \int_{|t|<A} |\sin th/2| \, |f(t)| dt$$

can be made smaller than $\varepsilon/2$ by choosing h sufficiently small. This completes the proof of the theorem.

§ 37. Helly's Theorems

In the sequel, we shall require two theorems of a purely analytical nature, the first and second theorems of Helly.

We shall agree to say that a sequence of non-decreasing functions

$$F_1(x), F_2(x), \ldots, F_n(x), \ldots$$

converges weakly to a non-decreasing function $F(x)$ if for $n \to \infty$ it converges to $F(x)$ at each of its points of continuity.

Henceforth, we shall always assume that the functions $F_n(x)$ satisfy the supplementary condition

$$F_n(-\infty) = 0$$

and we shall make no further mention of this.

We immediately observe that in order for a sequence of functions to be weakly convergent it is sufficient that it converge to the function $F(x)$ on some everywhere-dense set D. For let x be an arbitrary point, and x' and x'' any two points of the set D such that $x' \leq x \leq x''$. Then also

$$F_n(x') \leq F_n(x) \leq F_n(x'').$$

Therefore,

$$\lim_{n\to\infty} F_n(x') \leq \underline{\lim_{n\to\infty}} F_n(x) \leq \overline{\lim_{n\to\infty}} F_n(x) \leq \lim_{n\to\infty} F_n(x'').$$

And since by assumption

$$\lim_{n\to\infty} F_n(x') = F(x') \text{ and } \lim_{n\to\infty} F_n(x'') = F(x''),$$

it also follows that

$$F(x') \leq \underline{\lim_{n\to\infty}} F_n(x) \leq \overline{\lim_{n\to\infty}} F_n(x) \leq F(x'').$$

But the middle terms in these inequalities are independent of x' and x'', and hence

$$F(x-0) \leq \underline{\lim_{n\to\infty}} F_n(x) \leq \overline{\lim_{n\to\infty}} F_n(x) \leq F(x+0).$$

If the function $F(x)$ is continuous at the point x, then

$$F(x-0) = F(x) = F(x+0).$$

Consequently, at the points of continuity of the function $F(x)$ we have:

$$\lim_{n\to\infty} F_n(x) = F(x).$$

THE FIRST THEOREM OF HELLY: *Any sequence of non-decreasing functions*

$$F_1(x), F_2(x), \ldots, F_n(x), \ldots \tag{1}$$

which are uniformly bounded contains at least one subsequence

$$F_{n_1}(x), F_{n_2}(x), \ldots, F_{n_k}(x), \ldots$$

which converges weakly to some non-decreasing function $F(x)$.

Proof: Let D be some denumerable everywhere-dense set of points $x_1', x_2', \ldots, x_n', \ldots$. Let us consider the values of the functions of the sequence (1) at the point x_1':

$$F_1(x_1'), F_2(x_1'), \ldots, F_n(x_1'), \ldots$$

§ 37. HELLY'S THEOREM

Since the set consisting of these values is by assumption bounded, it contains at least one subsequence

$$F_{11}(x_1'), F_{12}(x_1'), \ldots, F_{1n}(x_1'), \ldots, \qquad (2)$$

which converges to some limiting value, which we shall denote by $G(x_1')$. Now consider the set of values

$$F_{11}(x_2'), F_{12}(x_2'), \ldots, F_{1n}(x_2'), \ldots.$$

Since this set is also bounded, it contains a subsequence which converges to some limiting value $G(x_2')$. Thus, we can pick out from the sequence (2) a subsequence

$$F_{21}(x), F_{22}(x), \ldots, F_{2n}(x), \ldots \qquad (3)$$

such that $\lim_{n \to \infty} F_{2n}(x_1') = G(x_1')$ and $\lim_{n \to \infty} F_{2n}(x_2') = G(x_2')$ hold simultaneously.

We continue to pick out such sequences

$$F_{k1}(x), F_{k2}(x), \ldots, F_{kn}(x), \ldots, \qquad (4)$$

for which the relations $\lim_{n \to \infty} F_{kn}(x_r') = G(x_r')$ hold simultaneously for all $r \leq k$. Now we consider the diagonal sequence

$$F_{11}(x), F_{22}(x), \ldots, F_{nn}(x), \ldots. \qquad (5)$$

In the final analysis, this entire sequence has been selected from the sequence (2) and therefore $\lim_{n \to \infty} F_{nn}(x_1') = G(x_1')$. Further, since the entire diagonal sequence, with the exception of the first term only, has been selected from the sequence (3), $\lim_{n \to \infty} F_{nn}(x_2') = G(x_2')$. And in general, the entire diagonal sequence with the exception of the first $k-1$ terms has been selected from the sequence (4); therefore, $\lim_{n \to \infty} F_{nn}(x_k') = G(x_k')$ for each value of k. The above result may be formulated as follows: The sequence (1) contains at least one subsequence converging at all the points x_k' of the set D to some function $G(x)$ defined on this set. In addition to this, since the functions $F_{nn}(x)$ are non-decreasing and uniformly bounded, it is evident that the function $G(x)$ will also be non-decreasing and bounded.

It is now clear that the function $G(x)$, which is defined on the set D, may be continued in such a way that it will be defined over the entire interval $-\infty < x < \infty$ while remaining non-decreasing and bounded.

266   VII. CHARACTERISTIC FUNCTIONS

The sequence (5) converges to this function on the everywhere-dense set D; consequently, as proved above, it converges weakly to this function, Q.E.D.

We note that the function obtained by continuing the function G may prove not to be continuous from the left. But we can alter its values at the points of discontinuity so as to re-establish this property. The subsequence F_{nn} will converge weakly to the function thus "corrected."

THE SECOND THEOREM OF HELLY: *Let $f(x)$ be a continuous function and let the sequence of non-decreasing uniformly bounded functions*

$$F_1(x), F_2(x), \ldots, F_n(x), \ldots$$

converge weakly to the function $F(x)$ on some finite interval $a \leq x \leq b$, where a and b are points of continuity of the function $F(x)$; then

$$\lim_{n \to \infty} \int_a^b f(x)\, dF_n(x) = \int_a^b f(x)\, dF(x).$$

Proof: From the continuity of the function $f(x)$, it follows that, given any positive constant ε no matter how small, we can find a subdivision of the interval $a \leq x \leq b$ by means of points $x_0 = a, x_1, \ldots, x_N = b$ into the subintervals (x_k, x_{k+1}) such that in each interval (x_k, x_{k+1}) the inequality $|f(x) - f(x_k)| < \varepsilon$ holds. Using this fact, we may introduce an auxiliary function $f_\varepsilon(x)$ which assumes only a finite number of values and is defined by means of the equations

$$f_\varepsilon(x) = f(x_k) \text{ for } x_k \leq x < x_{k+1}.$$

Clearly, the inequality

$$|f(x) - f_\varepsilon(x)| < \varepsilon$$

holds for all values of x in the interval $a \leq x \leq b$. Here, we may select the points of division $x_1, x_2, \ldots, x_{N-1}$ beforehand so that they are the points of continuity of the function $F(x)$. Since the sequence of functions $F_1(x), F_2(x), F_3(x), \ldots$ converges to the function $F(x)$, the inequality

$$|F(x_k) - F_n(x_k)| < \frac{\varepsilon}{MN} \tag{6}$$

will be satisfied at all the points of subdivision for n sufficiently large, where M is the maximum of the absolute value of $f(x)$ in the interval $a \leq x \leq b$.

§ 37. Helly's Theorem

As can readily be seen,

$$\left|\int_a^b f(x)\,dF(x) - \int_a^b f(x)\,dF_n(x)\right| \leq \left|\int_a^b f(x)\,dF(x) - \int_a^b f_\varepsilon(x)\,dF(x)\right| +$$

$$+ \left|\int_a^b f_\varepsilon(x)\,dF(x) - \int_a^b f_\varepsilon(x)\,dF_n(x)\right| +$$

$$+ \left|\int_a^b f_\varepsilon(x)\,dF_n(x) - \int_a^b f(x)\,dF_n(x)\right|.$$

It is not difficult to show that the first term on the right-hand side does not exceed $\varepsilon[F(z) - F(a)]$ and that the third term does not exceed $\varepsilon[F_n(b) - F_n(a)]$. As regards the second term, it is equal to

$$\left|\sum_{k=0}^{N-1} f(x_k)[F(x_{k+1}) - F(x_k)] - \sum_{k=0}^{N-1} f(x_k)[F_n(x_{k+1}) - F_n(x_k)]\right| =$$

$$= \left|\sum_{k=0}^{N-1} f(x_k)[F(x_{k+1}) - F_n(x_{k+1})] - \sum_{k=0}^{N-1} f(x_k)[F(x_k) - F_n(x_k)]\right|;$$

and therefore, for n sufficiently large, does not exceed 2ε, as follows from the inequality (6). Since the functions $F_n(x)$ are uniformly bounded, the sum

$$\varepsilon[F(b) - F(a)] + \varepsilon[F_n(b) - F_n(a)] + 2\varepsilon$$

can be made as small as we wish by choosing ε sufficiently small.

THE GENERALIZED SECOND THEOREM OF HELLY: *If the function $f(x)$ is continuous and bounded over the entire real line $-\infty < x < \infty$, the sequence of non-decreasing uniformly bounded functions*

$$F_1(x), F_2(x), \ldots, F_n(x), \ldots$$

converges weakly to the function $F(x)$, and

$$\lim_{n\to\infty} F_n(-\infty) = F(-\infty), \quad \lim_{n\to\infty} F_n(+\infty) = F(+\infty),$$

then

$$\lim_{n\to\infty} \int f(x)\,dF_n(x) = \int f(x)\,dF(x).$$

Proof: Let $A < 0$ and $B > 0$; set

$$J_1 = \left| \int_{-\infty}^{A} f(x) \, dF(x) - \int_{-\infty}^{A} f(x) \, dF_n(x) \right|,$$

$$J_2 = \left| \int_{A}^{B} f(x) \, dF(x) - \int_{A}^{B} f(x) \, dF_n(x) \right|,$$

$$J_3 = \left| \int_{B}^{\infty} f(x) \, dF(x) - \int_{B}^{\infty} f(x) \, dF_n(x) \right|.$$

Clearly,

$$\left| \int f(x) \, dF(x) - \int f(x) \, dF_n(x) \right| \leq J_1 + J_2 + J_3.$$

The quantities J_1 and J_3 can be made arbitrarily small by choosing A and B sufficiently large in absolute value and moreover such that A and B are points of continuity of the function $F(x)$ and by choosing n sufficiently large. In fact, if M is the least upper bound of $|f(x)|$ for $-\infty < x < \infty$, then

$$J_1 \leq M \left[F(A) + F_n(A) \right],$$

$$J_3 \leq M[F(+\infty) - F(B)] + M[F_n(+\infty) - F_n(B)].$$

But

$$\lim_{A \to -\infty} F(A) = 0, \quad \lim_{B \to \infty} F(B) = F(+\infty).$$

And since by assumption,

$$\lim_{n \to \infty} F_n(A) = F(A), \quad \lim_{n \to \infty} F_n(B) = F(B),$$

our assertion concerning J_1 and J_3 is proved. The quantity J_2 can be made arbitrarily small for n sufficently large by virtue of Helly's Theorem for a finite interval.

The theorem is thus proved.

§ 38. Limit Theorems for Characteristic Functions

Most important from the standpoint of applying characteristic functions in deriving asymptotic formulas in probability theory are two limit theorems, a direct theorem and its converse. These theorems establish that the correspondence existing between distribution functions and characteristic functions is not only one-to-one but also continuous.

§ 38. Limit Theorems for Characteristic Functions

THE DIRECT LIMIT THEOREM: *If the sequence of distribution functions*

$$F_1(x), F_2(x), \ldots, F_n(x), \ldots$$

converges weakly to the distribution function $F(x)$, then the sequence of characteristic functions

$$f_1(t), f_2(t), \ldots, f_n(t), \ldots$$

converges to the characteristic function $f(t)$. The convergence is uniform with respect to t in every finite interval $a \leq t \leq b$.

Proof: Since

$$f_n(t) = \int e^{itx} dF_n(x), \quad f(t) = \int e^{itx} dF(x),$$

and the function e^{itx} is continuous and bounded in the interval $-\infty < t < \infty$, then by the Generalized Helly Theorem, as $n \to \infty$

$$f_n(t) \to f(t).$$

The assertion that this convergence is uniform with respect to t in every finite interval can be verified by using word for word the same reasoning as in the proof of Helly's Second Theorem.

THE CONVERSE LIMIT THEOREM: *If the sequence of characteristic functions*

$$f_1(t), f_2(t), \ldots, f_n(t), \ldots \tag{1}$$

converges to a continuous function $f(t)$, then the sequence of distribution functions

$$F_1(x), F_2(x), \ldots, F_n(x), \ldots \tag{2}$$

converges weakly to some distribution function $F(x)$. (In virtue of the direct limit theorem, $f(t) = \int e^{itx} dF(x)$.)

Proof: By Helly's First Theorem, we can conclude that the sequence (2) certainly contains a subsequence

$$F_{n_1}(x), F_{n_2}(x), \ldots, F_{n_k}(x), \ldots, \tag{3}$$

which converges weakly to some non-decreasing function $F(x)$. It is clear that the function $F(x)$ may be considered to be continuous from the left:

$$\lim_{x' \to x-0} F(x') = F(x).$$

The function $F(x)$ need not, in general, be a distribution function, since for this to be the case, the conditions $F(-\infty) = 0$ and $F(+\infty) = 1$ also have to be satisfied. In fact, the limit of the sequence of functions

$$F_n(x) = \begin{cases} 0 & \text{for} \quad x \leq -n, \\ \frac{1}{2} & \text{for} \quad -n < x \leq n, \\ 1 & \text{for} \quad x > n, \end{cases}$$

is the function $F(x) \equiv 1/2$ and therefore both $F(-\infty)$ and $F(+\infty)$ are also equal to $1/2$. However, under the conditions of our theorem, as we shall now show, we are sure to have $F(-\infty) = 0$ and $F(+\infty) = 1$.

In fact, if this were not so, then taking into account that the relations $F(-\infty) \geq 0$ and $F(+\infty) \leq 1$ have to hold, we would have

$$\delta = F(+\infty) - F(-\infty) < 1.$$

Let us now take some positive number ε less than $1 - \delta$. Since by the hypothesis of the theorem, the sequence of characteristic functions (1) converges to the function $f(t)$, it follows that $f(0) = 1$. But since in addition to this the function $f(t)$ is continuous, it is possible to choose a positive number τ so small that the inequality

$$\frac{1}{2\tau} \left| \int_{-\tau}^{\tau} f(t)\, dt \right| > 1 - \frac{\varepsilon}{2} > \delta + \frac{\varepsilon}{2} \tag{4}$$

holds. But at the same time, we may choose $X > 4/\tau\varepsilon$ and K so large that for $k > K$

$$\delta_k = F_{n_k}(X) - F_{n_k}(-X) < \delta + \frac{\varepsilon}{4}.$$

Since $f_{n_k}(t)$ is a characteristic function, it follows that

$$\int_{-\tau}^{\tau} f_{n_k}(t)\, dt = \int \left[\int_{-\tau}^{\tau} e^{itx}\, dt \right] dF_{n_k}(x).$$

The integral on the right-hand side of this equation may be estimated in the following manner. On the one hand, since $|e^{itx}| = 1$, we have

$$\left| \int_{-\tau}^{\tau} e^{itx}\, dt \right| \leq 2\tau.$$

§ 38. Limit Theorems for Characteristic Functions

On the other hand,

$$\int_{-\tau}^{\tau} e^{itx} dt = \frac{2}{x} \sin \tau x,$$

and since $|\sin \tau x| \leq 1$, we have for $|x| > X$

$$\left|\int_{-\tau}^{\tau} e^{itx} dt\right| < \frac{2}{X}.$$

Hence, using the first estimate for $|x| \leq X$ and the second for $|x| > X$, we obtain:

$$\left|\int_{-\tau}^{\tau} f_{n_k}(t) dt\right| \leq \left|\int_{|x|\leq X} \left(\int_{-\tau}^{\tau} e^{itx} dt\right) dF_{n_k}(x) + \right.$$

$$\left. + \left|\int_{|x|> X} \left(\int_{-\tau}^{\tau} e^{itx} dt\right) dF_{n_k}(x)\right| < 2\tau \delta_k + \frac{2}{X},$$

and therefore

$$\frac{1}{2\tau}\left|\int_{-\tau}^{\tau} f_{n_k}(t) dt\right| < \delta + \frac{\varepsilon}{2}.$$

This inequality continues to hold in the limit:

$$\frac{1}{2\tau}\left|\int_{-\tau}^{\tau} f(t) dt\right| \leq \delta + \frac{\varepsilon}{2},$$

and this clearly contradicts inequality (4).

Thus, the function $F(x)$ to which the sequence of functions $F_{n_k}(x)$ converges weakly is a distribution function; by the direct limit theorem, its characteristic function is $f(t)$.

In order to complete the proof of the theorem, it remains to show that the sequence (2) itself converges weakly to the function $F(x)$. Let us suppose the contrary. Then one could find a subsequence of functions

$$F_{n'_1}(x), F_{n'_2}(x), \ldots, F_{n'_k}(x), \ldots, \qquad (5)$$

converging weakly to some function $F^*(x)$ which would differ from $F(x)$ in at least one of its points of continuity. According to what already has been proved, $F^*(x)$ has to be a distribution function with the characteristic function $f(t)$. By the uniqueness theorem, we must have

$$F^*(x) = F(x).$$

This contradicts our assumption.

We note that the conditions of the theorem are fulfilled in each of the following two cases:

1) The sequence of characteristic functions $f_n(t)$ converges uniformly to some function $f(t)$ in every finite interval $a \leq t \leq b$.

2) The sequence of characteristic functions $f_n(t)$ converges to a characteristic function $f(t)$.

Example. As an example of how the limit theorems are used, we shall discuss the proof of the DeMoivre-Laplace Theorem.

In Example 4 of § 35, we determined the characteristic function of the random variable $\eta = (\mu - np)/\sqrt{npq}$:

$$f_n(t) = \left(q e^{-it\sqrt{\frac{p}{nq}}} + p e^{it\sqrt{\frac{q}{np}}} \right)^n.$$

By utilizing the MacLaurin expansion for e^x, we find:

$$q e^{-it\sqrt{\frac{p}{nq}}} + p e^{it\sqrt{\frac{q}{np}}} = 1 - \frac{t^2}{2n}(1 + R_n),$$

where

$$R_n = 2 \sum_{k=3}^{\infty} \frac{1}{k!} \left(\frac{it}{\sqrt{n}}\right)^{k-2} \frac{pq^k + q(-p)^k}{\sqrt{(pq)^k}}.$$

Since $R_n \to 0$ as $n \to \infty$, we have

$$f_n(t) = \left[1 - \frac{t^2}{2n}(1 + R_n) \right]^n \to e^{-\frac{t^2}{2}}.$$

By virtue of the converse limit theorem, it follows from this that for any x

$$\mathsf{P}\left\{ \frac{\mu - np}{\sqrt{npq}} < x \right\} \to \frac{1}{\sqrt{2\pi}} \int_{-\infty}^{x} e^{-\frac{z^2}{2}} dz.$$

when $n \to \infty$.

From the continuity of the limit function, it is easy to deduce that the convergence is uniform in x.

§ 39. Positive-Semidefinite Functions

The purpose of the present section is to give an exhaustive description of the class of characteristic functions. The fundamental theorem which we quote below was simultaneously proved by Khintchine and Bochner and first published by Bochner.

To formulate and prove this theorem, we have to introduce a new notion. We shall say that a continuous function $f(t)$ of the real argument t is *positive-semidefinite** in the interval $-\infty < t < \infty$ if for any real numbers $t_1, t_2, \ldots, t_n$, complex numbers $\xi_1, \xi_2, \ldots, \xi_n$, and integer n

$$\sum_{k=1}^{n}\sum_{j=1}^{n} f(t_k - t_j)\, \bar{\xi}_j\, \xi_k \geq 0. \tag{1}$$

Let us enumerate a few of the simplest properties of positive-semidefinite functions.

1. $f(0) \geq 0$. In fact, if we set $n = 1$, $t_1 = 0$ and $\xi_1 = 1$, then from the condition for the positive-semidefiniteness of the function $f(t)$, we find:

$$\sum_{k=1}^{n}\sum_{j=1}^{n} f(t_k - t_j)\, \xi_k\, \bar{\xi}_j = f(0) \geq 0.$$

2. For any real value of t,

$$f(-t) = \overline{f(t)}.$$

In order to prove this, we insert in (1) the values $n = 2$, $t_1 = 0$, $t_2 = t$, and leave ξ_1 and ξ_2 arbitrary. We have by assumption

$$0 \leq \sum_{k=1}^{2}\sum_{j=1}^{2} f(t_k - t_j)\, \xi_k\, \bar{\xi}_j =$$
$$= f(0-0)\, \xi_1\, \bar{\xi}_1 + f(0-t)\, \xi_1\, \bar{\xi}_2 + f(t-0)\, \xi_2\, \bar{\xi}_1 + f(t-t)\, \xi_2\, \bar{\xi}_2$$
$$= f(0)\,(|\xi_1|^2 + |\xi_2|^2) + f(-t)\, \xi_1\, \bar{\xi}_2 + f(t)\, \bar{\xi}_1\, \xi_2, \tag{2}$$

and therefore, the quantity

$$f(-t)\, \xi_1\, \bar{\xi}_2 + f(t)\, \bar{\xi}_1\, \xi_2$$

---

* The terms *positive definite* and *non-negative definite* are also used. [*Trans.*]

has to be real. Thus, if we set $f(-t) = \alpha_1 + i\beta_1$, $f(t) = \alpha_2 + i\beta_2$, $\xi_1\bar{\xi}_2 = \gamma + i\delta$, and $\bar{\xi}_1\xi_2 = \gamma - i\delta$, then we must have

$$\alpha_1\delta + \beta_1\gamma - \alpha_2\delta + \beta_2\gamma = 0.$$

Since ξ_1 and ξ_2, and consequently γ and δ, are arbitrary, it follows that

$$\alpha_1 - \alpha_2 = 0, \ \beta_1 + \beta_1 = 0.$$

This implies our assertion.

3. For any real value of t

$$|f(t)| \leq f(0).$$

Let us suppose that in the inequality (2) $\xi_1 = f(t)$ and $\xi_2 = -|f(t)|$; then by the preceding result,

$$2 f(0) |f(t)|^2 - |f(t)|^2 |f(t)| - |f(t)|^2 |f(t)| \geq 0.$$

For $|f(t)| \neq 0$, we obtain from this that

$$f(0) \geq |f(t)|.$$

However, if $|f(t)| = 0$, then by virtue of property 1 we again have:

$$f(0) \geq |f(t)|.$$

In particular, it follows from property 3. that if a positive-semidefinite function is such that $f(0) = 0$ then $f(t) \equiv 0$.

THE BOCHNER-KHINTCHINE THEOREM: *In order for a continuous function $f(t)$ which satisfies the condition $f(0) = 1$ to be a characteristic function, it is necessary and sufficient that it be positive-semidefinite.*

Proof: In one direction, the theorem is trivial. In fact, if

$$f(t) = \int e^{itx} dF(x),$$

where $F(x)$ is some distribution function, then if n is any integer, $t_1, t_2, \ldots, t_n$ are arbitrary real numbers, and $\xi_1, \xi_2, \ldots, \xi_n$ arbitrary complex numbers, we have:

§ 39. POSITIVE-SEMIDEFINITE FUNCTIONS

$$\sum_{k=1}^{n}\sum_{j=1}^{n} f(t_k - t_j)\, \xi_k \bar{\xi}_j = \sum_{k=1}^{n}\sum_{j=1}^{n} \{\int e^{ix(t_k-t_j)}\, dF(x)\}\, \xi_k \bar{\xi}_j$$

$$= \int \sum_{k=1}^{n}\sum_{j=1}^{n} e^{ix(t_k-t_j)}\, \xi_k \bar{\xi}_j\, dF(x)$$

$$= \int \left(\sum_{k=1}^{n} e^{it_k x}\, \xi_k\right)\left(\sum_{j=1}^{n} e^{-it_j x}\, \bar{\xi}_j\right) dF(x)$$

$$= \int \left|\sum_{k=1}^{n} e^{it_k x}\, \xi_k\right|^2 dF(x) \geq 0.$$

The sufficiency proof requires a more complicated argument.

The proof given below is based on Theorem 3 of § 36, the limit theorems for characteristic functions, and the following lemma; it was suggested in the main by Linnik.

LEMMA: *If the function $f(t)$ is measurable, bounded, and summable in the interval $(-T, T)$ and*

$$p(x) = \int_{-T}^{T} e^{-itx} f(t)\, dt \geq 0 , \qquad (3)$$

then the function $p(x)$ is integrable over the entire line.

Proof: Since the function $p(x)$ is continuous, it is integrable in any finite interval. Let

$$G(x) = \int_{-x}^{x} p(z)\, dz.$$

The function $p(z)$ is non-negative and therefore $G(x)$ is a non-decreasing function. To prove the lemma it suffices to show that $G(x)$ is bounded. To this end, we consider the function

$$F(x) = \frac{1}{x} \int_{x}^{2x} G(u)\, du.$$

Clearly,

$$F(x) \geq \frac{G(x)}{x} \int_{x}^{2x} du = G(x).$$

It can easily be verified that

$$G(x) = 2 \int_{-T}^{T} \frac{\sin xt}{t}\, f(t)\, dt$$

and that

$$F(x) = \frac{4}{x}\int_{-T}^{T}\frac{\sin^2 xt}{t^2}f(t)dt - \frac{4}{x}\int_{-T}^{T}\frac{\sin^2(xt/2)}{t^2}f(t)dt.$$

Let $M = \sup |f(t)|$; then

$$\frac{4}{x}\int_{-T}^{T}\frac{\sin^2 xt}{t^2}f(t)dt < 4M\int\frac{\sin^2 u}{u^2}du.$$

Clearly,

$$\frac{4}{x}\int_{-T}^{T}\frac{\sin^2(xt/2)}{t^2}f(t)dt < 2M\int\frac{\sin^2 u}{u^2}du.$$

This proves the boundedness of the functions $F(x)$ and $G(x)$, and hence the lemma is also proved.

Let us now assume that the function $f(t)$ is positive-semidefinite and continuous and that $f(0) = 1$. Let $z > 0$ and consider the function

$$p_z(x) = \frac{1}{2\pi z}\int_0^z\int_0^z f(u-v)e^{-iux}e^{ivx}dudv.$$

By virtue of the positive-definiteness of $f(t)$, the function $p_z(x)$ is non-negative (the double integral is the limit of non-negative sums of the form (1)). If we make the change of variables $u = t+v$ in the inner integral and change the order of integration, then we find after a straightforward computation that

$$p_z(x) = \frac{1}{2\pi}\int_{-z}^{z}(1-|t|/z)f(t)e^{-itx}dt.$$

Since the function $p_z(x)$ is non-negative and representable in the form (3), the lemma just proved may be applied. Thus, $p_z(x)$ is integrable over the entire line. The function $f(t)$ is continuous, and therefore it follows from Theorem 3 of § 36 that

$$(1-|t|/z)f(t) = \int p_z(x)e^{itx}dx$$

for all t with $|t| \leq z$. In particular, when $t = 0$

$$\int p_z(x)dx = f(0) = 1.$$

This means that $p_z(x)$ is the density function of some probability distribution, and therefore $(1 - |t|/z) f(t)$ is the characteristic function corresponding to it. As $z \to \infty$, the function $(1 - |t|/z) f(t)$ converges uniformly to the function $f(t)$ in every finite interval $-T \leq t \leq T$. It follows from this that $f(t)$ is a characteristic function.

This completes the proof of the theorem.

§ 40. Characteristic Functions of Multi-Dimensional Random Variables

In the present section, we shall present without proof the basic properties of characteristic functions of multi-dimensional random variables.

The *characteristic function* of an n-dimensional random variable $(\xi_1, \xi_2, \ldots, \xi_n)$ is defined as the expectation of the variable

$$\exp(i(t_1\xi_1 + t_2\xi_2 + \ldots + t_n\xi_n)),$$

where $t_1, t_2, \ldots, t_n$ are real variables:

$$f(t_1, t_2, \ldots, t_n) = \mathrm{E} \exp\left(i \sum_{k=1}^{n} t_k \xi_k\right). \tag{1}$$

If $F(x_1, x_2, \ldots, x_n)$ is the distribution function of the variable $(\xi_1, \xi_2, \xi_3, \ldots, \xi_n)$, then as we know from our earlier discussions,[2]

$$f(t_1, t_2, \ldots, t_n) = \int \ldots \int \left(\exp i \sum_{k=1}^{n} t_k x_k\right) dF(x_1, \ldots, x_n). \tag{2}$$

As in the one-dimensional case, the characteristic function of an n-dimensional random variable is uniformly continuous over the entire space ($-\infty < t_j < \infty$; $1 \leq j \leq n$) and satisfies the following relations:

$$f(0, 0, \ldots, 0) = 1,$$
$$|f(t_1, t_2, \ldots, t_n)| \leq 1 \quad (-\infty < t_k < +\infty, \ k = 1, 2, \ldots n),$$
$$f(-t_1, -t_2, \ldots, -t_n) = \overline{f(t_1, t_2, \ldots, t_n)}.$$

---

[2] See Theorem 1 of § 27 and the remark concerning multi-dimensional Stieltjes integrals in § 26.

From the characteristic function $f(t_1, t_2, \ldots, t_n)$ of the random variable $(\xi_1, \xi_2, \ldots, \xi_n)$, it is easy to find the characteristic function of any k-dimensional $(k < n)$ variable $(\xi_{j_1}, \xi_{j_2}, \ldots, \xi_{j_k})$ whose components are the variables ξ_s $(1 \leq s \leq n)$. For this, it is necessary to set all the arguments t_s for $s \neq j_r$ $(1 \leq r \leq k)$ equal to zero in expression (2). Thus, for example, the characteristic function of the variable ξ_1 is

$$f_1(t_1) = f(t_1, 0, \ldots, 0).$$

If the components of the variable $(\xi_1, \xi_2, \ldots, \xi_n)$ are *independent* random variables, then from the definition it follows that its characteristic function is equal to the product of the characteristic functions of the components

$$f(t_1, t_2, \ldots, t_n) = f(t_1) \cdot f(t_2) \ldots f(t_n).$$

Just as in the one-dimensional case, the moments of different orders can readily be found from the characteristic function.

Thus, for example,

$$\mathsf{E}\, \xi_1^{k_1} \xi_2^{k_2} \ldots \xi_n^{k_n} = \int \int \ldots \int x_1^{k_1} x_2^{k_2} \ldots x_n^{k_n}\, dF(x_1, x_2, \ldots, x_n)$$

$$= (i)^{-\sum_{1}^{n} k_j} \left[\frac{\partial^{k_1+k_2+\ldots+k_n} f(t_1, t_2, \ldots, t_n)}{\partial t_1^{k_1} \partial t_2^{k_2} \ldots \partial t_n^{k_n}} \right]_{t_1 = t_2 = \ldots = t_n = 0}$$

In computing a characteristic function, it is useful to know the following theorem, which the reader will have no difficulty in proving.

THEOREM 1: *If the characteristic function of the variable* $(\xi_1, \xi_2, \xi_3, \ldots, \xi_n)$ *is* $f(t_1, t_2, \ldots, t_n)$, *then the characteristic function of the variable* $(\sigma_1 \xi_1 + \alpha_1, \sigma_2 \xi_2 + \alpha_2, \ldots, \sigma_n \xi_n + \alpha_n)$, *where* α_i *and* σ_i $(1 \leq i \leq n)$ *are real constants, is equal to*

$$\exp(i \sum_{k=1}^{n} \alpha_k t_k) \cdot f(\sigma_1 t_1, \sigma_2 t_2, \ldots, \sigma_n t_n).$$

Example 1. Let us compute the characteristic function of a two-dimensional random variable which is distributed according to the normal law:

$$p(x, y) = \frac{1}{2\pi\sqrt{1-r^2}} \exp\left\{ -\frac{1}{2(1-r^2)} [x^2 - 2rxy + y^2] \right\}. \tag{3}$$

§ 40. MULTI-DIMENSIONAL RANDOM VARIABLES

By formula (2),

$$f(t_1, t_2) = \int \int e^{i(t_1 x + t_2 y)} p(x, y) \, dx \, dy.$$

By a change of variables, we can reduce $f(t_1, t_2)$ to the form

$$f(t_1, t_2) = e^{-\frac{1}{2}(t_1^2 + 2r t_1 t_2 + t_2^2)} \frac{1}{2\pi} \int \int e^{-\frac{1}{2}(u^2 + v^2)} \, du \, dv = e^{-\frac{1}{2}(t_1^2 + 2r t_1 t_2 + t_2^2)}.$$

Example 2. By application of Theorem 1, we can find the characteristic function of the variable (η_1, η_2) distributed according to the normal law:

$$p(x, y) = \frac{1}{2\pi \sigma_1 \sigma_2 \sqrt{1 - r^2}} \times \qquad (4)$$

$$\times \exp\left\{ -\frac{1}{2(1-r^2)} \left[\frac{(x-a)^2}{\sigma_1^2} - 2r \frac{(x-a)(y-b)}{\sigma_1 \sigma_2} + \frac{(y-b)^2}{\sigma_2^2} \right] \right\}.$$

If we set $\eta_1 = \sigma_1 \xi_1 + a$ and $\eta_2 = \sigma_2 \xi_2 + b$, then the variable (ξ_1, ξ_2) will be distributed according to the normal law (3). By Theorem 1, the characteristic function of the variable (η_1, η_2) is

$$\varphi(t_1, t_2) = \exp\left[i a t_1 + i b t_2 - \frac{1}{2}(\sigma_1^2 t_1^2 + 2 \sigma_1 \sigma_2 r t_1 t_2 + \sigma_2^2 t_2^2) \right].$$

The following theorem follows from the definition of the characteristic function:

THEOREM 2: *If* $f(t_1, t_2, \ldots, t_n)$ *is the characteristic function of the variable* $(\xi_1, \xi_2, \ldots, \xi_n)$, *then the characteristic function of the sum* $\xi_1 + \xi_2 + \ldots + \xi_n$ *is equal to*

$$f_1(t) = f(t, t, \ldots, t).$$

Note: Observe that

$$f(t t_1, t t_2, \ldots, t t_n)$$

is the characteristic function of the sum $t_1 \xi_1 + t_2 \xi_2 + \ldots + t_n \xi_n$.

Example 3. Let us use Theorem 2 to determine the distribution function of the sum $\eta_1 + \eta_2$ if (η_1, η_2) is distributed according to the law (4).

By Theorem 2, the characteristic function of the sum $\eta_1 + \eta_2$ is

$$f(t) = \exp\left[i\,t\,(a+b) - \frac{t^2}{2}(\sigma_1^2 + 2r\,\sigma_1\,\sigma_2 + \sigma_2^2)\right].$$

We know (Example 1 of § 35) that this is the characteristic function for a normal law with expectation $a+b$ and variance $\sigma_1^2 + 2r\,\sigma_1\,\sigma_2 + \sigma_2^2$. We obtained this result earlier, directly (§ 24, Example 2).

At the beginning of the present chapter (§ 35), we saw that the characteristic function of the sum of two independent random variables is equal to the product of the characteristic functions of the summands. We shall now show that this property is only a necessary, and not a sufficient condition for the independence of random variables. With this purpose in mind, let us consider the two-dimensional random variable (ξ, η) having a density function representable in the form

$$p(x, y) = p_1(x)\,p_2(y) + \varphi(x)\,\psi(y) - \varphi(y)\,\psi(x),$$

where $p_1(x)$ and $p_2(x)$ are one-dimensional density functions and $\varphi(x)$ and $\psi(x)$ ($\varphi(x) \neq \psi(x)$) are odd integrable functions. It is easy to see that there exist such density functions. In fact, the function

$$p(x, y) = \frac{1}{4}e^{-|x|-|y|}\left\{1 + x\,y\,e^{-2|x|-|y|} - x\,y\,e^{-|x|-2|y|}\right\}$$

is an example of this very kind. For any x and y, it satisfies the inequality $p(x, y) > 0$ and moreover

$$\int_{-\infty}^{\infty} \int_{-\infty}^{\infty} p(x, y)\,dx\,dy = 1.$$

The random variables ξ and η are dependent, since their joint density function cannot be represented in the form of a product of two factors each depending on only one of the arguments x and y. The density function of the component ξ is

$$p_\xi(x) = \int_{-\infty}^{\infty} p(x, y)\,dy = p_1(x),$$

and that of the component η is

$$p_\eta(y) = \int_{-\infty}^{\infty} p(x, y)\,dx = p_2(y).$$

§ 40. Multi-Dimensional Random Variables

The two-dimensional characteristic function for the vector (ξ, η) is

$$f(t, \tau) = f_1(t) f_2(\tau) + $$

$$+ \int_{-\infty}^{\infty} \int_{-\infty}^{\infty} e^{itx+i\tau y} \varphi(x) \psi(y) \, dx \, dy - \int_{-\infty}^{\infty} \int_{-\infty}^{\infty} e^{itx+i\tau y} \psi(x) \varphi(y) \, dx \, dy,$$

where

$$f_1(t) = \int_{-\infty}^{\infty} e^{itx} p_1(x) \, dx, \quad f_2(\tau) = \int_{-\infty}^{\infty} e^{i\tau y} p_2(y) \, dy.$$

In our particular example, the characteristic function of the vector is

$$f(t, \tau) = \frac{1}{(1+t^2)(1+\tau^2)} + 24\, t\, \tau \left[\frac{1}{(4+t^2)(9+\tau^2)} - \frac{1}{(9+t^2)(4+\tau^2)} \right].$$

On the other hand, from the above general relation for $f(t, \tau)$, according to Theorem 2, the characteristic function of the sum $\xi + \eta$ is given by

$$f(t, t) = f_1(t) f_2(t),$$

i.e., it is equal to the product of the characteristic functions of the summands. Thus, we have shown that there exist dependent random variables the characteristic function of whose sum is equal to the product of the characteristic functions of the summands.

It is of importance to note that the following theorem also holds in the multi-dimensional case.

Theorem 3: *A distribution function $F(x_1, x_2, \ldots, x_n)$ is uniquely determined by its characteristic function.*

The basis for the proof of this proposition is the following *inversion formula*.

Theorem 4: *If $f(t_1, t_2, \ldots, t_n)$ is the characteristic function and $F(x_1, x_2, \ldots, x_n)$ the distribution function of the random variable $(\xi_1, \xi_2, \xi_3, \ldots, \xi_n)$, then*

$$\mathsf{P}\{a_k \leq \xi_k < b_k, k = 1, 2, \ldots, n\} = $$

$$= \lim_{T \to \infty} \frac{1}{(2\pi)^n} \int_{-T}^{T} \int_{-T}^{T} \cdots \int_{-T}^{T} \prod_{k=1}^{n} \frac{e^{-it_k a_k} - e^{it_k b_k}}{it_k} f(t_1, \ldots, t_n) \, dt_1 \, dt_2 \ldots dt_n$$

where a_k and b_k are any real numbers satisfying the single requirement: the probability of the point $(\xi_1, \xi_2, \ldots, \xi_n)$ falling on the surface of the parallelepiped $a_k \leq \xi_k < b_k$ $(k = 1, 2, \ldots, n)$ is zero.

Just as in the one-dimensional case, the direct limit theorem for characteristic functions and its converse both hold. We shall not stop to consider them.

Example 4. An n-dimensional random variable $(\xi_1, \xi_2, \ldots, \xi_n)$ is said to have a *non-degenerate* (proper) *n-dimensional normal distribution* if its density function is given by

$$p(x_1, x_2, \ldots, x_n) = C e^{-\frac{1}{2} Q(x_1, x_2, \ldots, x_n)},$$

where

$$Q(x_1, x_2, \ldots, x_n) = \sum_{i,j} b_{ij}(x_i - a_i)(x_j - a_j)$$

is a positive-definite quadratic form, and C, a_i, and b_{ij} are real constants.

By a straight-forward computation, we can show[3] that

$$C = (\sqrt{2\pi})^{-n} \sqrt{D},$$

where

$$D = \begin{vmatrix} b_{11} & b_{12} & \ldots & b_{1n} \\ b_{21} & b_{22} & \ldots & b_{2n} \\ \cdots & \cdots & \cdots & \cdots \\ b_{n1} & b_{n2} & \ldots & b_{nn} \end{vmatrix}.$$

Let D_{ij} denote the minor of D corresponding to the element b_{ij}; then

$$\mathsf{E}\,\xi_j = a_j, \quad \sigma_j^2 = \mathsf{D}\,\xi_j = \frac{D_{jj}}{D} \qquad (j = 1, 2, \ldots, n),$$

$$r_{ij} = \frac{\mathsf{E}\,(\xi_i - a_i)(\xi_j - a_j)}{\sigma_i \sigma_j} = \frac{D_{ij}}{\sqrt{D_{ii} D_{jj}}} \quad (i, j = 1, 2, \ldots, n).$$

The determinant D and all its principal minors are positive.

By the usual operations, it is easily verified that the characteristic function of the variable $(\xi_1, \xi_2, \ldots, \xi_n)$ is

$$f(t_1, t_2, \ldots, t_n) = e^{i \sum_{j=1}^{n} a_j t_j - \frac{1}{2} \sum_{k=1}^{n} \sum_{j=1}^{n} \sigma_j \sigma_k r_{jk} t_j t_k}.$$

Thus, *the n-dimensional normal distribution is completely determined by assignment of its expectation and variance.*

---

[3] The usual procedure in such computations is to make a change of variables which reduces the form Q to a sum of squares and then to carry out the computations in the new variables.

From this expression for the characteristic function of an n-dimensional normal distribution, we see that the distribution of the variable

$$(\xi_{i_1}, \xi_{i_2}, \ldots, \xi_{i_k})$$

for any $1 \leq i_1 < i_2 < \ldots < i_k \leq n$ is a k-dimensional normal distribution.

EXERCISES

1. Prove that the functions

$$f_1(t) = \sum_{k=0}^{\infty} a_k \cos kt, \quad f_2(t) = \sum_{k=0}^{\infty} a_k e^{i\lambda_k t},$$

where $a_k \geq 0$ and $\sum_{k=0}^{\infty} a_k = 1$, are characteristic functions; determine the corresponding probability distributions.

2. Find the characteristic function corresponding to each of the probability density functions:

a) $p(x) = \dfrac{a}{2} e^{-a|x|}$;

b) $p(x) = \dfrac{a}{\pi(a^2 + x^2)}$;

c) $p(x) = \begin{cases} 0 & \text{for } |x| \geq a, \\ \dfrac{a - |x|}{a^2} & \text{for } |x| \leq a; \end{cases}$

d) $p(x) = \dfrac{2 \sin^2 \dfrac{a x}{2}}{\pi a x^2}$.

Note: The alert reader will notice that Examples a) and b), and also c) and d) are, so to speak, converses.

3. Prove that

$$\varphi_1(t) = \frac{1}{\cosh t}, \quad \varphi_2(t) = \frac{t}{\sinh t}, \quad \varphi_3(t) = \frac{1}{\cosh^2 t}$$

are the characteristic functions corresponding to the density functions

$$p_1(x) = \frac{1}{2 \cosh \dfrac{\pi x}{2}}, \quad p_2(x) = \frac{\pi}{4 \cosh^2 \dfrac{\pi x}{2}}, \quad p_3(x) = \frac{x}{2 \sinh \dfrac{\pi x}{2}}.$$

4. Find the probability distribution of each of the random variables whose characteristic function is equal to

a) $\cos t$, b) $\cos^2 t$, c) $\dfrac{1}{1+it}$, d) $\dfrac{\sin at}{at}$.

5. Prove that the function defined by the relations
$$f(t) = f(-t),\ f(t+2a) = f(t),\ f(t) = 1 - t/a \text{ for } 0 \leq t \leq a$$
is a characteristic function.

Note: The characteristic functions in Examples 2d) and 5 have the following striking property:
$$f_2(t) = f_5(t) \quad \text{for } |t| \leq a,$$
$$f_2(t) \neq f_5(t) \quad \text{for } |t| \geq a \text{ and } t \neq \pm 2a, \ldots$$

Thus, there exist characteristic functions whose values coincide on an interval $(-a, a)$ as large as we please and which are not identically equal. The first of these two examples of such characteristic functions was pointed out by the author; Krein later gave necessary and sufficient conditions for two characteristic functions that coincide in some interval $(-a, a)$ to be identical.

6. Prove that it is possible to find independent random variables ξ_1, ξ_2, and ξ_3 such that the probability distributions of ξ_2 and ξ_3 are different but that the distribution functions of the sums $\xi_1 + \xi_2$ and $\xi_1 + \xi_3$ are identical.

Hint: Use the results of Examples 2a) and 5.

7. Prove that if $f(t)$ is a characteristic function which is zero for $|t| \geq a$, then the function defined by
$$\varphi(t) = \begin{cases} f(t) & \text{for } |t| \leq a, \\ f(t+2a) & \text{for } -\infty < t < \infty, \end{cases}$$
is also a characteristic function.

Hint: Use the Bochner-Khintchine Theorem.

8. Prove that if $f(t)$ is a characteristic function, then
$$\varphi(t) = e^{f(t)-1}$$
is also a characteristic function.

9. Prove that if $f(t)$ is a characteristic function, then
$$\varphi(t) = \frac{1}{t} \int_0^t f(z)\, dz$$
is also a characteristic function.

10. Prove that the inequality
$$1 - \varphi(2t) \leq 4\{1 - \varphi(t)\}$$

holds for any real characteristic function $\varphi(t)$ (which implies that the inequality

$$1 - |f(2t)|^2 \leq 4\{1 - |f(t)|^2\}$$

holds for any characteristic function).

11. Prove that the inequality

$$1 + \varphi(2t) \geq 2\{\varphi(t)\}^2$$

holds for any real characteristic function.

12. Prove that if $F(x)$ is a distribution function and $f(t)$ its corresponding characteristic function, then the relation

$$\lim_{T \to \infty} \frac{1}{2T} \int_{-T}^{T} f(t) e^{-itx} dt = F(x+0) - F(x-0)$$

holds for any value of x.

13. Prove that if $F(x)$ is a distribution function and $f(t)$ its corresponding characteristic function, and if x_ν are the abscissas of the jumps in the function $F(x)$, then

$$\lim_{T \to \infty} \frac{1}{2T} \int_{-T}^{T} |f(t)|^2 dt = \sum_{\nu} \{F(x_\nu + 0) - F(x_\nu - 0)\}^2 .$$

14. Prove that if a random variable has a density function, then its characteristic function tends to zero as $t \to \infty$.

15. A random variable ξ is distributed according to the Poisson law; $E\xi = \lambda$. Prove that for $\lambda \to \infty$, the distribution of the variable $(\xi - \lambda)/\sqrt{\lambda}$ tends to the normal distribution for which the parameters a and σ are $a = 0$ and $\sigma = 1$.

16. The random variable ξ has the density function

$$p(x) = \begin{cases} 0 & \text{for } x \leq 0, \\ \dfrac{\beta^\alpha}{\Gamma(\alpha)} x^{\alpha-1} e^{-\beta x} & \text{for } x > 0. \end{cases}$$

Prove that the distribution of the variable $(\beta\xi - \alpha)/\sqrt{\alpha}$ tends to the normal distribution with parameters $a = 0$ and $\sigma = 1$ for $\alpha \to \infty$.

Note: The results of Exercises 15 and 16 enable us to use normal distribution tables to compute the probabilities $P\{a \leq \xi < b\}$ for large values of λ (or α). In particular, it turns out that the indicated limiting relation gives excellent accuracy for the χ^2 distribution when $n \geq 30$. This last observation is used constantly in statistics.

17. Prove that if $\varphi(t)$ is a characteristic function and $\psi(t)$ is a function such that for some sequence $\{h_n\}$ ($h_n \to \infty$ as $n \to \infty$) the product

$$\varphi(t) \psi(h_n t) = f_n(t)$$

is also a characteristic function, then the function $\psi(t)$ is a characteristic function.

CHAPTER VIII

THE CLASSICAL LIMIT THEOREM

§ 41. Statement of the Problem

The DeMoivre-Laplace Integral Limit Theorem, which we proved in Chapter II, has served as a basis for a large group of investigations of fundamental significance both in the theory of probability and in its numerous applications to the natural sciences, engineering and economics. In order to get an idea of the course of these investigations, we shall put the DeMoivre-Laplace Theorem in a slightly different form. Specifically, if we denote the number of times an event A occurs in the k-th trial by μ_k, as we have done many times before, then the number of times the event A occurs in n successive trials is $\sum_{k=1}^{n}\mu_k$. Furthermore, in Example 5 of § 28, we showed that $\mathsf{E}\sum_{k=1}^{n}\mu_k = np$ and $\mathsf{D}\sum_{k=1}^{n}\mu_k = npq$. Therefore, the DeMoivre-Laplace Theorem can be expressed in the following form: As $n \to \infty$,

$$\mathsf{P}\left\{a \leq \frac{\sum_{k=1}^{n}(\mu_k - \mathsf{E}\,\mu_k)}{\sqrt{\sum_{k=1}^{n}\mathsf{D}\,\mu_k}} < b\right\} \to \frac{1}{\sqrt{2\pi}}\int_{a}^{b} e^{-\frac{z^2}{2}}\,dz, \qquad (1)$$

which stated in words, is: *If there exist independent random variables each assuming only two values 0 and 1 with probabilities q and $p = 1 - q$, respectively $(0 < p < 1)$, then the probability that the sum of the deviations of these variables from their expectations divided by the square root of the sum of their respective variances lies between a and b approaches the integral*

$$\frac{1}{\sqrt{2\pi}}\int_{a}^{b} e^{-\frac{z^2}{2}}\,dz$$

uniformly in a and b as the number of summands is increased indefinitely.

§ 41. STATEMENT OF THE PROBLEM

The following question naturally arises: In so far as relation (1) is closely connected with the special choice of the variables μ_k, would it not also hold if weaker restrictions were imposed on the distribution functions of the summands? The formulation of this problem, as well as its solution, are essentially due to Tchebychev and his students, Markov and Liapounov. Their investigations have shown that it is merely necessary to impose a very general restriction on the variables, the significance of which is that the individual terms have a negligible effect on the sum. In the next section, we shall give a precise formulation of this condition. The reasons for the great importance of these results in applications lie in the very essence of mass phenomena; and, as we have said earlier, the laws governing mass phenomena are what constitute the subject matter of probability theory.

One of the most important schemes serving as a basis for applying the results of probability theory to natural science and engineering is the following. Suppose that a process takes place under the influence of a large number of independently acting random factors each of which only changes the course of the phenomenon or process insignificantly. The research worker, being interested in studying the whole process rather than the effect of individual factors, merely looks at the summary effect of these factors.

We give two typical examples.

Example 1. Let a certain measurement be made. The outcome is inevitably affected by a great number of factors that produce errors in the measurement. Relevant here are the errors due to the state of the measuring device which can change imperceptibly owing to diverse atmospheric and mechanical causes. Relevant here are the human errors of the observer which are due to peculiarities in his eyesight or hearing and which may also change slightly with changes in his psychological or physical state, etc. Each of these factors may introduce an insignificant error. But all of these errors tell immediately upon the measurement; there is a "cumulative" error observed. In other words, the error of measurement actually observed will be a random variable which is the sum of a great number of mutually independent random variables of insignificant magnitude. And although these random variables, as well as their distribution functions, are not known, their effect on the outcome of the measurement is noticeable and must therefore be made the subject of study.

VIII. The Classical Limit Theorem

Example 2. Large batches of identical items are produced by mass production in many branches of industry. Let us focus our attention on some numerical characteristic of such a product. Inasmuch as this article is manufactured in accordance with industrial norms, the numerical characteristic we have chosen has a certain normal value. Actually, however, there is always some deviation from this normal value. In a well-organized manufacturing process, such deviations can only be caused by chance factors, each of which has only an imperceptible effect. The cumulative effect, however, brings about a noticeable deviation from the norm.

One could give any number of similar examples.

The problem thus arises of investigating the laws that are characteristic of sums of a larger number of independent random variables each of which has only a small effect on the sum. Below, we shall give this last requirement a more precise meaning. In order to study such sums of a very large but finite number of terms, we shall instead consider a sequence of sums having an ever-increasing number of terms and we shall take as the solution of our problem the distribution function which is the limit of the sequence of distribution functions for the sums. This sort of passing to the limit from a finite formulation of a problem is very usual in modern mathematics and in many branches of the natural sciences.

Thus, we have arrived at the consideration of the following problem: Let there be given a sequence of mutually independent random variables $\xi_1, \xi_2, \ldots, \xi_n, \ldots$ each having a finite expectation and variance. In the following, we shall adhere to the notation:

$$a_k = \mathsf{E}\,\xi_k,\ \ b_k^2 = \mathsf{D}\,\xi_k,\ \ B_n^2 = \sum_{k=1}^n b_k^2 = \mathsf{D} \sum_{k=1}^n \xi_k.$$

The question arises as to what conditions must be imposed on the variables ξ_k so that the distribution functions of the sums

$$\frac{1}{B_n} \sum_{k=1}^n (\xi_k - a_k) \tag{2}$$

converge to the normal distribution law.[1] In the following section, we

---

[1] The answer to this question is contained in a number of theorems which go by the generic name of *Central Limit Theorem*. A few will be considered in this chapter.

§ 41. STATEMENT OF THE PROBLEM

shall see that it is sufficient that the *Lindeberg condition* be satisfied, which is as follows: *For any $\tau > 0$*

$$\lim_{n \to \infty} \frac{1}{B_n^2} \sum_{k=1}^n \int_{|x-a_k|>\tau B_n} (x-a_k)^2 \, dF_k(x) = 0,$$

where $F_k(x)$ is the distribution function of the variable ξ_k.

Let us make clear what this condition means.

Let A_k denote the event that

$$|\xi_k - a_k| > \tau B_n \quad (k = 1, 2, \ldots, n),$$

and let us estimate the probability

$$P\{\max_{1 \le k \le n} |\xi_k - a_k| > \tau B_n\}.$$

Since

$$P\{\max_{1 \le k \le n} |\xi_k - a_k| > \tau B_n\} = P\{A_1 + A_2 + \ldots + A_n\}$$

and

$$P\{A_1 + A_2 + \ldots + A_n\} \le \sum_{k=1}^n P\{A_k\},$$

by noting that

$$P\{A_k\} = \int_{|x-a_k|>\tau B_n} dF_k(x) \le \frac{1}{(\tau B_n)^2} \int_{|x-a_k|>\tau B_n} (x-a_k)^2 \, dF_k(x),$$

we obtain the inequality

$$P\{\max_{1 \le k \le n} |\xi_k - a_k| \ge \tau B_n\} \le \frac{1}{\tau^2 B_n^2} \sum_{k=1}^n \int_{|x-a_k|>\tau B_n} (x-a_k)^2 \, dF_k(x).$$

By virtue of the Lindeberg condition, the last sum tends to zero as $n \to \infty$ for any positive constant τ. Thus, the Lindeberg condition represents the distinctive requirement that the terms $(\xi_k - a_k)/B_n$ in the sum (2) be uniformly small.

We note once more that the investigations of Markov and Liapounov had already made quite clear the meaning of the conditions sufficient for the distribution functions of the sums (2) to converge to the normal law.

VIII. THE CLASSICAL LIMIT THEOREM

§ 42. Liapounov's Theorem

We begin with the proof of the sufficiency of the Lindeberg condition.

THEOREM: *If a sequence of mutually independent random variables* $\xi_1, \xi_2, \ldots, \xi_n, \ldots$ *satisfies the Lindeberg condition*

$$\lim_{n \to \infty} \frac{1}{B_n^2} \sum_{k=1}^{n} \int_{|x-a_k| > \tau B_n} (x - a_k)^2 \, dF_k(x) = 0 \tag{1}$$

for any positive constant τ, *then as* $n \to \infty$,

$$\mathsf{P}\left\{ \frac{1}{B_n} \sum_{k=1}^{n} (\xi_k - a_k) < x \right\} \to \frac{1}{\sqrt{2\pi}} \int_{-\infty}^{x} e^{-\frac{z^2}{2}} dz \tag{2}$$

uniformly in x.

Proof: For brevity, we introduce the notation

$$\xi_{nk} = \frac{\xi_k - a_k}{B_n}, \quad F_{nk}(x) = \mathsf{P}\{\xi_{nk} < x\}.$$

Clearly,

$$\mathsf{E}\,\xi_{nk} = 0, \quad \mathsf{D}\,\xi_{nk} = \frac{1}{B_n^2} \mathsf{D}\,\xi_k,$$

and therefore

$$\sum_{k=1}^{n} \mathsf{D}\,\xi_{nk} = 1. \tag{2'}$$

It is easy to satisfy oneself that in this notation the Lindeberg condition becomes the following:

$$\lim_{n \to \infty} \sum_{k=1}^{n} \int_{|x| > \tau} x^2 \, dF_{nk}(x) = 0. \tag{1'}$$

The characteristic function of the sum

$$\frac{1}{B_n} \sum_{k=1}^{n} (\xi_k - a_k) = \sum_{k=1}^{n} \xi_{nk}$$

is

$$\varphi_n(t) = \prod_{k=1}^{n} f_{nk}(t).$$

§ 42. Liapounov's Theorem

We now must prove that

$$\lim_{n\to\infty} \varphi_n(t) = e^{-\frac{t^2}{2}}.$$

With this in mind, we shall first of all establish that the factors $f_{nk}(t)$ tend to 1 as $n \to \infty$ uniformly in k ($1 \leq k \leq n$). In fact, taking into account that $\mathsf{E}\xi_{nk} = 0$, we find:

$$f_{nk}(t) - 1 = \int (e^{itx} - 1 - itx)\, dF_{nk}(x).$$

Since, for any real a,[2]

$$\left|e^{ia} - 1 - ia\right| \leq \frac{a^2}{2}, \tag{3}$$

we have

$$\left|f_{nk}(t) - 1\right| \leq \frac{t^2}{2} \int x^2\, dF_{nk}(x).$$

Let ε be an arbitrary positive number; then it is clear that

$$\int x^2\, dF_{nk}(x) = \int_{|x|\leq\varepsilon} x^2\, dF_{nk}(x) + \int_{|x|>\varepsilon} x^2\, dF_{nk}(x) \leq \varepsilon^2 + \int_{|x|>\varepsilon} x^2\, dF_{nk}(x).$$

According to (1'), the last term can be made smaller than ε^2 for all sufficiently large values of n. Thus, for all sufficiently large n and for t in any finite interval $|t| \leq T$,

$$\left|f_{nk}(t) - 1\right| \leq \varepsilon^2 T^2$$

uniformly in k ($1 \leq k \leq n$). From this we conclude that

---

[2] This inequality, and an entire series of similar ones may be derived, for example, as follows. From the fact that

$$\left|e^{i\alpha} - 1\right| = \left|\int_0^\alpha e^{ix}\, dx\right| < \alpha, \qquad (\alpha > 0)$$

there follows the inequality

$$\left|e^{i\alpha} - 1 - i\alpha\right| = \left|\int_0^\alpha (e^{ix} - 1)\, dx\right| \leq \frac{\alpha^2}{2}.$$

It then follows from this inequality that

$$\left|e^{i\alpha} - 1 - i\alpha + \frac{\alpha^2}{2}\right| = \left|\int_0^\alpha (e^{ix} - 1 - ix)\, dx\right| \leq \int_0^\alpha \left|e^{ix} - 1 - ix\right| dx \leq \int_0^\alpha \frac{\alpha^2}{2}\, dx = \frac{\alpha^3}{6},$$

etc.

$$\lim_{n \to \infty} f_{nk}(t) = 1 \tag{4}$$

uniformly in k $(1 \leq k \leq n)$ and that for all sufficiently large n, the inequality

$$|f_{nk}(t) - 1| < \frac{1}{2} \tag{5}$$

is satisfied for t lying in an arbitrary finite interval $|t| \leq T$. Therefore, in the interval $|t| \leq T$ we can write the expansion (log representing the *principal value* of the logarithm)

$$\log \varphi_n(t) = \sum_{k=1}^{n} \log f_{nk}(t) = \sum_{k=1}^{n} \log [1 + (f_{nk}(t) - 1)] = \sum_{k=1}^{n} (f_{nk}(t) - 1) + R_n, \tag{6}$$

where

$$R_n = \sum_{k=1}^{n} \sum_{s=2}^{\infty} \frac{(-1)^s}{s} (f_{nk}(t) - 1)^s.$$

In view of (5),

$$|R_n| \leq \sum_{k=1}^{n} \sum_{s=2}^{\infty} \frac{1}{2} |f_{nk}(t) - 1|^s = \frac{1}{2} \sum_{k=1}^{n} \frac{|f_{nk}(t) - 1|^2}{1 - |f_{nk}(t) - 1|} \leq \sum_{k=1}^{n} |f_{nk}(t) - 1|^2.$$

Since

$$\sum_{k=1}^{n} |f_{nk}(t) - 1| = \sum_{k=1}^{n} \left| \int (e^{itx} - 1 - itx) \, dF_{nk}(x) \right| \leq$$

$$\leq \frac{t^2}{2} \sum_{k=1}^{n} \int x^2 \, dF_{nk}(x) = \frac{t^2}{2},$$

we have:

$$|R_n| \leq \frac{t^2}{2} \max_{1 \leq k \leq n} |f_{nk}(t) - 1|.$$

Thus it follows from (4) that, as $n \to \infty$, $R_n \to 0$ uniformly for t in an arbitrary finite interval $|t| \leq T$. But

$$\sum_{k=1}^{n} (f_{nk}(t) - 1) = -\frac{t^2}{2} + \varrho_n, \tag{7}$$

where

$$\varrho_n = \frac{t^2}{2} + \sum_{k=1}^{n} \int (e^{itx} - 1 - itx) \, dF_{nk}(x).$$

Let ε be an arbitrary positive number; then by virtue of (2')

$$\varrho_n = \sum_{k=1}^{n} \int_{|x|\leq\varepsilon} \left(e^{itx} - 1 - itx - \frac{(itx)^2}{2}\right) dF_{nk}(x) +$$
$$+ \sum_{k=1}^{n} \int_{|x|>\varepsilon} \left(\frac{t^2 x^2}{2} + e^{itx} - 1 - itx\right) dF_{nk}(x).$$

The inequalities (3) and (3') enable us to obtain the following estimate:

$$|\varrho_n| \leq \frac{|t|^3}{6} \sum_{k=1}^{n} \int_{|x|\leq\varepsilon} |x|^3 dF_{nk}(x) + t^2 \sum_{k=1}^{n} \int_{|x|>\varepsilon} x^2 dF_{nh}(x)$$
$$\leq \frac{|t|^3}{6} \cdot \varepsilon \sum_{k=1}^{n} \int_{|x|\leq\varepsilon} x^2 dF_{nk}(x) + t^2 \sum_{k=1}^{n} \int_{|x|>\varepsilon} x^2 dF_{nk}(x)$$
$$= \frac{|t|^3}{6} \cdot \varepsilon + t^2 \left(1 - \frac{|t|}{6}\varepsilon\right) \sum_{k=1}^{n} \int_{|x|>\varepsilon} x^2 dF_{nk}(x).$$

According to condition (1'), given any $\varepsilon > 0$, the second term can be made less than any $\eta > 0$ by making n sufficiently large. And since ε is an arbitrary positive quantity, we can select it so small that for any T and any $\eta > 0$, the inequality

$$|\varrho_n| < 2\eta \quad (n \geq n_0(\varepsilon, \eta, T)),$$

is satisfied for all t in the interval $|t| \leq T$. This inequality shows that

$$\lim_{n\to\infty} \varrho_n = 0 \tag{8}$$

uniformly in every finite interval $|t| \leq T$. Collecting together (6), (7), (8), and the relation $R_n \to 0$, we finally have that

$$\lim_{n\to\infty} \log \varphi_n(t) = -\frac{t^2}{2}$$

uniformly in every finite interval $|t| \leq T$. The theorem is proved.

COROLLARY: *If the independent random variables $\xi_1, \xi_2, \ldots, \xi_n, \ldots$ are identically distributed and have non-vanishing finite variances, then as $n \to \infty$,*

$$P\left\{\frac{1}{B_n} \sum_{k=1}^{n} (\xi_k - \mathsf{E}\,\xi_k) < x\right\} \to \frac{1}{\sqrt{2\pi}} \int_{-\infty}^{x} e^{-\frac{z^2}{2}} dz$$

uniformly in x.

VIII. THE CLASSICAL LIMIT THEOREM

Proof: It suffices to verify that the Lindeberg condition is satisfied under the given assumptions. With this in mind, we observe that in our case

$$B_n = b\sqrt{n},$$

where b is the variance of the individual terms. By putting $\mathsf{E}\xi_k = a$, we may write the following obvious equations:

$$\sum_{k=1}^n \frac{1}{B_n^2} \int\limits_{|x-a|>\tau B_n} (x-a)^2 \, dF_k(x) =$$

$$= \frac{1}{nb^2} n \int\limits_{|x-a|>\tau B_n} (x-a)^2 \, dF_1(x) = \frac{1}{b^2} \int\limits_{|x-a|>\tau B_n} (x-a)^2 \, dF_1(x).$$

By the assumption that the variance exists and is non-vanishing, we conclude that the integral on the right-hand side of the last equation tends to zero as $n \to \infty$.

LIAPOUNOV'S THEOREM: *If for a sequence of mutually independent random variables $\xi_1, \xi_2, \ldots, \xi_n, \ldots$ a positive number δ can be found such that as $n \to \infty$*

$$\frac{1}{B_n^{2+\delta}} \sum_{k=1}^n \mathsf{E}\,|\xi_k - a_k|^{2+\delta} \to 0, \tag{10}$$

then as $n \to \infty$

$$\mathsf{P}\left\{\frac{1}{B_n} \sum_{k=1}^n (\xi_k - a_k) < x\right\} \to \frac{1}{\sqrt{2\pi}} \int_{-\infty}^x e^{-\frac{z^2}{2}} \, dz$$

uniformly in x.

Proof: Again, it suffices to verify that the *Liapounov condition* (condition (10)) implies the Lindeberg condition. But this is obvious from the following chain of inequalities:

$$\frac{1}{B_n^2} \sum_{k=1}^n \int\limits_{|x-a_k|>\tau B_n} (x-a_k)^2 \, dF_k(x) \leq$$

$$\leq \frac{1}{B_n^2 (\tau B_n)^\delta} \sum_{k=1}^n \int\limits_{|x-a_k|>\tau B_n} |x-a_k|^{2+\delta} \, dF_k(x) \leq \frac{1}{\tau^\delta} \frac{\sum\limits_{k=1}^n \int |x-a_k|^{2+\delta} \, dF_k(x)}{B_n^{2+\delta}}.$$

§ 43. The Local Limit Theorem

We shall now give some sufficient conditions for the applicability of another classical limit theorem, the *Local Limit Theorem*. In doing so, we shall confine ourselves to the case of mutually independent summands that have identical distributions.

Let us agree to say that a discrete random variable ξ has a *lattice distribution* if there exist numbers a and $h > 0$ such that all possible values of ξ are representable in the form $a + kh$, where k may take on any integral value ($-\infty < k < \infty$).

The Poisson and the Bernoulli distributions are among the examples of lattice distributions.

We now express the condition for a lattice distribution in terms of characteristic functions. For this purpose, we prove the following lemma.

LEMMA: *A random variable ξ has a lattice distribution if and only if for some $t \neq 0$ the absolute value of its characteristic function is equal to one.*

Proof: If ξ has a lattice distribution and p_k is the probability of the equality $\xi = a + kh$, then the characteristic function of the variable ξ is

$$f(t) = \sum_{k=-\infty}^{\infty} p_k \, e^{it(a+kh)} = e^{iat} \sum_{k=-\infty}^{\infty} p_k \, e^{itkh}.$$

From this, we find:

$$f\left(\frac{2\pi}{h}\right) = e^{2\pi i \frac{a}{h}} \sum_{k=-\infty}^{\infty} p_k \, e^{2\pi i k} = e^{2\pi i \frac{a}{h}}.$$

We thus see that

$$\left| f\left(\frac{2\pi}{h}\right) \right| = 1$$

for every lattice distribution.

Let us now assume that for some $t_1 \neq 0$

$$|f(t_1)| = 1,$$

and let us show that ξ then has a lattice distribution. The last equation implies that for some θ

$$f(t_1) = e^{i\theta}.$$

Thus
$$\int e^{it_1 x} dF(x) = e^{i\theta},$$
and therefore
$$\int e^{i(t_1 x - \theta)} dF(x) = 1.$$
From this, it follows that
$$\int \cos(t_1 x - \theta) dF(x) = 1.$$

In order that this relation be possible, it is necessary that the function $F(x)$ increase only at those values of x for which
$$\cos(t_1 x - \theta) = 1.$$
This implies that all the possible values of ξ must be of the form
$$x = \frac{\theta}{t_1} + k\frac{2\pi}{t_1},$$

Q.E.D.

We call the number h the *distribution span*. The distribution span is *maximal* if for any choice of b ($-\infty < b < \infty$) and $h_1 > h$, all of the possible values of ξ cannot be represented in the form $b + kh_1$.

To illustrate the distinction between the notions of distribution span and maximal distribution span, we consider the following example. Suppose that ξ takes on merely odd integral values. Clearly, all the values of ξ may be expressed in the form $a + kh$, where $a = 0$ and $h = 1$. However, the span h is not maximal, since all the possible values of ξ may also be expressed in the form $b + kh_1$, with $b = 1$ and $h_1 = 2$.

The conditions for a distribution span to be maximal can be expressed in other ways.

First: A distribution span h is maximal if and only if the greatest common divisor of the differences of every pair of possible values of the variable ξ divided by h is equal to one.

Second: A distribution span h is maximal if and only if the absolute value of the characteristic function is equal to one for $t = 2\pi/h$ and less than one in the interval $0 < |t| < 2\pi/h$.

The last assertion follows immediately from the lemma just proved. In fact, if for $0 < t_1 < 2\pi/h$
$$|f_1(t)| = 1,$$

§ 43. THE LOCAL LIMIT THEOREM

then by what was shown above, the quantity $2\pi/t_1$ must be a distribution span, and since

$$h < 2\pi/t_1,$$

the span h cannot be maximal.

From this, we may draw the following conclusion: If h is a maximal distribution span, then for any $\varepsilon > 0$ we can find a number $c_0 > 0$ such that the inequality

$$|f(t)| \leq e^{-c_0} \qquad (1)$$

holds for all t in the interval $\varepsilon \leq |t| \leq 2\pi/h - \varepsilon$.

Suppose now that $\xi_1, \xi_2, \ldots, \xi_n, \ldots$ are mutually independent, lattice random variables having the same distribution function $F(x)$. Let us consider the sum

$$\zeta_n = \xi_1 + \xi_2 + \ldots + \xi_n.$$

Clearly, it is also a lattice random variable and its possible values are representable in the form $na + kh$. Let $P_n(k)$ denote the probability of the equality

$$\zeta_n = na + kh;$$

in particular, $P_1(k) = \mathsf{P}\{\xi_1 = a + kh\} = p_k$.

Next, let

$$z_{nk} = \frac{an + kh - A_n}{B_n},$$

where $A_n = \mathsf{E}\zeta_n$, $B_n^2 = \mathsf{D}\zeta_n = n\mathsf{D}\xi_1$.

We can now prove the following statement, which is evidently a generalization of the DeMoivre-Laplace Local Limit Theorem.

THEOREM:[3] *Let $\xi_1, \xi_2, \ldots, \xi_n, \ldots$ be independent lattice random variables each having the same distribution function $F(x)$ and a finite expectation and variance. Then the relation*

$$\frac{B_n}{h} P_n(k) - \frac{1}{\sqrt{2\pi}} e^{-\frac{z_{nk}^2}{2}} \to 0$$

holds uniformly in k ($-\infty < k < \infty$) for $n \to \infty$ if and only if the distribution span h is maximal.

---

[3] This theorem is due to B. V. Gnedenko. [*Trans.*]

Proof: The necessity of the condition is almost obvious. In fact, if the span h were not maximal, then there would be systematic omissions among the possible values of the sum $\zeta_n = \sum_{k=1}^{n} \xi_n$: the difference between the closest possible values of the sum could not be less than dh, where d is the greatest common divisor of the differences of the possible values of ξ_n divided by h. If h were not the maximal span, then $d > 1$ for all values of n.

The proof of the sufficiency of the condition of the theorem requires a slightly more complicated argument.

The characteristic function of the variable ξ_k ($k = 1, 2, 3, \ldots$) is given by

$$f(t) = \sum_{k=-\infty}^{\infty} p_k\, e^{iat + itkh} = e^{iat} \sum_{k=-\infty}^{\infty} p_k\, e^{itkh},$$

and that of the sum ζ_n by

$$f^n(t) = e^{iant} \sum_{k=-\infty}^{\infty} P_n(k)\, e^{itkh}.$$

Multiplying both sides of this equation by $e^{-iant - itkh}$ and integrating from $-\pi/h$ to π/h, we obtain:

$$\frac{2\pi}{h} P_n(k) = \int_{-\frac{\pi}{h}}^{\frac{\pi}{h}} f^n(t)\, e^{-iant - itkh}\, dt.$$

By noting that

$$h\,k = B_n\, z_{nk} + A_n - a\,n$$

(in the following, we shall write z instead of z_{nk}), we may write:

$$\frac{2\pi}{h} P_n(k) = \int_{-\frac{\pi}{h}}^{\frac{\pi}{h}} f^{*n}(t)\, e^{-itzB_n}\, dt,$$

where

$$f^*(t) = e^{-\frac{itA_n}{n}} f(t).$$

§ 43. The Local Limit Theorem

Finally, by setting $x = tB_n$, we obtain:

$$\frac{2\pi B_n}{h} P_n(k) = \int_{-\frac{\pi B_n}{h}}^{\frac{\pi B_n}{h}} e^{-izx} f^{\cdot n}\left(\frac{x}{B_n}\right) dx.$$

It is easily shown that

$$\frac{1}{\sqrt{2\pi}} e^{-\frac{z^2}{2}} = \frac{1}{2\pi} \int e^{-izx - \frac{x^2}{2}} dx.$$

Let us represent the difference

$$R_n = 2\pi \left[\frac{B_n}{h} P_n(k) - \frac{1}{\sqrt{2\pi}} e^{-\frac{z^2}{2}}\right]$$

as the sum of four integrals

$$R_n = J_1 + J_2 + J_3 + J_4,$$

where

$$J_1 = \int_{-A}^{A} e^{-izx} \left[f^{\cdot n}\left(\frac{x}{B_n}\right) - e^{-\frac{x^2}{2}}\right] dx,$$

$$J_2 = -\int_{|x|>A} e^{-izx - \frac{x^2}{2}} dx,$$

$$J_3 = \int_{\varepsilon B_n \leq |x| \leq \frac{\pi B_n}{h}} e^{-izx} f^{\cdot n}\left(\frac{x}{B_n}\right) dx,$$

$$J_4 = \int_{A \leq |x| < \varepsilon B_n} e^{-izx} f^{\cdot n}\left(\frac{x}{B_n}\right) dx,$$

and where A is a sufficiently large positive quantity and ε a sufficiently small positive quantity, more precise values of which will be chosen later.

By virtue of the corollary of the theorem proved in the preceding section, the relation

$$f^{\cdot n}\left(\frac{t}{B_n}\right) \to e^{-\frac{t^2}{2}} \quad (n \to \infty)$$

is satisfied uniformly for t in any finite interval. But from this it

follows that
$$J_1 \to 0 \quad (n \to \infty)$$
no matter what the value of the constant A. The integral J_2 can be estimated by means of the inequality

$$|J_2| \leq \int_{|x|>A} e^{-\frac{x^2}{2}} dx \leq \frac{2}{A} \int_A^\infty x e^{-\frac{x^2}{2}} dx = \frac{2}{A} e^{-\frac{A^2}{2}}.$$

By choosing A sufficiently large, we can make J_2 arbitrarily small.

According to the inequality (1), we have:

$$|J_3| \leq \int_{\varepsilon B_n \leq |x| \leq \frac{\pi B_n}{h}} \left|f^*\left(\frac{x}{B_n}\right)\right|^n dx \leq e^{-nc_0} 2 B_n\left(\frac{\pi}{h} - \varepsilon\right).$$

From this it is clear that as $n \to \infty$,

$$J_3 \to 0.$$

To estimate the integral J_4, we note that the existence of the variance implies the existence of the second derivative of the function $f^*(t)$. We can therefore use (3) of § 35 to obtain the expansion

$$f^*(t) = 1 - \frac{\sigma^2 t^2}{2} + o(t^2),$$

where $\sigma^2 = B_n^2/n$, which is valid in the neighborhood of the point $t = 0$, and if ε is sufficiently small, we obtain

$$|f^*(t)| < 1 - \frac{\sigma^2 t^2}{4} < e^{-\frac{\sigma^2 t^2}{4}}$$

for $|t| \leq \varepsilon$. Then, for $|x| \leq \varepsilon B_n$,

$$\left|f^*\left(\frac{x}{B_n}\right)\right|^n < e^{-\frac{n\sigma^2 x^2}{4B_n^2}} = e^{-\frac{x^2}{4}}.$$

Therefore,

$$|J_4| \leq 2 \int_A^{\varepsilon B_n} e^{-\frac{x^2}{4}} dt < 2 \int_A^\infty e^{-\frac{x^2}{4}} dt.$$

By choosing A sufficiently large, we can make the integral J_4 arbitrarily small. This proves the theorem.

There still exists one other case in which it is natural to ask about the local behavior of distribution functions of sums. This is the case of continuous distributions.

The problem which can be posed here is, When do the density functions of normalized sums converge to the normal density function if the corresponding distribution functions converge to the normal distribution? The following theorem gives a comprehensive solution to this problem.

THEOREM: *Let* $\xi_1, \xi_2, \ldots, \xi_n, \ldots$ *be independent random variables each having the same distribution function* $F(x)$, *let their expectation and variance be finite and, beginning with a certain* n_0, *let the random variable*

$$s_n = \frac{1}{\sqrt{n\,D\,\xi_1}} \sum_{k=1}^{n} (\xi_k - \mathsf{E}\,\xi_k)$$

have a density function $p_n(x)$. *Then*

$$p_n(x) - \frac{1}{\sqrt{2\pi}} e^{-\frac{x^2}{2}} \to 0$$

uniformly in x ($-\infty < x < \infty$) *as* $n \to \infty$ *if and only if there exists a number* n_1 *such that the function* $p_{n_1}(x)$ *is bounded.*

We shall not carry out the proof of this theorem, since to a considerable extent it duplicates the above argument.

EXERCISES

1. Prove that as $n \to \infty$

$$\frac{\sqrt{n^n}}{\sqrt{2^n}\,\Gamma\left(\frac{n}{2}\right)} \int_0^{1+t\sqrt{\frac{2}{n}}} z^{\frac{n}{2}-1} e^{-\frac{nz}{2}}\,dz \to \frac{1}{\sqrt{2\pi}} \int_{-\infty}^{t} e^{-\frac{z^2}{2}}\,dz.$$

Hint: Apply the Liapounov Theorem to the χ^2-distribution.

2. The random variables

$$\xi_n = \begin{cases} -n^\alpha, \text{ with probability } 1/2, \\ +n^\alpha, \text{ with probability } 1/2 \end{cases}$$

are independent. Prove that for $\alpha > -1/2$ the Liapounov Theorem is applicable to them.

VIII. The Classical Limit Theorem

3. Prove that as $n \to \infty$

$$e^{-n} \sum_{k=0}^{n} \frac{n^k}{k!} \to \frac{1}{2}.$$

Hint: Apply the Liapounov Theorem to a sum of random variables distributed according to the Poisson law with parameter $\lambda = 1$.

4. Let p_i denote the probability of occurrence of the event A in the i-th trial and μ the number of times A occurs in n independent trials. Prove that

$$P\left\{ \frac{\mu - \sum_{k=1}^{n} p_k}{\sqrt{\sum_{k=1}^{n} p_i q_i}} < x \right\} \to \frac{1}{\sqrt{2\pi}} \int_{-\infty}^{x} e^{-\frac{z^2}{2}} dz$$

if and only if $\sum_{i=1}^{\infty} p_i q_i = \infty$.

5. Prove that under the conditions of the preceding example the requirement $\sum_{i=1}^{\infty} p_i q_i = \infty$ is sufficient not only for the Integral Theorem but also for the Local Theorem.

CHAPTER IX

THE THEORY OF INFINITELY DIVISIBLE DISTRIBUTION LAWS

For a long time, the central problem of the theory of probability was considered to be the discovery of conditions under which distributions for sums of independent random variables would converge to the normal law. Very general sufficient conditions for this convergence were found by Liapounov (see Chap. VIII).

Attempts to broaden the Liapounov conditions met with success only in very recent years, with the determination of conditions that were not only sufficient but, under quite natural restrictions, also necessary.

At the same time as the solutions to the classical problems were being completed, there was created and developed a new theoretical approach to limit theorems for sums of independent random variables, closely connected with the introduction and development of the theory of stochastic (random) processes. In the first place, the question arose as to what laws, besides the normal law, could be the limits for sums of independent random variables.

It was found that the normal law did not exhaust by far the class of limit laws. The problem thereupon arose of determining the conditions that had to be imposed on the summands in order that the distribution functions for sums might converge to this or some other limit law.

In the present chapter, we make it our object to present some of the investigations of recent years which are devoted to limit theorems for sums of independent random variables. In doing so, we confine ourselves to the case in which the summands have finite variances. To consider this problem without this restriction would require more formidable computation; for an account of the solution to this problem, we refer the reader to the monograph already mentioned of Gnedenko and Kolmogrov. As a simple consequence of the general theorems presented, we shall obtain the conditions, referred to above, which are necessary and sufficient for the distribution functions of sums to converge to the normal law.

§ 44. Infinitely Divisible Laws and Their Fundamental Properties

A distribution law $\Phi(x)$ is called *infinitely divisible* if, for any natural number n, the random variable which is distributed according to the law $\Phi(x)$ is the sum of n independent random variables $\xi_1, \xi_2, \ldots, \xi_n$ having a common distribution (depending on the number of summands n).

Clearly, this definition is equivalent to the following: The law $\Phi(x)$ is infinitely divisible if for any n its characteristic function is the n-th power of some other characteristic function.

The investigations of recent years have shown that infinitely divisible laws play an important role in diverse problems in probability theory. In particular, the class of limit laws for sums of independent random variables has been found to coincide with the class of infinitely divisible laws.

We now proceed to a presentation of the properties of infinitely divisible laws that we shall need in the sequel. We begin our discussion with a proof of the fact that the normal and the Poisson laws are infinitely divisible. Let us consider the characteristic function for the normal law having expectation a and variance σ^2:

$$\varphi(t) = e^{iat - \frac{1}{2}\sigma^2 t^2}.$$

For any n, the n-th root of $\varphi(t)$ is again the characteristic function for a normal law, but with expectation a/n and variance σ^2/n.

We next slightly generalize the concept of the Poisson law which we encountered earlier, and we shall say that the random variable ξ is distributed according to the Poisson law if it can assume only the values $ak + b$, where a and b are real constants and $k = 0, 1, 2, \ldots$, and if

$$\mathsf{P}\{\xi = ak + b\} = \frac{e^{-\lambda}\lambda^k}{k!}, \tag{1}$$

where λ is a positive constant. As can easily be shown, the characteristic function for the law (1) is given by

$$\varphi(t) = e^{\lambda(e^{iat} - 1) + ibt}.$$

We see that, for any n, the n-th root of $\varphi(t)$ is again the characteristic function for a Poisson law, but with different parameters: a, λ/n, and b/n.

§ 44. Infinitely Divisible Laws

THEOREM 1: *The characteristic function for an infinitely divisible law never vanishes.*

Proof: Let $\Phi(x)$ be an infinitely divisible law and $\varphi(t)$ its characteristic function. Then for any n, we have by definition

$$\varphi(t) = \{\varphi_n(t)\}^n, \qquad (2)$$

where $\varphi_n(t)$ is some characteristic function. By virtue of the continuity of the function $\varphi(t)$, there exists an interval $|t| \leq a$ in which $\varphi(t) \neq 0$; clearly, $\varphi_n(t) \neq 0$ also holds in this interval. For sufficiently large n, we can make the quantity $|\varphi_n(t)| = \sqrt[n]{|\varphi(t)|}$ arbitrarily close to one uniformly in the interval $|t| \leq a$.

Let us now take two mutually independent random variables η_1 and η_2 each of which is distributed according to some law $F(x)$, and let us consider their difference $\eta = \eta_1 - \eta_2$. The characteristic function of the variable η is

$$f^*(t) = \mathsf{E}\, e^{it(\eta_1 - \eta_2)} = \left|\mathsf{E}\, e^{it\eta_1}\right|^2 = |f(t)|^2.$$

We thus see that the square of the absolute value of any characteristic function is a characteristic function.

Further, since a real characteristic function is of the form

$$f(t) = \int \cos x\, t\, dF(x),$$

we may consequently write the inequality

$$1 - f(2t) = \int (1 - \cos 2xt)\, dF(x) =$$
$$= 2\int \sin^2 xt\, dF(x) = 2\int (1 - \cos xt)(1 + \cos xt)\, dF(x) \leq$$
$$\leq 4\int (1 - \cos xt)\, dF(x) = 4(1 - f(t)).$$

From this we see that the function $|\varphi_n(t)|^2$ satisfies the inequality

$$1 - |\varphi_n(2t)|^2 \leq 4(1 - |\varphi_n(t)|^2).$$

If n is sufficiently large, so that $1 - |\varphi_n(t)| < \varepsilon$ for $|t| \leq a$, it follows from the last inequality that in this interval

$$1 - |\varphi_n(2t)| \leq 1 - |\varphi_n(2t)|^2 \leq 4(1 - |\varphi_n(t)|^2) \leq 8(1 - |\varphi_n(t)|) < 8\varepsilon.$$

Thus, in the interval $|t| \leq 2a$,

$$1 - |\varphi_n(t)| < 8\varepsilon.$$

Hence, for sufficiently large values of n, $\varphi_n(t)$ does not vanish in the interval $|t| \leq 2a$, and therefore, neither does $\varphi(t)$.

Similarly, we can show that $\varphi(t) \neq 0$ in the interval $|t| \leq 4a$, etc.

This proves our theorem.

THEOREM 2: *The distribution function for a sum of independent random variables having infinitely divisible distribution functions is itself infinitely divisible.*

Proof: It obviously suffices to prove the theorem for the case of two summands. If $\varphi(t)$ and $\psi(t)$ are the characteristic functions for the summands then, by hypothesis, we have for any n:

$$\varphi(t) = \{\varphi_n(t)\}^n, \quad \psi(t) = \{\psi_n(t)\}^n,$$

where $\varphi_n(t)$ and $\psi_n(t)$ are characteristic functions. Therefore, the characteristic function for the sum satisfies the relation

$$\chi(t) = \varphi(t) \cdot \psi(t) = \{\varphi_n(t) \cdot \psi_n(t)\}^n$$

for any value of n.

THEOREM 3: *The limit distribution function (in the sense of weak convergence) of a sequence of infinitely divisible distribution functions is itself infinitely divisible.*

Proof: Let the sequence $\Phi^{(k)}(x)$ of infinitely divisible distribution functions converge weakly to the distribution function $\Phi(x)$. Then

$$\lim_{k \to \infty} \varphi^{(k)}(t) = \varphi(t) \tag{3}$$

uniformly in any finite interval $t_1 \leq t \leq t_2$. By hypothesis, the functions

$$\varphi_n^{(k)}(t) = \sqrt[n]{\varphi^{(k)}(t)} \tag{4}$$

($\sqrt[n]{}$ is the principal value of the root) are characteristic functions for any n. From (3), we conclude that

$$\lim_{k \to \infty} \varphi_n^{(k)}(t) = \varphi_n(t) \tag{5}$$

for every value of n. From the continuity of $\varphi_n^{(k)}(t)$ follows the continuity of $\varphi_n(t)$. By virtue of the limit theorem for characteristic functions, $\varphi_n(t)$ is a characteristic function. From (3), (4), and (5), we find that the equality

$$\varphi(t) = \{\varphi_n(t)\}^n$$

holds for every n, Q.E.D.

§ 45. Canonical Representation of Infinitely Divisible Laws

In the following, we shall confine ourselves to an investigation of infinitely divisible laws with finite variance. The purpose of the present section is to give the proof of the following theorem, which was discovered in 1932 by Kolmogorov, and which completely characterizes the class of distribution laws with which we are concerned.

THEOREM: *In order that a distribution function $\Phi(x)$ with a finite variance be infinitely divisible it is necessary and sufficient that the logarithm of its characteristic function be given by*

$$\log \varphi(t) = i\gamma t + \int \{e^{itx} - 1 - itx\} \frac{1}{x^2} dG(x), \qquad (1)$$

where γ is a real constant and $G(x)$ is a non-decreasing function of bounded variation.

Proof: Let us suppose first of all that $\Phi(x)$ is an infinitely divisible law and that $\varphi(t)$ is its characteristic function. Then for any n

$$\varphi(t) = \{\varphi_n(t)\}^n,$$

where $\varphi_n(t)$ is some characteristic function. Since $\varphi(t) \neq 0$, this equation is equivalent to the following:[1]

$$\log \varphi(t) = n \log \varphi_n(t) = n \log [1 + (\varphi_n(t) - 1)].$$

For any T,

$$\varphi_n(t) \to 1$$

as $n \to \infty$, uniformly in the interval $|t| < T$, and therefore, in any finite interval of the t-axis the quantity $|\varphi_n(t) - 1|$ can be made smaller than any given number if n is taken sufficiently large. Consequently, we can make use of the relation

$$\log [1 + (\varphi_n(t) - 1)] = (\varphi_n(t) - 1)(1 + o(1)),$$

---

[1] It is understood that the principal value of the logarithm is to be taken here.

which yields:

$$\log \varphi(t) = \lim_{n \to \infty} n(\varphi_n(t) - 1) = \lim_{n \to \infty} n \int (e^{itx} - 1) \, d\Phi_n(x), \qquad (2)$$

where $\Phi_n(x)$ is the distribution function having $\varphi_n(t)$ as its characteristic function. From the definition of expectation and the relation between the functions $\Phi_n(x)$ and $\Phi(x)$, it follows that

$$n \int x \, d\Phi_n(x) = \int x \, d\Phi(x).$$

Let γ denote this quantity; then equation (2) may be written in the following form:

$$\log \varphi(t) = i \gamma t + \lim_{n \to \infty} n \int \{e^{itx} - 1 - i t x\} \, d\Phi_n(x).$$

Let us now define

$$G_n(x) = n \int_{-\infty}^{x} u^2 \, d\Phi_n(u).$$

Clearly, the functions $G_n(x)$ do not decrease with increasing argument, and $G_n(-\infty) = 0$. Moreover, the functions $G_n(x)$ are uniformly bounded. The last assertion follows from the properties of the variance and the relation between the functions $\Phi(x)$ and $\Phi_n(x)$. In fact,

$$G_n(+\infty) = n \int u^2 \, d\Phi_n(u) \qquad (3)$$
$$= n \left[\int u^2 \, d\Phi_n(u) - \left(\int u \, d\Phi_n(u) \right)^2 \right] + n \left(\int u \, d\Phi_n(u) \right)^2 = \sigma^2 + \frac{1}{n} \gamma^2,$$

where σ^2 is the variance of the distribution $\Phi(x)$.

In terms of $G_n(x)$, we have (see property 6 of the Stieltjes integral in § 25):

$$\log \varphi(t) = i \gamma t + \lim_{n \to \infty} \int (e^{itx} - 1 - i t x) \frac{1}{x^2} \, dG_n(x).$$

According to Helly's First Theorem, we may select from the sequence of functions $G_n(x)$ a subsequence that converges to some limit function $G(x)$. If $A < 0$ and $B > 0$ are points of continuity of the function $G(x)$, then by virtue of Helly's Second Theorem,

§ 45. Canonical Representation of Infinitely Divisible Laws 309

$$\int_A^B (e^{itx} - 1 - itx) \frac{1}{x^2} dG_{n_k}(x) \to \int_A^B (e^{itx} - 1 - itx) \frac{1}{x^2} dG(x) \quad (4)$$

as $k \to \infty$. We know that

$$|e^{itx} - 1 - itx| \leq |e^{itx} - 1| + |tx| \leq |tx| + |tx| = 2|t| \cdot |x|$$

and therefore

$$\left| \int_{-\infty}^A + \int_B^\infty (e^{itx} - 1 - itx) \frac{1}{x^2} dG_{n_k}(x) \right| \leq \int_{-\infty}^A + \int_B^\infty \frac{|e^{itx} - 1 - itx|}{x^2} dG_{n_k}(x) \leq$$

$$\leq 2|t| \left(\int_{-\infty}^A + \int_B^\infty \frac{1}{|x|} dG_{n_k}(x) \right) \leq \frac{2|t|}{\Gamma} \left(\int_{-\infty}^A + \int_B^\infty dG_{n_k}(x) \right) \leq$$

$$\leq \frac{2|t|}{\Gamma} \max_{1 \leq k < \infty} \int dG_{n_k}(x),$$

where $\Gamma = \min(|A|, B)$. Since the variations of the functions $G_{n_k}(u)$ are uniformly bounded, for any $\varepsilon > 0$ we can obtain the inequality

$$\left| \int_{-\infty}^A + \int_B^\infty (e^{itx} - 1 - itx) \frac{1}{x^2} dG_{n_k}(x) \right| < \frac{\varepsilon}{2} \quad (5)$$

for all t in some finite interval and for all k, by choosing $|A|$ and B sufficiently large.

It follows from (4) and (5) that for any $\varepsilon > 0$ and for all t in an arbitrary finite interval the inequality

$$\left| \int (e^{itx} - 1 - itx) \frac{1}{x^2} dG_{n_k}(x) - \int (e^{itx} - 1 - itx) \frac{1}{x^2} dG(x) \right| < \varepsilon$$

holds for sufficiently large values of n; that is to say,

$$\lim_{k \to \infty} \int (e^{itx} - 1 - itx) \frac{1}{x^2} dG_{n_k}(x) = \int (e^{itx} - 1 - itx) \frac{1}{x^2} dG(x).$$

We have thus shown that the logarithm of the characteristic function of any infinitely divisible distribution is expressible in the form (1). We now have to prove the converse statement, that any function whose logarithm is representable by formula (1) is the characteristic function of some infinitely divisible distribution.

For any ε $(0 < \varepsilon < 1)$, the integral

$$\int_{\varepsilon}^{\frac{1}{\varepsilon}} (e^{itx} - 1 - itx) \frac{1}{x^2} dG(x), \qquad (6)$$

by the definition of the Stieltjes integral, is the limit of the sum

$$\sum_{s=1}^{n} (e^{it\bar{x}_s} - 1 - it\bar{x}_s) \frac{1}{\bar{x}_s^2} (G(x_{s+1}) - G(x_s)),$$

where $x_1 = \varepsilon$, $x_{n+1} = 1/\varepsilon$, $x_s \leq \bar{x}_s \leq x_{s+1}$, and $\max (x_{s+1} - x_s) \to 0$. Each term in this sum is the logarithm of the characteristic function for some Poisson law. By Theorems 2 and 3 of § 44, the integral (6) is the logarithm of the characteristic function of some infinitely divisible distribution. By taking the limit as $\varepsilon \to 0$, we can satisfy ourselves that the same is true of the integral

$$\int_{x>0} (e^{itx} - 1 - itx) \frac{1}{x^2} dG(x). \qquad (7)$$

In an exactly analogous way, we can show that the integral

$$\int_{x<0} (e^{itx} - 1 - itx) \frac{1}{x^2} dG(x) \qquad (8)$$

is the logarithm of the characteristic function of some infinitely divisible law. The integral on the right-hand side of the expression (1) is the sum of the integrals (7) and (8) and the quantity

$$i\gamma t - \frac{1}{2} t^2 (G(+0) - G(-0)).$$

This last term is the logarithm of the characteristic function of the normal law. It follows from Theorem 2 of § 44 that the function $\varphi(t)$ given by the expression (1) is the characteristic function of some infinitely divisible law.[2] It now remains to show that the representation of $\log \varphi(t)$ in

---

[2] We have just proved that every infinitely divisible law is either the composition of a finite number of Poisson and normal laws or the limit of a uniformly convergent sequence of such laws. We thus see that the normal and the Poisson laws are those fundamental elements that go to make up any given infinitely divisible law.

§ 45. Canonical Representation of Infinitely Divisible Laws

expression (1) is unique, i.e., that the function $G(x)$ and the constant γ are uniquely determined by specifying $\varphi(t)$.

By differentiation of (1), we find:

$$\frac{d^2}{dt^2} \log \varphi(t) = -\int e^{itx} \, dG(x). \tag{9}$$

From the theory of characteristic functions, we know that the function $G(x)$ in this integral is uniquely determined by $d^2\{\log \varphi(t)\}/dt^2$. In the course of proving the theorem, we saw that the constant γ was the expectation and hence it too is uniquely determined by the function $\varphi(t)$.

Finally, we shall indicate the probabilistic significance of the total variation of the function $G(x)$. We know that if a random variable ξ is distributed according to the law $\Phi(x)$, then (see (5) of § 35)

$$\mathsf{D}\,\xi = -\left[\frac{d^2}{dt^2} \log \varphi(t)\right]_{t=0};$$

it therefore follows from (9) that

$$\mathsf{D}\,\xi = \int dG(x) = G(+\infty).$$

By way of example, we shall give the canonical representations of the normal and the Poisson laws.

For the normal law with variance σ^2 and expectation a,

$$\gamma = a \quad \text{and} \quad G(x) = \begin{cases} 0 & \text{for } x < 0, \\ \sigma^2 & \text{for } x > 0. \end{cases}$$

In fact, this function and the constant γ lead to the law in question, since

$$\int \{e^{itx} - 1 - itx\} \frac{1}{x^2} dG(x) =$$

$$= \lim_{u \to 0} \frac{e^{itu} - 1 - itu}{u^2} [G(+0) - G(-0)] = -\frac{t^2 \sigma^2}{2},$$

and by virtue of the uniqueness of the canonical representation, no other function $G(x)$ can yield the normal law.

In an analogous way, it is easy to show that the $G(x)$ corresponding to the Poisson law with the characteristic function

$$\varphi(t) = e^{\lambda(e^{ita}-1)+ibt}$$

is the function having a single jump at the point a:

$$G(x) = \begin{cases} 0 & \text{for } x < a, \\ a^2 \lambda & \text{for } x > a \end{cases}$$

and $\gamma = b + a\lambda$.

§ 46. A Limit Theorem for Infinitely Divisible Laws

We already know that if a sequence of infinitely divisible distribution laws converges to a limit distribution law, then this limit distribution is itself infinitely divisible. We shall now indicate conditions which are sufficient for the convergence of a given sequence of infinitely divisible distribution functions.

THEOREM: *In order for a sequence $\Phi_n(x)$ of infinitely divisible distribution functions to converge to some distribution function $\Phi(x)$ as $n \to \infty$ and for their respective variances to converge to the variance of the limit distribution, it is necessary and sufficient that there exist a constant γ and function $G(x)$ such that as $n \to \infty$*

1) *$G_n(x)$ converges weakly to $G(x)$,*
2) *$G_n(\infty) - G_n(-\infty) \to G(\infty) - G(-\infty)$,*
3) *$\gamma_n \to \gamma$,*

where γ_n and $G_n(x)$ are defined by formula (1) of § 45 for the law $\Phi_n(x)$ and where the constant γ and the function $G(x)$ define, by the same formula, the limit distribution $\Phi(x)$.

Proof: The sufficiency of the conditions of the theorem is an immediate consequence of Helly's Second Theorem. In fact, from these assumptions and formula (1) of § 45, it follows that

$$\log \varphi_n(t) \to \log \varphi(t)$$

as $n \to \infty$ uniformly in every finite interval $t_1 \leq t \leq t_2$.

In the preceding section, we saw that the integrals

$$\int dG_n(u) \quad \text{and} \quad \int dG(u)$$

were the respective variances of the laws $\Phi_n(x)$ and $\Phi(x)$; therefore the second condition of the theorem is nothing other than a requirement that the variances converge.

§ 46. A Limit Theorem

Now suppose that we know that

$$\Phi_n(x) \to \Phi(x) \tag{1}$$

as $n \to \infty$ and that the variances of the laws $\Phi_n(x)$ converge to the variance of the limit distribution $\Phi(x)$. We shall prove that these requirements imply the three conditions of the theorem. As we have just noted, no additional arguments are needed as regards condition 2. From this condition it follows that the total variations of the functions $G_n(u)$ are uniformly bounded. We can therefore apply Helly's First Theorem and from the sequence of functions $G_n(u)$ select a subsequence $G_{n_k}(u)$ which converges to some limit $G_\infty(u)$ as $k \to \infty$. Our purpose is to prove that

$$G_\infty(u) = G(u) .$$

To do this, we shall first establish that

$$J_k = \int \{e^{itu} - 1 - itu\} \frac{1}{u^2} dG_{n_k}(u) \to J_\infty =$$

$$= \int \{e^{itu} - 1 - itu\} \frac{1}{u^2} dG_\infty(u) \tag{2}$$

as $k \to \infty$. Let $A < 0$ and $B > 0$ be points of continuity of the function $G_\infty(u)$; then, by Helly's Second Theorem,

$$\int_A^B \{e^{itu} - 1 - itu\} \frac{1}{u^2} dG_{n_k}(u) \to \int_A^B \{e^{itu} - 1 - itu\} \frac{1}{u^2} dG_\infty(u), \tag{3}$$

as $k \to \infty$. On the other hand, from the inequality

$$|e^{itx} - 1 - itx| \leq 2|tx|,$$

we see that

$$L_k = \left| \int_{-\infty}^A + \int_B^\infty \{e^{itu} - 1 - itu\} \frac{1}{u^2} dG_{n_k}(u) \right| \leq 2|t| \left| \int_{-\infty}^A + \int_B^\infty \frac{1}{|u|} dG_{n_k}(u) \right|$$

$$\leq \frac{2|t|}{\Gamma} \left(\int_{-\infty}^A + \int_B^\infty dG_{n_k}(u) \right) \leq \frac{2|t|}{\Gamma} \int dG_{n_k}(u),$$

where $\Gamma = \min(-A, B)$. By virtue of the uniform boundedness of the total variations of the functions $G_n(u)$, for any $\varepsilon > 0$ it is possible to choose $|A|$ and B sufficiently large that

Similarly, the inequality

$$\left| \int_{-\infty}^{A} + \int_{B}^{\infty} \{e^{itu} - 1 - itu\} \frac{1}{u^2} dG_\infty(u) \right| < \varepsilon. \qquad (5)$$

holds for any given positive ε if B and $|A|$ are sufficiently large. From (3), (4), and (5), we conclude that for any $\varepsilon > 0$

$$|J_k - J_\infty| < 3\varepsilon$$

for sufficiently large values of k. This proves (2). From (1), we see that

$$\lim_{n \to \infty} \log \varphi_n(t) = \lim_{n \to \infty} \left(i\gamma_n t + \int \{e^{itu} - 1 - itu\} \frac{1}{u^2} dG_n(u) \right)$$
$$= \log \varphi(t) = i\gamma t + \int \{e^{itu} - 1 - itu\} \frac{1}{u^2} dG(u),$$

or

$$\lim_{k \to \infty} \left(i\gamma_{n_k} + \int \{e^{itu} - 1 - itu\} \frac{1}{tu^2} dG_{n_k}(u) \right) =$$
$$= i\gamma + \int \{e^{itu} - 1 - itu\} \frac{1}{tu^2} dG(u). \qquad (6)$$

From the inequality

$$|e^{itu} - 1 - itu| \leq \frac{t^2 u^2}{2}$$

and the uniform boundedness of the total variations of the functions $G_{n_k}(u)$, we conclude that as $t \to 0$

$$\left| \int (e^{itu} - 1 - itu) \frac{1}{tu^2} dG_{n_k}(u) \right| \leq \left| t \int dG_{n_k}(u) \right| \to 0$$

uniformly in n. Therefore, for $t \to 0$, (6) yields:

$$\lim_{k \to \infty} \gamma_{n_k} = \gamma, \qquad (7)$$

but on the other hand, by (2) and (7),

$$\log \varphi(t) = i\gamma t + \int \{e^{iut} - 1 - iut\} \frac{1}{u^2} dG_\infty(u).$$

§ 47. Limit Theorems for Sums: Formulation of the Problem 315

Since formula (1) of § 45 gives a unique representation for infinitely divisible distributions, we conclude that $G_\infty(u) = G(u)$.

Thus, *any* convergent subsequence of functions $G_{n_k}(u)$ converges to the function $G(u)$, and at the same time the sequence of constants γ_{n_k} converges to γ.

It is now easy to show that the sequence $G_n(u)$ itself also converges to $G(u)$, which implies that, at the same time, $\lim_{n\to\infty} \gamma_n = \gamma$. If this were not the case, then we could find a point of continuity of the function $G(u)$, say $u = c$, and a subsequence of functions $G_{n_k}(u)$ which converges at the point $u = c$ to a value distinct from $G(c)$. By Helly's First Theorem, we can select from this subsequence a convergent subsequence $G_{n_{k_r}}(u)$.

From the above, it follows that

$$\lim_{r \to \infty} G_{n_{k_r}}(u) = G(u)$$

at all points of continuity of the function $G(u)$. This contradicts our assumption. Thus, at all points of continuity of the function $G(u)$,

$$\lim_{n \to \infty} G_n(u) = G(u);$$

and, as we have seen, it immediately follows from this that

$$\lim_{n \to \infty} \gamma_n = \gamma .$$

This completes the proof.

§ 47. Limit Theorems for Sums: Formulation of the Problem

Let there be given the double sequence

$$\left.\begin{array}{l} \xi_{11}, \xi_{12}, \ldots, \xi_{1k_1}, \\ \xi_{21}, \xi_{22}, \ldots, \xi_{2k_2}, \\ \cdots\cdots\cdots\cdots \\ \xi_{n1}, \xi_{n2}, \ldots, \xi_{nk_n}, \\ \cdots\cdots\cdots\cdots \end{array}\right\} \quad (1)$$

in which the random variables in each row are independent. The question is: To what limit distribution function can the distribution functions of the sequence of sums

$$\zeta_n = \xi_{n1} + \xi_{n2} + \cdots + \xi_{nk_n}$$

converge as $n \to \infty$, and under what conditions?

In what follows, we shall confine ourselves to the study of *elementary systems*, i.e., to double sequences (1) satisfying the following conditions:

1) The variables ξ_{nk} have finite variances;
2) The variances of the sums ζ_n are bounded from above by a constant C which is independent of n;
3) $\beta_n = \max\limits_{1 \leq k \leq k_n} D\xi_{nk} \to 0$ as $n \to \infty$. The last requirement means that the effect of the individual terms in the sum becomes smaller and smaller with increasing n.

The limit theorems for sums considered earlier are clearly contained in this general scheme. Thus, in the DeMoivre-Laplace and Liapounov Theorems we had the following double sequence:

$$\xi_{n1}, \xi_{n2}, \ldots, \xi_{nn},$$

where

$$\xi_{nk} = \frac{\xi_k - E\,\xi_k}{\sqrt{\sum\limits_{k=1}^{n} D\,\xi_k}} \quad (1 \leq k \leq n, \; n = 1, 2, \ldots).$$

We also dealt with such a scheme in the theorems concerning the law of large numbers of Bernoulli, Tchebychev, and Markov, where the ξ_{nk} are the quantities

$$\xi_{nk} = \frac{\xi_k - E\,\xi_k}{n}.$$

§ 48. Limit Theorems for Sums

Let there be given an elementary system; let $F_{nk}(x)$ denote the distribution function of the random variable ξ_{nk} and $\bar{F}_{nk}(x)$ that of the random variable $\bar{\xi}_{nk} = \xi_{nk} - E\xi_{nk}$; clearly,

$$\bar{F}_{nk}(x) = F_{nk}(x + E\xi_{nk}).$$

§ 48. Limit Theorems for Sums

Theorem 1: *The distribution functions of the sequence of sums*

$$\zeta_n = \xi_{n1} + \xi_{n2} + \cdots + \xi_{nk_n} \tag{1}$$

converge to a limit distribution function as $n \to \infty$ if and only if the sequence of infinitely divisible laws whose characteristic functions have logarithms given by the formula

$$\psi_n(t) = \sum_{k=1}^{k_n} \{i t \, \mathsf{E}\, \xi_{nk} + \int (e^{itx} - 1) \, d\bar{F}_{nk}(x)\} \tag{2}$$

converge to a limit law.

The limit laws for both sequences coincide.[3]

Proof: The characteristic function of the sum (1) is

$$f_n(t) = \prod_{k=1}^{k_n} f_{nk}(t) = e^{it \sum_{k=1}^{k_n} \mathsf{E}\, \xi_{nk}} \prod_{k=1}^{k_n} \bar{f}_{nk}(t),$$

where $f_{nk}(t)$ is the characteristic function of the random variable ξ_{nk} and $\bar{f}_{nk}(t)$ is that of the random variable $\bar{\xi}_{nk}$.

We know that in order for the distribution functions of the sequence of sums (1) to converge to a limit $\Phi(x)$, it is necessary and sufficient that $f_n(t)$ converge to $\varphi(t)$ as $n \to \infty$, where $\varphi(t)$ is a continuous function; then $\varphi(t)$ turns out to be the characteristic function of the law $\Phi(x)$.

Let us set

$$\alpha_{nk} = \bar{f}_{nk}(t) - 1 .$$

---

[3] If we introduce the notation

$$\gamma_n = \sum_{k=1}^{k_n} \mathsf{E}\, \xi_{nk}, \qquad G_n(u) = \sum_{k=1}^{k_n} \int_{-\infty}^{u} x^2 \, d\bar{F}_{nk}(x)$$

and observe that $\int x d\bar{F}_{nk}(x) = 0$, then the function $\psi_n(t)$ can be written in the form

$$\psi_n(t) = i \gamma_n t + \int \{e^{itu} - 1 - itu\} \frac{1}{u^2} dG_n(u). \tag{3}$$

We already know that this means that $\psi_n(t)$ is the logarithm of the characteristic function of some infinitely divisible law.

The variance of ζ_n coincides with that of the infinitely divisible law (2).

For the quantities α_{nk} we have that

$$\alpha_n = \max_{1 \leq k \leq k_n} |\alpha_{nk}| \to 0. \tag{4}$$

uniformly in every finite interval $t_1 \leq t \leq t_2$. In fact,

$$\alpha_{nk} = \int (e^{itx} - 1)\, d\overline{F}_{nk}(x) = \int (e^{itx} - 1 - itx)\, d\overline{F}_{nk}(x),$$

since

$$\mathsf{E}\,\xi_{nk} = \int x\, d\overline{F}_{nk}(x) = 0.$$

We know that for all real values of α

$$|e^{i\alpha} - 1 - i\alpha| \leq \frac{\alpha^2}{2};$$

therefore

$$|\alpha_{nk}| \leq \frac{t^2}{2} \int x^2\, d\overline{F}_{nk}(x) = \frac{t^2}{2}\,\mathsf{D}\,\xi_{nk}. \tag{5}$$

From (5) and the third condition for an elementary system there follows (4).

From (4), we first of all conclude that for arbitrary T we may take

$$|\alpha_{nk}| < \frac{1}{2} \tag{6}$$

for n sufficiently large and $|t| \leq T$. By virtue of this, we can apply the series expansion for the logarithm:

$$\log \overline{f}_{nk}(t) = \log(1 + \alpha_{nk}) = \alpha_{nk} - \frac{\alpha_{nk}^2}{2} + \frac{\alpha_{nk}^2}{3} - \cdots.$$

Clearly,

$$R_n = \left| \log f_n(t) - \sum_{k=1}^{k_n} (it\,\mathsf{E}\,\xi_{nk} + \alpha_{nk}) \right| =$$

$$= \left| \sum_{k=1}^{k_n} (\log \overline{f}_{nk}(t) - \alpha_{nk}) \right| \leq \sum_{k=1}^{k_n} \sum_{s=2}^{\infty} \frac{|\alpha_{nk}|^s}{s} \leq \frac{1}{2} \sum_{k=1}^{k_n} \frac{|\alpha_{nk}|^2}{1 - |\alpha_{nk}|}. \tag{7}$$

Formulas (5) and (6) and the second condition for an elementary system lead to the inequality

$$R_n \leq \max_{1 \leq k \leq k_n} |\alpha_{nk}| \sum_{k=1}^{k_n} |\alpha_{nk}| \leq \frac{t^2}{2}\, C \max_{1 \leq k \leq k_n} |\alpha_{nk}|.$$

§ 48. LIMIT THEOREMS FOR SUMS

By virtue of (4), we can conclude that

$$|\log f_n(t) - \psi_n(t)| \to 0 \qquad (8)$$

uniformly in every finite interval $|t| \leq T$.

We have thus established that *the distribution functions of the sums ζ_n in any elementary system and the infinitely divisible distributions defined by* (2) *are asymptotic as* $n \to \infty$, by virtue of which Theorem 1 has been proved.

This theorem allows one to replace the investigation of sums of random variables (1) having, generally speaking, arbitrary distribution functions by the investigation of infinitely divisible laws. The latter turns out, as we shall see, to be quite simple in many cases.

THEOREM 2: *Every limit distribution of the sequence of distribution functions of the sums in an elementary system is infinitely divisible and has a finite variance and, conversely, every infinitely divisible law with a finite variance is the limit of the distribution functions of the sums in some elementary system.*

Proof: From Theorem 1, we know that the limit law for the distribution functions of the sums (1) is the limit of infinitely divisible laws and by Theorem 3 of § 44, this means that the limit law is infinitely divisible; its variance is finite, since the variances of the sums are uniformly bounded according to the second condition for an elementary system. The converse statement, that every infinitely divisible law with a finite variance is the limit of distribution functions for sums follows immediately from the definition of infinitely divisible laws.

THEOREM 3: *The distribution functions of the sequence of sums* (1) *converge to some limit distribution function as* $n \to \infty$ *and their variances converge to the variance of the limit distribution if and only if there exist a function $G(u)$ and constant γ such that as* $n \to \infty$

1. $\sum_{k=1}^{k_n} \int_{-\infty}^{u} x^2 \, d\overline{F}_{nk}(x) \to G(u)$

at the points of continuity of $G(u)$,

2. $\sum_{k=1}^{k_n} \int x^2 \, d\overline{F}_{nk}(x) \to G(+\infty)$,

3. $\sum_{k=1}^{k_n} \int x \, dF_{nk}(x) \to \gamma$.

The logarithm of the characteristic function of the limit distribution is determined by expression (1) of § 45, where $G(u)$ and γ are the function and constant just defined.

Proof: If we introduce the notation

$$G_n(u) = \sum_{k=1}^{k_n} \int_{-\infty}^{u} x^2 \, d\overline{F}_{nk}(x)$$

and

$$\gamma_n = \sum_{k=1}^{k_n} \int x \, dF_{nk}(x),$$

then we have the conditions of the theorem of § 46. This proves the theorem.

By slightly modifying the formulation of Theorem 3, we can obtain not only conditions for the existence of the limit distribution but also conditions for convergence to any given limit distribution.

THEOREM 4: *The distribution functions of the sequence of sums* (1) *converge to a given distribution function $\Phi(x)$ as $n \to \infty$ and the variances of the sums converge to the variance of the limit distribution if and only if the following conditions are satisfied for $n \to \infty$:*

1. $\sum_{k=1}^{k_n} \int_{-\infty}^{u} x^2 \, d\overline{F}_{nk}(x) \to G(u)$

at the points of continuity of the function $G(u)$;

2. $\sum_{k=1}^{k_n} \int x^2 \, d\overline{F}_{nk}(x) \to G(\infty)$;

3. $\sum_{k=1}^{k_n} \int x \, dF_{nk}(x) \to \gamma$,

where the function $G(u)$ and the constant γ are determined by formula (1) *of § 45 for the function $\Phi(x)$.*

§ 49. Conditions for Convergence to the Normal and Poisson Laws

We shall now apply the results of § 48 in deriving conditions for the convergence of sequences of distribution functions of sums to the normal and Poisson laws.

§ 49. Conditions for Convergence to Normal and Poisson Laws

Theorem 1: *Let there be given an elementary system of independent random variables. The sequence of distribution functions for the sums*

$$\zeta_n = \xi_{n1} + \xi_{n2} + \cdots + \xi_{nk_n} \tag{1}$$

converges to the law

$$\Phi(x) = \frac{1}{\sqrt{2\pi}} \int_{-\infty}^{x} e^{-\frac{x^2}{2}} dx$$

as $n \to \infty$ if and only if the conditions

1. $\sum_{k=1}^{k_n} \int x \, dF_{nk}(x) \to 0$,

2. $\sum_{k=1}^{k_n} \int_{|x|>\tau} x^2 \, d\bar{F}_{nk}(x) \to 0$,

3. $\sum_{k=1}^{k_n} \int_{|x|<\tau} x^2 \, d\bar{F}_{nk}(x) \to 1$,

are satisfied for $n \to \infty$, where τ is an arbitrary positive constant.

Proof: From Theorem 4 of § 48, it follows that the required conditions consist in the relations

$$\sum_{k=1}^{k_n} \int x \, dF_{nk}(x) \to 0,$$

$$\sum_{k=1}^{k_n} \int_{-\infty}^{u} x^2 \, d\bar{F}_{nk}(x) \to \begin{cases} 0 & \text{for } u < 0, \\ 1 & \text{for } u > 0, \end{cases}$$

$$\sum_{k=1}^{k_n} \int x^2 \, d\bar{F}_{nk}(x) \to 1$$

being satisfied for $n \to \infty$. The first of these conditions coincides with the first hypothesis of the theorem and the remaining two are clearly equivalent to the second and third hypotheses of the theorem.

This theorem assumes a particularly simple form if the elementary system in question is normalized beforehand by the conditions

$$\sum_{k=1}^{k_n} \int x^2 \, dF_{nk}(x) = 1,$$

$$\int x \, dF_{nk}(x) = 0 \qquad (1 \leq k \leq k_n, \qquad n = 1, 2, \ldots). \tag{2}$$

THEOREM 2: *If the elementary system is normalized by relations* (2), *then the sequence of distribution functions of the sums* (1) *converges to the normal law if and only if for all* $\tau > 0$

$$\sum_{k=1}^{k_n} \int_{|x|>\tau} x^2 \, dF_{nk}(x) \to 0 \tag{3}$$

for $n \to \infty$.

The proof of this theorem is obvious.

Condition (3) is called the *Lindeberg condition*, since Lindeberg proved, in 1923, that it is sufficient for the distribution functions of sums to converge to the normal law. In 1935, Feller showed the necessity of this condition.

As another illustration of how the general limit theorems of the preceding section can be used, we consider the convergence of distribution functions for elementary systems to the Poisson law

$$P(x) = \begin{cases} 0 & \text{for } x \leq 0, \\ \sum_{0 \leq k < x} e^{-\lambda} \frac{\lambda^k}{k!} & \text{for } x > 0. \end{cases} \tag{4}$$

If ξ is a random variable which is distributed according to the law (4), then we know that $\mathsf{E}\xi = \mathsf{D}\xi = \lambda$.

We shall confine ourselves to elementary systems for which

$$\left. \begin{array}{c} \sum_{k=1}^{k_n} \mathsf{E}\,\xi_{nk} \to \lambda, \\ \sum_{k=1}^{k_n} \mathsf{D}\,\xi_{nk} \to \lambda. \end{array} \right\} \tag{5}$$

THEOREM 3: *Let there be given an elementary system satisfying the conditions* (5). *The distribution functions of the sums*

$$\zeta_n = \xi_{n1} + \xi_{n2} + \cdots + \xi_{nk_n}$$

converge to the law (4) *if and only if for any* $\tau > 0$

$$\sum_{k=1}^{k_n} \int_{|x-1|>\tau} x^2 \, dF_{nk}(x + \mathsf{E}\,\xi_{nk}) \to 0 \quad (n \to \infty).$$

We leave the proof of this theorem to the reader.

In § 15, we proved the Poisson Theorem. One can easily satisfy oneself that for $np_n = \lambda$ it is a special case of the statement just proved. In fact, let ξ_{nk} $(1 \leq k \leq n)$ be a random variable which assumes the values 0 or 1 depending on whether the event A that is being observed does not or does occur in the k-th trial of the n-th series of trials. Here

$$P\{\xi_{nk} = 1\} = \frac{\lambda}{n} \quad \text{and} \quad P\{\xi_{nk} = 0\} = 1 - \frac{\lambda}{n}.$$

Clearly, the sum

$$\mu_n = \xi_{n1} + \xi_{n2} + \cdots + \xi_{nn}$$

is the number of times the event A occurs in the n-th series of trials.

According to the Poisson Theorem, the distribution functions of the variables μ_n converge to the Poisson law (4) as $n \to \infty$. This result also follows from the theorem just formulated, since all the conditions of the theorem are satisfied in the case under consideration.

General theorems concerning the approach of distribution functions of the sums (1) to some infinitely divisible distribution function, proved under broader assumptions than ours, enable us also to obtain necessary and sufficient conditions for the law of large numbers (in the case of independent summands). For this, see the monograph by Gnedenko and Kolmogorov, referred to above.

EXERCISES

1. Prove that the distributions of
a) Pascal (Exercise 1a) of Chapter V),
b) Polya (Exercise 1b) of Chapter V),
c) Cauchy (Example 5 of § 24)
are infinitely divisible.

2. Prove that the random variable having the density function

$$p(x) = \begin{cases} 0 & \text{for } x \leq 0, \\ \dfrac{\beta^\alpha}{\Gamma(\alpha)} x^{\alpha-1} e^{-\beta x} & \text{for } x > 0, \end{cases}$$

where α and β are positive constants, is infinitely divisible.

Note: In particular, it follows from this that the Maxwell distribution and the χ^2-distribution (for any value of n) are infinitely divisible.

3. Prove that for any positive constants α and β

$$\varphi(t) = \left(1 + \frac{t^2}{\beta^2}\right)^{-\alpha}$$

is an infinitely divisible characteristic function.

Note: In particular, it follows from this that the Laplace distribution is infinitely divisible.

4. Determine the function $G(x)$ and the parameter γ in Kolmogorov's formula for the logarithm of an infinitely divisible characteristic function for the following distributions:

a) That of Example 2;
b) The Laplace distribution.

5. Prove that if the sum of two independent infinitely divisible random variables is distributed according to

a) the Poisson law,
b) the normal law,

then each term is distributed

in case a), according to the Poisson law,
in case b) according to the normal law.

6. Determine conditions under which the distribution functions of sums of random variables comprising an elementary system converge to

a) the distribution of Example 2,
b) that of Laplace.

CHAPTER X

THE THEORY OF STOCHASTIC PROCESSES

§ 50. Introductory Remarks

The perfecting of physical statistics as well as of a number of fields of engineering has posed a great number of new problems for probability theory that do not fit into the framework of the classical theory. Whereas the physicist and engineer were interested in studying a *process*, i.e., a phenomenon that changed with time, probability theory either lacked general methods or had not worked out special schemes for handling the problems that arose in the investigation of such phenomena. There appeared a pressing need for the development of a general theory of random processes, i.e., a theory that could be used in the study of random variables that depend on one or several continuously varying parameters.

Earlier we considered several simple problems of an applied nature in which real random processes were studied (see Chapter I, § 10, Examples 2, 3, 4; Chapter I, Exercises 21, 22; and Chapter II, Exercises 10, 11). We now enumerate several problems that illustrate the necessity for constructing a theory of random processes.

Let us imagine that our object is to follow the motion of some molecule of gas or fluid. At random moments of time, the molecule collides with other molecules, resulting in a change of its velocity and position. The state of the molecule is thus subject to random change at each moment of time. Many physical phenomena require for their analysis that we be able to compute the probability that some specified number of molecules will be displaced a certain distance in a certain interval of time. Thus, for example, if two gases or fluids are brought into contact, there begins a mutual permeation of the molecules of one fluid by those of the other, i.e., diffusion occurs. How fast does the process of diffusion take place? according to what laws? and when does the mixture that is formed become practically homogeneous? The answers to these and many other questions are given by the statistical theory of diffusion, the basis for which is the theory of random processes or, as it is now customary to say, the *theory of stochastic* (or: *probabilistic*) *processes*. An analogous problem obviously arises in chemistry when the process of a chemical reaction is studied. What proportion of the molecules has already entered into the reaction? how does the reaction proceed with time? and when is the reaction practically complete?

A very important class of phenomena takes place in accordance with the principle of radioactive decay. The course of these phenomena is as follows: Atoms of a radioactive substance disintegrate and are converted into atoms of some other element. The decay of each atom occurs instantaneously, like an explosion, with the liberation of a certain amount of energy. Numerous observations show that the decay of the various atoms occurs at moments of time which, to the observer, are random. Moreover, these moments of time are stochastically independent of one another. To study the process of radioactive decay, it is necessary to determine the probability that a certain number of atoms will disintegrate in a given interval of time. From a formal point of view, if we are concerned solely with the mathematical structure of a phenomenon, then there are other phenomena that also proceed in precisely this way: the number of calls that are received at a telephone exchange in a given interval of time (the traffic at the telephone exchange), the breaking of threads on a ring spinning frame (a type of loom) or, in Brownian motion, the fluctuation in the number of particles appearing in a given region of space at any moment of time. In this chapter, we shall give a simple solution to the mathematical problems to which these phenomena lead.

The general theory of stochastic processes had its origin in the fundamental papers of the Soviet mathematicians Kolmogorov and Khintchine in the early thirties. In Kolmogorov's article "On analytical methods in the theory of probability," [in Russian] there is given a systematical and rigorous presentation of the foundations of the theory of *stochastic processes without aftereffect* or *non-hereditary stochastic processes* or, as they are often called, *Markov processes*. The so-called theory of *stationary processes* originated in a series of papers by Khintchine.

Before some natural phenomenon or technical process can be studied mathematically, it must be schematized. The reason for this lies in the fact that mathematical analysis can be used to investigate the process of change in some system only if it is supposed that every possible state of the system can be fully defined by means of some definite mathematical machinery. Of course, such a mathematically definable system is not

§ 50. INTRODUCTORY REMARKS

reality itself but merely a scheme—a mathematical model—that is suitable for describing it. We encounter such a picture in, say, mechanics where it is supposed that the actual motion of a system of particles is completely describable at any moment of time whatever by specifying the moment of time as well as the state of the system at any earlier moment of time t_0. In other words, the scheme that is used in theoretical mechanics to describe the motion is the following: It is supposed that the state of the system y at any moment of time t is completely determined by its state x at any preceding moment of time t_0. The state of a mechanical system is understood to be the position and the velocities of the particles in the system.

Outside of classical mechanics, and strictly speaking, in all of modern physics, one has to deal with more complicated situations, wherein a knowledge of the state of a system at some moment of time t_0 no longer uniquely determines the states of the system at succeeding moments of time but only determines the probability that the system will be in one of the states belonging to some class of states of the system. If x denotes the state of the system at time t_0 and E some class of states of the system, then for the processes just described there is defined the probability

$$\mathsf{P}\{t_0, x; t, E\}$$

that the system, which is in the state x at time t_0, passes from x into one of the states of the class E at the instant of time t.

If any additional knowledge concerning the states of the system at the times $t < t_0$ does not affect this probability, then it is natural to refer to the class of stochastic processes that we have singled out as *processes without aftereffect*, or *non-hereditary processes*, or, because of their analogy to Markov chains, *Markov processes*.

The general notion of a stochastic process is based on the axiomatics presented earlier and can be introduced in the following way. Let U be the set of elementary events and t a continuous parameter. The function of two arguments

$$\xi(t) = \varphi(e, t) \qquad (e \in U)$$

is called a *stochastic process*.

For each value of the parameter t, the function $\varphi(e, t)$ is a function of e alone and consequently is a random variable. For each fixed value of the argument e (i.e., for each elementary event), $\varphi(e, t)$ depends only on t and is thus an ordinary function of one real variable. Each such func-

X. THE THEORY OF STOCHASTIC PROCESSES

tion is called a *realization* of the stochastic process $\xi(t)$. A random process may be interpreted either as a collection of random variables $\xi(t)$ depending on the parameter t or as a collection of realizations of the process $\xi(t)$. Of course, in order to define a process it is necessary to assign a probability measure in the function space of its realizations.

The entire present chapter will be devoted to the study of processes without aftereffect and stationary processes.

In the introduction to the book, it was stated that the first sketches of the theory of stochastic processes were made by certain outstanding physicists at the beginning of the present century on various particular occasions. We now give a brief account of the way in which they obtained the differential equations of diffusion theory, starting from the problem of a random walk. An outline of the reasoning is as follows. Suppose that a particle, located initially at the origin, undergoes independent random impacts at the times $k\tau$ ($k = 1, 2, 3, \ldots$). As a result, it is displaced each time a distance h to the right, with probability p, or a distance h to the left, with probability $q = 1 - p$. Let $f(x, t)$ denote the probability that the particle will be at the position x as the result of n impacts (obviously, for an even number of impacts the quantity x must be an even multiple of h and for an odd number, an odd multiple of h). We know that

$$f(x, t) = \binom{n}{m} p^m q^{n-m},$$

where $t = n\tau$ and m and $n - m$ are, respectively, the number of steps the particle has taken to the right and to the left. Clearly, the quantities m, n, x, and h are related by the equation

$$m - (n - m) = \frac{x}{h}.$$

We can easily satisfy ourselves, either by direct computation or by the simpler means of applying the formula on total probability (as we repeatedly did in the theory of Markov chains) that $f(x, t)$ satisfies the following difference equation

§ 50. Introductory Remarks

$$f(x, t + \tau) = pf(x - h, t) + qf(x + h, t) \tag{1}$$

as well as the initial conditions

$$f(0, 0) = 1, \quad f(x, 0) = 0 \quad \text{for} \quad x \neq 0.$$

Let us see what happens to this difference equation when we let both h and τ tend to zero. In so doing, we are compelled by the physical nature of the problem to impose certain conditions on h and τ. Similarly, the quantities p and q cannot be arbitrary; otherwise, the particle could go off to infinity with probability one in a finite interval of time. It turns out that if one is to obtain a meaningful result, it is necessary to require that

$$x = nh, \; t = n\tau, \; h^2/\tau = 2D, \; (p - q)/h \to c/D \tag{2}$$

as $n \to \infty$, where c and D are certain constants. The quantity c is called the *drift* and D the *diffusion coefficient*.

Subtracting the quantity $f(x, t)$ from both sides of equation (1), we find that

$$f(x, t + \tau) - f(x, t) =$$
$$= p[f(x - h, t) - f(x, t)] + q[f(x + h, t) - f(x, t)]. \tag{3}$$

Suppose now that $f(x, t)$ has a continuous first derivative with respect to t and a continuous second derivative with respect to x. Then

$$f(x, t + \tau) - f(x, t) = \tau \frac{\partial f(x, t)}{\partial t} + o(\tau),$$

$$f(x - h, t) - f(x, t) = -h \frac{\partial f(x, t)}{\partial x} + \frac{h^2}{2} \frac{\partial^2 f(x, t)}{\partial x^2} + o(h^2),$$

$$f(x + h, t) - f(x, t) = h \frac{\partial f(x, t)}{\partial x} + \frac{h^2}{2} \frac{\partial^2 f(x, t)}{\partial x^2} + o(h^2).$$

After substituting these expressions into (3), we obtain

$$\tau \frac{\partial f(x,t)}{\partial t} + o(\tau) = -(p-q)h \frac{\partial f(x,t)}{\partial x} + \frac{h^2}{2} \frac{\partial^2 f(x,t)}{\partial x^2} + o(h^2).$$

Hence, by virtue of relations (2), we find in the limit that

$$\frac{\partial f(x,t)}{\partial t} = -2c \frac{\partial f(x,t)}{\partial x} + D \frac{\partial^2 f(x,t)}{\partial x^2}.$$

We have thus derived the equation which is called in diffusion theory the *Fokker-Planck equation*.

It is interesting to note that we have obtained a physically meaningful and definitive result, which reflects the true picture of diffusion well, for a problem whose formulation was very artificial. Later on, we shall derive the general equations that are satisfied by the distributions of stochastic processes under very broad assumptions concerning the nature of the processes.

§ 51. The Poisson Process

Before presenting some of the general results that have now already become classical, we shall study in detail an example of a stochastic process without aftereffect that plays an important role both in the theory and in a variety of applications.

Let us suppose that a certain event occurs at random instants of time. We are interested in the number of occurrences of this event in the interval of time from 0 to t. Let us denote this number by $\xi(t)$. We shall assume that the process characterizing the occurrence of the event is (1) stationary, (2) without aftereffect, or non-hereditary, and (3) ordinary. By these three terms, we mean the following.

A process is called *stationary* if for any finite set of non-overlapping intervals of time, the probability of occurrence of a specified number of events during each of these intervals depends only on these numbers and on the length of the intervals; the probability is unaffected when all the intervals are shifted by the same amount. In particular, the probability of occurrence of k events during the interval of time from T to $T+t$ is a function only of t and k and not of T.

§ 51. THE POISSON PROCESS

By *no aftereffect*, or *non-hereditary*, we mean that the probability of occurrence of k events during the interval of time $(T, T + t)$ is independent of past history, i.e., of how many events have occurred earlier or how they occurred. This means that the conditional probability of occurrence of k events in the interval $(T, T + t)$ under any assumption about the occurrences of the events up to the time T, coincides with the unconditional probability. In particular, no aftereffect means that the occurrence of n events and m events, respectively, in each of two non-overlapping intervals of time are mutually independent.

Ordinariness expresses the condition that the occurrence of two or more events in a small interval of time Δt is practically impossible. Let $P_{>1}(\Delta t)$ denote the probability of occurrence of more than one event in the interval of time Δt. Then the precise formulation of the condition of ordinariness is as follows:

$$P_{>1}(\Delta t) = o(\Delta t).$$

Our immediate problem is that of determining the probability $P_k(\Delta t)$ that k events will occur in the interval of time Δt. By stationarity, this probability does not depend on where the interval of time Δt is located. To solve the problem, we first show that for small Δt, we have

$$P_1(\Delta t) = \lambda \Delta t + o(\Delta t),$$

where λ is a constant. Consider an interval of time of duration one, and let p denote the probability that no event occurs during this period. We now split up our interval of time into n equal non-overlapping parts. By the stationarity and absence of aftereffect, we have

$$p = [P_0(1/n)]^n,$$

or

$$P_0(1/n) = p^{1/n}.$$

From this, it follows that for any integer k

$$P_0(k/n) = p^{k/n}.$$

Now let t be a non-negative number. For any n, we can find a k such that

$$(k - 1)/n \leqq t < k/n.$$

Since the probability $P_0(t)$ is a decreasing function of time,

$$P_0\big((k-1)/n\big) \geq P_0(t) \geq P_0(k/n).$$

Thus $P_0(t)$ satisfies the inequality

$$p^{(k-1)/n} \geq P_0(t) \geq p^{k/n}.$$

We now let k and n tend to infinity in such a way that

$$\lim_{n \to \infty} k/n = t.$$

From the above inequality, it is clear that

$$P_0(t) = p^t.$$

Since p, being a probability, satisfies the inequality

$$0 \leq p \leq 1,$$

any of the following three cases can occur: $p = 0$, $p = 1$, or $0 < p < 1$. The first two cases are of little interest. In the first, we have $P_0(t) = 0$ for any t, which means that the probability is 1 that at least one event will occur during an interval of time of arbitrary length. In other words, an infinite number of events will occur in an interval of time of arbitrary length with probability one. In the second case, $P_0(t) = 1$, and therefore no events occur. The only interesting case is the third one, in which we shall put $p = e^{-\lambda}$, where λ is a positive quantity ($\lambda = -\log p$).

Thus, from the assumptions of stationarity and no aftereffect, we have found that for any $t \geq 0$,

$$P_0(t) = e^{-\lambda t} \tag{1}$$

(the assumption of ordinariness still not having been used).

Of course, for any value of t the following relation holds:

$$P_0(t) + P_1(t) + P_{>1}(t) = 1.$$

§ 51. THE POISSON PROCESS

From equation (1), it follows that for small values of t

$$P_0(t) = 1 - \lambda t + o(t).$$

Therefore, for small t, we have

$$P_1(t) = \lambda t + o(t). \tag{2}$$

We can now proceed to derive the formula for $P_k(t)$ for $k \geq 1$. To this end, let us determine the probability that the event will occur exactly k times in the time $t + \Delta t$. This may happen in $k+1$ different ways, namely:

1) All the k events occur in the interval of time t, and none in the interval of time Δt;

2) $k-1$ events occur in the interval of length t and one in the time Δt;

. . . .

$k+1$) The event does not occur in the interval of length t, but occurs k times during Δt.

By the formula on total probability, we have

$$P_k(t + \Delta t) = \sum_{j=0}^{k} P_{k-j}(t) \, P_j(\Delta t)$$

(here, we have made use of the condition of stationarity and the condition of no aftereffect). Let us set

$$R_k = \sum_{j=2}^{k} P_{k-j}(t) \, P_j(\Delta t).$$

It is obvious that

$$R_k \leq \sum_{j=2}^{k} P_j(\Delta t) \leq \sum_{j=2}^{\infty} P_j(\Delta t) = P_{>1}(\Delta t) = o(\Delta t),$$

by the condition of ordinariness.

Thus,

$$P_k(t + \Delta t) = P_k(t) P_0(\Delta t) + P_{k-1}(t) P_1(\Delta t) + o(\Delta t).$$

But, from the above,

$$P_0(\Delta t) = e^{-\lambda \Delta t} = 1 - \lambda \Delta t + o(\Delta t).$$

Furthermore, by (2),

$$P_1(\Delta t) = \lambda \Delta t + o(\Delta t),$$

and therefore

$$P_k(t + \Delta t) = (1 - \lambda \Delta t) P_k(t) + \lambda \Delta t\, P_{k-1}(t) + o(\Delta t).$$

From this, we have

$$\frac{P_k(t + \Delta t) - P_k(t)}{\Delta t} = -\lambda P_k(t) + \lambda P_{k-1}(t) + o(1).$$

Since the right-hand side of this equation has a limit for $\Delta t \to 0$, so does the left-hand side. As a consequence, we obtain the equation

$$\frac{dP_k(t)}{dt} = -\lambda P_k(t) + \lambda P_{k-1}(t) \tag{3}$$

for the determination of $P_k(t)$. Clearly, the condition of ordinariness and the expression for $P_0(t)$ we obtained above imply the following initial conditions:

$$P_0(0) = 1, \; P_k(0) = 0 \text{ for } k \geq 1. \tag{4}$$

The solution of equation (3) can be obtained very simply by making the substitution

$$P_k(t) = e^{-\lambda t} v_k(t), \tag{5}$$

where $v_k(t)$ is the new unknown function. Note that, by virtue of (1), $v_0(t) = 1$, and that the relations (4) imply the following initial conditions:

$$v_0(0) = 1 \text{ and } v_k(0) = 0 \text{ for } k \geq 1.$$

Substituting (5) in (3), we arrive at the equation

$$v_k'(t) = \lambda v_{k-1}(t). \tag{6}$$

Now, in particular,

$$v_1'(t) = \lambda. \tag{7}$$

Successively solving (7) and (6) and taking into account the initial conditions, we arrive at the relations

$$v_1(t) = \lambda t, \quad v_2(t) = \frac{(\lambda t)^2}{2!}, \quad v_3(t) = \frac{(\lambda t)^3}{3!}$$

and, in general,

$$v_k(t) = \frac{(\lambda t)^k}{k!}.$$

Thus, for any $k \geq 0$, we have

$$P_k(t) = \frac{(\lambda t)^k}{k!} e^{-\lambda t},$$

and our problem is solved.

The conditions imposed on the process characterizing the occurrence of the event are satisfied with surprising accuracy in many scientific and technological processes. By way of example, we mention (1) the number of spontaneously disintegrating atoms of a radioactive substance in a given interval of time and (2) the number of cosmic particles falling in a specified area in an interval of time t. If we were concerned with a complex communications system consisting of a large number of components each having small probability of failing to work in unit time, independently of the other components, then the number of components that fail in some interval of time $(0, t)$ is a stochastic process. In many instances, this process is described well by the Poisson process. If we so desired, we could go on adding to the number of such examples, literally, indefinitely.

We shall pause here to give two simple properties of the Poisson process.

The interval of time between two successive occurrences of some event in which we are interested is a random variable which we shall denote by τ. Let us find the probability distribution of τ. Since the event $\tau \geq t$ is obviously equivalent to the non-occurrence of our event in the interval of time t, we can say that

$$\mathsf{P}\{\tau \geq t\} = e^{-\lambda t}.$$

The required distribution function is thus given by the formula

$$\mathsf{P}\{\tau < t\} = 1 - e^{-\lambda t}.$$

The result obtained can be interpreted physically in a variety of ways. For example, we can look upon it as the time distribution of the free motion of a molecule or as the time distribution between two failures of the components in a complex communications system.

Suppose we know that n ($n > 0$) events of our process have occurred in an interval of duration t. The question we ask is, How are the occurrences of these events distributed within the interval? It turns out, as we shall next show, that the conditional distribution of the instants at which the events occur is uniform over this interval of time. Furthermore, the instants at which the n events occur are mutually independent.

Let B denote the event consisting in the occurrence of n events of the process in the interval of time $(0, t)$. From the above discussion, we know that the probability of B is

$$\mathsf{P}(B) = \frac{(\lambda t)^n}{n!} e^{-\lambda t}.$$

Since the events have already occurred in the interval of time $(0, t)$, we can distinguish between them and fix our attention on a particular one of them. We let A denote the event that our particular event has occurred in an interval (a, b) lying within the interval $(0, t)$. We should like to determine the probability $\mathsf{P}(A/B)$. By the Multiplication Theorem, we have

$$\mathsf{P}(A/B) = \frac{\mathsf{P}(AB)}{\mathsf{P}(A)}.$$

We must now determine the probability of the combined occurrence of the events A and B. To this end, consider C_{rs}, the event consisting in (a) r events of the process, excluding our particular event, falling in $(0, a)$; (b) s events, including our particular event, falling in the interval (a, b); (c) the other $n - r - s$ events falling in the interval (b, t). Obviously, for distinct pairs of values of (r, s), the events C_{rs} will be mutually exclusive, and therefore

$$AB = \sum_{r=0}^{n-1} \sum_{s=1}^{n-r} C_{rs}.$$

The probability of having r events of the process in $(0, a)$, s events in (a, b), and $n - r - s$ in the interval (b, t), by the assumption of no aftereffect, is

$$\frac{(\lambda a)^r}{r!} e^{-\lambda a} \frac{[\lambda(b-a)]^s}{s!} e^{-\lambda(b-a)} \frac{[\lambda(t-b)]^{n-r-s}}{(n-r-s)!} e^{-\lambda(t-b)}. \tag{8}$$

This expression, however, does not quite give the probability of the event C_{rs}, since we have not accounted for the fact that our particular event has to occur in the interval (a, b). To take this into account, we must still multiply (8) by the probability that our particular event will be among those falling in (a, b). This probability is the quotient of the number of combinations of $n - 1$ things taken $s - 1$ at a time and the number of combinations of n things taken s at a time, i.e., it is equal to

§ 51. The Poisson Process

$$\frac{\binom{n-1}{s-1}}{\binom{n}{s}} = \frac{s}{n}.$$

Thus

$$P(AB) = e^{-\lambda t} \lambda^n \sum_{r=0}^{n-1} \sum_{s=1}^{n-r} \frac{s}{n} \frac{a^r}{r!} \frac{(b-a)^s}{s!} \frac{(t-b)^{n-r-s}}{(n-r-s)!}.$$

Simple algebraic manipulations then lead to the formula

$$P(AB) = \frac{b-a}{t} \frac{(\lambda t)^n}{n!} e^{-\lambda t}.$$

Collecting our results, we find that

$$P(A/B) = \frac{b-a}{t}. \tag{9}$$

This formula thus substantiates the statements made above.

We note that the theory developed in this section is applicable even when the parameter t does not represent time. To bear out this statement, we consider the following example.

Example. Points are distributed in space in accordance with the following conditions:

1) The probability of k points being in a region R depends only on the volume v of the region R and not on its shape nor its position in space. We denote this probability by $p_k(v)$;

2) The numbers of points falling in non-overlapping regions are independent random variables;

3) $\sum_{k=2}^{\infty} p_k(\Delta v) = o(\Delta v).$

The conditions imposed are nothing other than those of stationarity, no aftereffect, and ordinariness. Hence, we have

$$p_n(v) = \frac{(av)^n}{n!} e^{-av}.$$

If minute particles of a substance are suspended in a fluid, these particles are found to be in a continuous state of chaotic motion (Brownian motion) as a consequence of collisions with the surrounding molecules.

As a result, at every instant of time we have a random distribution of the particles in space of the kind discussed above. By the theory of the present example, we may consider that the distribution of the particles in some given region will obey the Poisson law.

Table 12 compares (1) the empirical results obtained with particles of gold suspended in water, which we have adopted from an article by Smolukhovsky, and (2) the results computed according to the Poisson law.

TABLE 12

Number of Particles	Number of Cases Observed	Relative Frequency $\frac{m}{518}$	$\frac{\lambda^n e^{-\lambda}}{n!}$	Number of Cases Computed
0	112	0.216	0.213	110
1	168	0.325	0.328	173
2	130	0.251	0.253	131
3	69	0.133	0.130	67
4	32	0.062	0.050	26
5	5	0.010	0.016	8
6	1	0.002	0.004	2
7	1	0.002	0.001	1

The constant $\lambda = av$, which defines the Poisson law, has been chosen to be the arithmetic average of the number of particles observed, i.e.,

$$\lambda = \frac{0 \cdot 112 + 1 \cdot 168 + 2 \cdot 130 + 3 \cdot 69 + 4 \cdot 32 + 5 \cdot 5 + 6 \cdot 1 + 7 \cdot 1}{518} \approx 1.54.$$

§ 52. Conditional Distribution Functions and Bayes' Formula

For the further development of the theory, we need to generalize the concept of conditional probability, introduced in the first chapter, to the case of an infinite number of hypotheses. In particular, we shall have to introduce the notion of a conditional distribution function with respect to some random variable.

§ 52. Conditional Distribution Functions and Bayes' Formula

Consider some event B and a random variable ξ with a distribution function $F(x)$. Let $A_{\alpha\beta}$ denote the event that

$$x - \alpha \leq \xi < x + \beta.$$

By virtue of the definitions of the first chapter,

$$\mathsf{P}\{BA_{\alpha\beta}\} = \mathsf{P}\{A_{\alpha\beta}\} \cdot \mathsf{P}\{B/A_{\alpha\beta}\} = [F(x+\beta) - F(x-\alpha)]\, \mathsf{P}\{B/A_{\alpha\beta}\},$$

and hence

$$\mathsf{P}\{B/A_{\alpha\beta}\} = \frac{\mathsf{P}\{BA_{\alpha\beta}\}}{F(x+\beta) - F(x-\alpha)}.$$

If the limit

$$\lim_{\alpha,\beta \to 0} \frac{\mathsf{P}\{BA_{\alpha\beta}\}}{F(x+\beta) - F(x-\alpha)}$$

exists,[1] it is called the *conditional probability of the event B under the condition that $\xi = x$* and it is denoted by the symbol $\mathsf{P}\{B/x\}$. It is clear that for fixed x, $\mathsf{P}\{B/x\}$ will be a finitely additive function of the event B defined over some field of events.

Under certain conditions, which practically always prove to be fulfilled, $\mathsf{P}\{B/x\}$ has all the properties of an ordinary probability corresponding to Axioms 1-3 of § 8.

If η is a random variable and B denotes the event that $\eta < y$, then the function $\Phi(y/x) = \mathsf{P}\{\eta < y/x\}$, which, as can easily be seen, is a distribution function, is called the *conditional distribution function of the variable η under the condition that $\xi = x$*.

It is obvious that if $F(x, y)$ is the distribution function of the pair of random variables ξ and η then

$$\Phi(y/x) = \lim_{\alpha,\beta \to 0} \frac{F(x+\beta, y) - F(x-\alpha, y)}{F(x+\beta, \infty) - F(x-\alpha, \infty)},$$

provided this limit exists.

If the function $\mathsf{P}\{B/x\}$ is integrable with respect to $F(x)$, then the *formula for total probability* holds:

$$\mathsf{P}\{B\} = \int \mathsf{P}\{B/x\}\, dF(x).$$

---

[1] This limit exists for almost all values of x—almost all, in the sense of the measure defined by the function $F(x)$.

To derive this formula, we divide the interval of definition of the variable ξ into the subintervals $x_i \leq \xi < x_{i+1}$ by means of the points x_i ($i = 0, \pm 1, \pm 2, \ldots$) and we let A_i denote the event $x_i \leq \xi < x_{i+1}$. By virtue of the extended addition axiom, we have:

$$\mathsf{P}\{B\} = \sum_{i=-\infty}^{\infty} \mathsf{P}\{BA_i\} = \sum_{i=-\infty}^{\infty} \mathsf{P}\{B/A_i\}\,[F(x_{i+1}) - F(x_i)].$$

We now proceed to subdivide the intervals (x_i, x_{i+1}) into subintervals, in such a way that the length of the largest subinterval approaches zero. By the definition of conditional probability and of the Stieltjes integral, we thus obtain:

$$\mathsf{P}\{B\} = \int \mathsf{P}\{B/x\}\,dF(x).$$

In particular,

$$\Phi(y) = \mathsf{P}\{\eta < y\} = \int \Phi(y/x)\,dF(x). \tag{1}$$

If a probability density function exists for the variable η, then

$$\varphi(y) = \int \varphi(y/x)\,dF(x), \tag{1'}$$

where $\varphi(y/x)$ is the *conditional density function* of the variable η.

Example. As an example illustrative of formula (1), we consider the following problem in the theory of ballistics. In firing at a certain target, two kinds of errors are possible: 1) an error in determining the position of the target; and 2) errors in shooting which occur for a large number of diverse reasons (fluctuations in the amount of charge in the shells, irregularities in the machining of the shell casing, errors due to sighting, small variations in atmospheric conditions, and so on). Errors of the second kind are referred to as the technical dispersion.

Let n independent shots be fired at a single fix of the target position. It is required to find the probability of at least one target hit.

For the sake of simplicity, we shall restrict our discussion to a one-dimensional target of length $2a$, and the shell will be assumed to be a point. Let $f(x)$ denote the density function of the position of the target and $\varphi_i(x)$ the density function of the point of impact of the i-th shell.

If the center of the target is at the point z, then the probability of hitting the target in the i-th round is equal to the probability of a hit in the interval $(z - a, z + a)$, i.e. is equal to[2]

---

[2] It is understood here that the determination of the target position and the technical dispersion are independent.

§ 52. Conditional Distribution Functions and Bayes' Formula

$$\int_{z-\alpha}^{z+\alpha} \varphi_i(x)\, dx.$$

The conditional probability of a miss in the i-th round given that the target center is located at the point z is

$$1 - \int_{z-\alpha}^{z+\alpha} \varphi_i(x)\, dx.$$

The conditional probability of a miss in all n rounds (given the same condition) is

$$\prod_{i=1}^{n}\left(1 - \int_{z-\alpha}^{z+\alpha} \varphi_i(x)\, dx\right).$$

From this we conclude that the probability of at least one hit under the condition that the target center is at the point z is

$$1 - \prod_{i=1}^{n}\left(1 - \int_{z-\alpha}^{z+\alpha} \varphi_i(x)\, dx\right).$$

The unconditional probability of at least one target hit (according to formula (1)) is thus

$$P = \int f(z)\left[1 - \prod_{i=1}^{n}\left(1 - \int_{z-\alpha}^{z+\alpha} \varphi_i(x)\, dx\right)\right] dz.$$

If the firing conditions do not change from round to round, then $\varphi_i(x) = \varphi(x)$ ($i = 1, 2, \ldots, n$) and therefore

$$P = \int f(z)\left[1 - \left(1 - \int_{z-\alpha}^{z+\alpha} \varphi(x)\, dx\right)^n\right] dz.$$

As before, let A_i denote the event that $x_i \leq \xi < x_{i+1}$. According to the classical Theorem of Bayes,

$$\mathsf{P}\{A_i/B\} = \frac{\mathsf{P}\{A_i\}\, \mathsf{P}\{B/A_i\}}{\mathsf{P}\{B\}}.$$

If $F(x) = \mathsf{P}\{\xi < x\}$ and $\mathsf{P}\{\xi < x/B\}$ have continuous derivatives with respect to x, then by making use of the mean-value theorem, we obtain:

$$\mathsf{P}\{A_i/B\} = p_\xi(\bar{x}_i/B)(x_{i+1} - x_i) = \frac{F'(\bar{x}_i')\, \mathsf{P}\{B/A_i\}}{\mathsf{P}\{B\}}(x_{i+1} - x_i),$$

where
$$x_i < \bar{x}_i < x_{i+1} \text{ and } x_i < \bar{x}_i' < x_{i+1}.$$

In the limit, as $x_i \to x$ and $x_{i+1} \to x$, we obtain:
$$p_\xi(x/B) = \frac{p(x)\,\mathsf{P}\{B/x\}}{\mathsf{P}\{B\}},$$
or
$$p_\xi(x/B) = \frac{p(x)\,\mathsf{P}\{B/x\}}{\int \mathsf{P}\{B/x\}\,p(x)\,dx}. \tag{2}$$

It is natural to call this relation *Bayes' formula*.

Suppose now that B is the event that some random variable η assumes a value lying between $y - a$ and $y + \beta$ and that the conditional distribution function $\Phi(y/x)$ of the variable η has a continuous density $p_\eta(y/x)$ for each value of x. Then, as follows from equation (2), if $\mathsf{P}\{B/x\}/(a+\beta)$ approaches $p_\eta(y/x)$ uniformly in x as a and β tend to zero, then the relation
$$p_\xi(x/y) = \frac{p(x)\,p_\eta(y/x)}{\int p_\eta(y/x)\,p(x)\,dx}.$$
holds. This expression will be used extensively in the next chapter.

§ 53. The Generalized Markov Equation

We now pass to an investigation of stochastic processes without aftereffect, and in doing so we confine ourselves to the *simplest* problems only. In particular, we shall assume that the class of possible states of a system is the set of real numbers. Thus, for us, a *stochastic process* will be a collection of random variables $\xi(t)$ depending on a single real parameter t. We shall call the parameter t time, and we shall speak of the state of a system at a given moment of time.

We obtain a complete probabilistic characterization of a process without aftereffect by giving a function $F(t, x; \tau, y)$, the probability that the random variable will assume a value less than y at time τ if it is known that at time t ($t < \tau$) the equation $\xi(t) = x$ holds. Additional knowledge concerning the states of a system at instants of time earlier than t for a process without aftereffect does not change the function $F(t, x; \tau, y)$.

§ 53. GENERALIZED MARKOV EQUATION

Let us now note some conditions that have to be satisfied by the function $F(t, x; \tau, y)$. In the first place, like any distribution function, it has to satisfy the relations

$$\lim_{y \to -\infty} F(t, x; \tau, y) = 0 \quad \text{and} \quad \lim_{y \to +\infty} F(t, x; \tau, y) = 1$$

for arbitrary[3] values of x, t, and τ;

Second, the function $F(t, x; \tau y)$ is continuous from the left in the argument y.

We shall now assume that the function $F(t, x; \tau, y)$ is continuous with respect to t, τ, and x.

Let us consider the instants of time t, s, and τ, with $t < s < \tau$. Since the system passes from the state x at time t to one of the states in the interval $(z, z + dz)$ with probability $d_z F(t, x; s, z)$ and from the state z at time s to a state less than y at time τ with a probability $F(s, z; \tau, y)$, then according to formula (1) of § 52, we find:

$$F(t, x; \tau, y) = \int F(s, z; \tau, y) \, d_z F(t, x; s, z) \,.$$

It is natural to call this equation the *generalized Markov equation*, since it extends identity (1), § 17 in the theory of Markov chains to the theory of stochastic processes; and in this theory, it plays just as important a part as the identity mentioned does in the theory of Markov chains.

Until now, the probability $F(t, x; \tau, y)$ has only been defined for $\tau > t$. We complete the definition by taking

$$\lim_{\tau \to t+0} F(t, x; \tau, y) = \lim_{t \to \tau - 0} F(t, x; \tau, y) = E(x, y) = \begin{cases} 0 & \text{for } y \leq x, \\ 1 & \text{for } y > x. \end{cases}$$

If a density

$$f(t, x; \tau, y) = \frac{\partial}{\partial y} F(t, x; \tau, y)$$

exists, then it satisfies the following obvious equalities:

$$\int_{-\infty}^{y} f(t, x; \tau, z) \, dz = F(t, x; \tau, y) \,,$$

$$\int f(t, x; \tau, z) \, dz = 1 \,.$$

---

[3] We note that the values of the parameter t (time) are usually assigned on the half-line $t \geq t_0$.

In this case, the generalized Markov equation is expressible in the following form:

$$f(t, x; \tau, y) = \int f(s, z; \tau, y) f(t, x; s, z) \, dz.$$

§ 54. Continuous Stochastic Processes. Kolmogorov's Equations

We say that a stochastic process $\xi(t)$ is *continuous* if there is only a small probability that $\xi(t)$ will take on an appreciable increment in a short interval of time. Here, we shall require $\xi(t)$ to be strongly continuous. That is, for any positive constant δ, the relation

$$\lim_{\Delta t \to 0} \frac{1}{\Delta t} \int_{|y-x| \geq \delta} d_y F(t - \Delta t, x; t, y) = 0 \tag{1}$$

holds.

Our next problem is to derive the differential equations which, under certain conditions, are satisfied by the function $F(t, x; \tau, y)$ governing a continuous stochastic process without aftereffect. These equations were first rigorously derived by Kolmogorov (although the second equation was to be found before this in works on Physics), and they are referred to as *Kolmogorov's equations*.

We assume the following:

1) The partial derivatives

$$\frac{\partial F(t, x; \tau, y)}{\partial x} \quad \text{and} \quad \frac{\partial^2 F(t, x; \tau, y)}{\partial x^2}$$

exist and are continuous for arbitrary values of t, x, y and $\tau > t$;

2) For any $\delta > 0$, the limits

$$\lim_{\Delta t \to 0} \frac{1}{\Delta t} \int_{|y-x| < \delta} (y - x) \, d_y F(t - \Delta t, x; t, y) = a(t, x) \tag{2}$$

and

$$\lim_{\Delta t \to 0} \frac{1}{\Delta t} \int_{|y-x| < \delta} (y - x)^2 \, d_y F(t - \Delta t, x; t, y) = b(t, x) \tag{3}$$

§ 54. CONTINUOUS STOCHASTIC PROCESSES. KOLMOGOROV'S EQUATIONS

exist[4] and the convergence is uniform in x.

The left-hand sides of equations (2) and (3) depend on δ. However, in view of the definition of continuity of a process (i.e., in view of (1)), this dependence is merely apparent.

THE FIRST KOLMOGOROV EQUATION: *If conditions* 1) *and* 2) *above are satisfied, then the function* $F(t, x; \tau, y)$ *is a solution of the equation*

$$\frac{\partial F(t, x; \tau, y)}{\partial t} = - a(t, x) \frac{\partial F(t, x; \tau, y)}{\partial x} - \frac{b(t, x)}{2} \frac{\partial^2 F(t, x; \tau, y)}{\partial x^2}. \quad (4)$$

Proof: According to the generalized Markov equation

$$F(t - \Delta t, x; \tau, y) = \int F(t, z; \tau, y) \, d_z F(t - \Delta t, x; t, z).$$

Moreover, by the properties of a distribution function,

$$F(t, x; \tau, y) = \int F(t, x; \tau, y) \, d_z F(t - \Delta t, x; t, z).$$

From these equations, we conclude that

$$\frac{F(t - \Delta t, x; \tau, y) - F(t, x; \tau, y)}{\Delta t} =$$

$$= \frac{1}{\Delta t} \int [F(t, z; \tau, y) - F(t, x; \tau, y)] \, d_z F(t - \Delta t, x; t, z).$$

---

[4] Kolmogorov has shown that the limits $a(t, x)$ and $b(t, x)$ exist under the assumption that the determinant

$$\begin{vmatrix} \frac{\partial}{\partial x} f(s, x, t', y'), & \frac{\partial}{\partial x} f(s, x, t'', y'') \\ \frac{\partial^2}{\partial x^2} f(s, x, t', y'), & \frac{\partial^2}{\partial x^2} f(s, x, t'', y'') \end{vmatrix}$$

does not vanish identically for given values of x and s and arbitrary values of t', t'', y', and y''. By literally repeating Kolmogorov's argument, word for word, we can prove that the existence of the limits $a(t, x)$ and $b(t, x)$ follows from (1) and the assumption that the determinant

$$\begin{vmatrix} \frac{\partial}{\partial x} F(s, x, t', y'), & \frac{\partial}{\partial x} F(s, x, t'', y'') \\ \frac{\partial^2}{\partial x^2} F(s, x, t', y'), & \frac{\partial^2}{\partial x^2} F(s, x, t'', y'') \end{vmatrix}$$

does not vanish identically for given x and s and arbitrary t', t'', y' and y''.

We shall make clear what the physical meaning is of the functions a and b, at the end of this section.

By Taylor's Theorem, the expansion

$$F(t, z; \tau, y) = F(t, x; \tau, y) + (z - x)\frac{\partial F(t, x; \tau, y)}{\partial x}$$
$$+ \frac{1}{2}(z-x)^2 \frac{\partial^2 F(t, x; \tau, y)}{\partial x^2} + o\left((z-x)^2\right)$$

holds under our assumptions. The following analytical transformations require no explanation:

$$\frac{F(t - \Delta t, x; \tau, y) - F(t, x; \tau, y)}{\Delta t} =$$
$$= \frac{1}{\Delta t} \int_{|z-x| \geq \delta} [F(t, z; \tau, y) - F(t, x; \tau, y)] \, d_z F(t - \Delta t, x; t, z) +$$
$$+ \frac{1}{\Delta t} \int_{|z-x| < \delta} [F(t, z; \tau, y) - F(t, x; \tau, y)] \, d_z F(t - \Delta t, x; t, z) =$$
$$= \frac{1}{\Delta t} \int_{|z-x| \geq \delta} [F(t, z; \tau, y) - F(t, x; \tau, y)] \, d_z F(t - \Delta t, x; t, z) +$$
$$+ \frac{\partial F(t, x; \tau, y)}{\partial x} \cdot \frac{1}{\Delta t} \int_{|z-x| < \delta} (z-x) \, d_z F(t - \Delta t, x; t, z) +$$
$$+ \frac{1}{2} \frac{\partial^2 F(t, x; \tau, y)}{\partial x^2} \cdot \frac{1}{\Delta t} \int_{|z-x| < \delta} [(z-x)^2 + o((z-x)^2)] \, d_z F(t - \Delta t, x; t, z). \quad (5)$$

We now pass to the limit, letting $\Delta t \to 0$. The first term on the right-hand side has the limit 0, by virtue of (1). The second term, by (2) is equal to $a(t, x) \, \partial F/\partial x$ in the limit. Finally, the third term can only differ from $2^{-1} b(t, x) \, \partial^2 F/\partial x^2$ by a term which approaches zero as $\delta \to 0$. But since the left-hand side of the last equation is independent of δ, as are the limiting values just indicated, it follows that the limit of the right-hand side exists and is equal to

$$a(t, x) \frac{\partial F(t, x; \tau, y)}{\partial x} + \frac{1}{2} b(t, x) \frac{\partial^2 F(t, x; \tau, y)}{\partial x^2}.$$

Hence, we conclude that the limit

$$\lim_{\Delta t \to 0} \frac{F(t - \Delta t, x; \tau, y) - F(t, x; \tau, y)}{\Delta t} = -\frac{\partial F(t, x; \tau, y)}{\partial t}$$

exists.

Equation (5) thus leads to equation (4).

§ 54. Continuous Stochastic Processes. Kolmogorov's Equations

If it is assumed that a density function

$$f(t, x; \tau, y) = \frac{\partial}{\partial y} F(t, x; \tau, y)$$

exists, then a straightforward differentiation of (4) shows that the density $f(t, x; \tau, y)$ satisfies the equation

$$\frac{\partial f(t, x; \tau, y)}{\partial t} + a(t, x) \frac{\partial f(t, x; \tau, y)}{\partial x} + \frac{1}{2} b(t, x) \frac{\partial^2 f(t, x; \tau, y)}{\partial x^2} = 0. \quad (4')$$

We now proceed to derive the second Kolmogorov equation. In so doing, we shall not strive for the greatest possible generality, and we shall make assumptions that are not mandatory for our argument. Besides the assumptions already made, we impose the following further restrictions on the function $F(t, x; \tau, y)$:

3) A probability density function

$$f(t, x; \tau, y) = \frac{\partial F(t, x; \tau, y)}{\partial y}$$

exists;

4) The derivatives

$$\frac{\partial f(t, x; \tau, y)}{\partial \tau}, \quad \frac{\partial}{\partial y} [a(\tau, y) f(t, x; \tau, y)], \quad \frac{\partial^2}{\partial y^2} [b(\tau, y) f(t, x; \tau, y)]$$

exist and are continuous.

THE SECOND KOLMOGOROV EQUATION:[5] *For any continuous stochastic process without aftereffect satisfying conditions* 1)-4), *the density* $f(t, x; \tau, y)$ *is a solution of the equation*

$$\frac{\partial f(t, x; \tau, y)}{\partial \tau} = -\frac{\partial}{\partial y} [a(\tau, y) f(t, x; \tau, y)] + \frac{1}{2} \frac{\partial^2}{\partial y^2} [b(\tau, y) f(t, x; \tau, y)]. \quad (6)$$

Proof: Let a and b ($a < b$) be certain numbers and $R(y)$ a nonnegative continuous function having continuous derivatives up to and including those of the second order. Moreover, let it be required that

$$R(y) = 0 \quad \text{for} \quad y < a \quad \text{and} \quad y > b.$$

---

[5] The second Kolmogorov equation was obtained earlier by the physicists Fokker and Planck in connection with the development of diffusion theory (see § 50).

X. The Theory of Stochastic Processes

From the condition that the function $R(y)$ and its derivatives are continuous, we conclude that

$$R(a) = R(b) = R'(a) = R'(b) = R''(a) = R''(b) = 0. \qquad (7)$$

We first note that

$$\int_a^b \frac{\partial f(t, x; \tau, y)}{\partial \tau} R(y)\, dy = \frac{\partial}{\partial \tau} \int_a^b f(t, x; \tau, y)\, R(y)\, dy =$$

$$= \lim_{\Delta\tau \to 0} \int \frac{f(t, x; \tau + \Delta\tau, y) - f(t, x; \tau, y)}{\Delta\tau} R(y)\, dy.$$

According to the generalized Markov equation,

$$f(t, x; \tau + \Delta\tau, y) = \int f(t, x; \tau, z)\, f(\tau, z; \tau + \Delta\tau, y)\, dz,$$

and therefore

$$\int_a^b \frac{\partial f(t, x; \tau, y)}{\partial \tau} R(y)\, dy =$$

$$= \lim_{\Delta\tau \to 0} \frac{1}{\Delta\tau} \Bigl[\int\int f(t, x; \tau, z)\, f(\tau, z; \tau + \Delta\tau, y)\, R(y)\, dz\, dy -$$

$$- \int f(t, x; \tau, y)\, R(y)\, dy\Bigr]$$

$$= \lim_{\Delta\tau \to 0} \frac{1}{\Delta\tau} \Bigl[\int f(t, x; \tau, z) \int f(\tau, z; \tau + \Delta\tau, y)\, R(y)\, dy\, dz -$$

$$- \int f(t, x; \tau, y)\, R(y)\, dy\Bigr]$$

$$= \lim_{\Delta\tau \to 0} \frac{1}{\Delta\tau} \int f(t, x; \tau, y) \Bigl[\int f(\tau, y; \tau + \Delta\tau, z)\, R(z)\, dz - R(y)\Bigr] dy.$$

The transformations that have been carried out are obvious: the first time, we interchanged the order of integration, and the second time, we changed the notation for the variables of integration (we replaced y by z and z by y).

By Taylor's Theorem,

$$R(z) = R(y) + (z - y)\, R'(y) + \frac{1}{2}(z - y)^2\, R''(y) + o[(z - y)^2].$$

Since, by the boundedness of the function $R(z)$ and the condition (1),

$$\int_{|y - z| \geq \delta} f(\tau, y; \tau + \Delta\tau, z)\, R(z)\, dz = o(\Delta\tau)$$

and

§ 54. Continuous Stochastic Processes. Kolmogorov's Equations

$$\int_{|y-z|\leq \delta} f(\tau, y; \tau + \varDelta\tau, z) \, dz = 1 + o(\varDelta\tau),$$

it follows that

$$\int f(\tau, y; \tau + \varDelta\tau, z) R(z) \, dz - R(y) =$$

$$= R'(y) \int_{|y-z|<\delta} (z-y) f(\tau, y; \tau + \varDelta\tau, z) \, dz +$$

$$+ \frac{1}{2} R''(y) \int_{|y-z|<\delta} [(z-y)^2 + o((z-y)^2)] f(\tau, y; \tau + \varDelta\tau, z) \, dz + o(\varDelta\tau).$$

Thus,

$$\int_a^b \frac{\partial f(t, x; \tau, y)}{\partial \tau} R(y) \, dy =$$

$$= \lim_{\varDelta\tau \to 0} \frac{1}{\varDelta\tau} \int f(t, x; \tau, y) \Big\{ R'(y) \int_{|y-z|<\delta} (z-y) f(\tau, y; \tau + \varDelta\tau, z) \, dz +$$

$$+ \frac{1}{2} R''(y) \int_{|y-z|<\delta} [(z-y)^2 + o(z-y)^2] f(\tau, y; \tau + \varDelta\tau, z) \, dz + o(\varDelta\tau) \Big\} dy.$$

We now pass to the limit, letting $\varDelta t \to 0$. By the assumption that the limits in (2) and (3) are uniform in x, we conclude that the limit equation is expressible in the form

$$\int_a^b \frac{\partial f(t, x; \tau, y)}{\partial \tau} R(y) \, dy = \int f(t, x; \tau, y) \Big[a(\tau, y) R'(y) + \frac{1}{2} b(\tau, y) R''(y) \Big] dy.$$

Since $R'(y) = R''(y) = 0$ for $y \leq a$ and $y \geq b$, we have

$$\int_a^b \frac{\partial f(t, x; \tau, y)}{\partial \tau} R(y) \, dy = \qquad\qquad\qquad\qquad (8)$$

$$= \int_a^b f(t, x; \tau, y) \Big[a(\tau, y) R'(y) + \frac{1}{2} b(\tau, y) R''(y) \Big] dy.$$

By making use of integration by parts and equations (7), we find:

$$\int_a^b f(t, x; \tau, y) a(\tau, y) R'(y) \, dy = - \int_a^b R(y) \frac{\partial}{\partial y} [a(\tau, y) f(t, x; \tau, y)] \, dy,$$

$$\int_a^b f(t, x; \tau, y) b(\tau, y) R''(y) dy = \int_a^b R(y) \frac{\partial^2}{\partial y^2} [b(\tau, y) f(t, x; \tau, y)] dy.$$

If we now substitute these expressions into (8), we obtain:

$$\int_a^b \frac{\partial f(t, x; \tau, y)}{\partial \tau} R(y) dy =$$

$$= \int_a^b \left\{ -\frac{\partial}{\partial y} [a(\tau, y) f(t, x; \tau, y)] + \frac{1}{2} \frac{\partial^2}{\partial y^2} [b(\tau, y) f(t, x; \tau, y)] \right\} R(y) dy.$$

This equation is clearly expressible in the following form:

$$\int_a^b \left\{ \frac{\partial f(t, x; \tau, y)}{\partial \tau} + \frac{\partial}{\partial y} [a(\tau, y) f(t, x; \tau, y)] - \right.$$

$$\left. - \frac{1}{2} \frac{\partial^2}{\partial y^2} [b(\tau, y) f(t, x; \tau, y)] \right\} R(y) dy = 0. \quad (9)$$

Since the function $R(y)$ is arbitrary, equation (6) follows from this last identity. For, suppose that this were not so. Then there exists a quadruplet of numbers (t, x, τ, y) such that the expression within the braces in (9) is different from zero. Under our assumptions, this expression is a continuous function; therefore, an interval $\alpha < y < \beta$ can be found in which its algebraic sign stays the same. If $a \leq \alpha$ and $b \geq \beta$, then we set $R(y) = 0$ for $y \leq \alpha$ and $y \geq \beta$ and $R(y) > 0$ for $\alpha < y < \beta$. With this choice of $R(y)$, the integral on the left-hand side of equation (9) must differ from zero. Thus, our assumption is false, and equation (6) follows from (9).

Of course, our basic purpose is not to show that some given function $f(t, x; \tau, y)$ satisfies the Kolmogorov equations but rather to seek an unknown function $f(t, x; \tau, y)$ which satisfies these equations for given coefficients $a(t, x)$ and $b(t, x)$. And, in doing this, we do not look for just any solution of the Kolmogorov equations but only for those that satisfy the following conditions:

§ 54. Continuous Stochastic Processes. Kolmogorov's Equations

1. $f(t, x; \tau, y) \geq 0$ for all t, x, τ, y;
2. $\int f(t, x; \tau, y) dy = 1$;

and for any $\delta > 0$,

3. $\lim_{\tau \to t} \int_{|y-x| \geq \delta} f(t, x; \tau, y) dy = 0$.

We shall not stop to set up the conditions that have to be imposed on the functions $a(t, x)$ and $b(t, x)$ in order for a solution of the Kolmogorov equations to exist which would satisfy the above requirements and which would, in addition, be unique.

We now strengthen the continuity conditions slightly in order to give a physical interpretation of the coefficients $a(t, x)$ and $b(t, x)$. Instead of (1), we assume that for any $\delta > 0$ the relation

$$\lim_{\Delta t \to 0} \frac{1}{\Delta t} \int_{|y-x| > \delta} (y-x)^2 \, d_y F(t - \Delta \tau, x; t, y) = 0 \tag{1'}$$

holds. It is easy to see that (1) follows from (1'). Conditions 2. and 3. are now expressible as follows:

$$\lim_{\Delta t \to 0} \frac{1}{\Delta t} \int (y-x) \, d_y F(t - \Delta t, x; t, y) = a(t, x) \tag{2'}$$

and

$$\lim_{\Delta t \to 0} \frac{1}{\Delta t} \int (y-x)^2 \, d_y F(t - \Delta t, x; t, y) = b(t, x). \tag{3'}$$

The remaining requirements, as well as our final conclusions, are not changed by the replacement of (1) by (1'). Since

$$\int (y-x) \, d_y F(t - \Delta t, x; t, y) = \mathsf{E}\left[\xi(t) - \xi(t - \Delta t)\right]$$

is the expectation of the increment in $\xi(t)$ in time Δt and

$$\int (y-x)^2 \, d_y F(t - \Delta t, x; t, y) = \mathsf{E}\left[\xi(t) - \xi(t - \Delta t)\right]^2$$

is the expectation of the square of this increment, and is therefore proportional to the kinetic energy (assuming that $\xi(t)$ is the coordinate of the position of a point moving about as a result of chance factors), it is clear from (2') and (3') that $a(t, x)$ is the average rate of change of $\xi(t)$ and that $b(t, x)$ is proportional to the average kinetic energy of our system.

We conclude this section with a discussion of a special case of the Kolmogorov equations, the case where the function $f(t, x; \tau, y)$ depends on t, τ, and $y - x$, but not on x and y themselves. Physically, this means that the process is spatially homogeneous: the probability of the increment $\Delta = y - x$ is independent of what position x the system was in at the instant t. In this case, the functions $a(t, x)$ and $b(t, x)$ are clearly independent of x and are functions of the single variable t:

$$a(t) = a(t, x) \quad \text{and} \quad b(t) = b(t, x).$$

For the case in question, the Kolmogorov equations can be rewritten in the following form:

$$\left.\begin{aligned}\frac{\partial f}{\partial t} &= -a(t)\frac{\partial f}{\partial x} - \frac{1}{2}b(t)\frac{\partial^2 f}{\partial x^2}, \\ \frac{\partial f}{\partial \tau} &= -a(\tau)\frac{\partial f}{\partial y} + \frac{1}{2}b(\tau)\frac{\partial^2 f}{\partial y^2}.\end{aligned}\right\} \quad (11)$$

Let us first consider the special case $a(t) = 0$, $b(t) = 1$. For this, equations (11) reduce to the heat equation

$$\frac{\partial f}{\partial \tau} = \frac{1}{2}\frac{\partial^2 f}{\partial y^2}$$

and its adjoint (12)

$$\frac{\partial f}{\partial t} = -\frac{1}{2}\frac{\partial^2 f}{\partial x^2}.$$

From the general theory of the equation of heat conduction, it is known that the unique solution of these equations satisfying conditions (10) is the function

$$f(t, x; \tau, y) = \frac{1}{\sqrt{2\pi(\tau - t)}} e^{-\frac{(y-x)^2}{2(\tau-t)}}.$$

By the change of variables

$$x' = x - \int_a^t a(z)\,dz, \quad y' = y - \int_a^\tau a(z)\,dz,$$

$$t' = \int_a^t b(z)\,dz, \quad \tau' = \int_a^\tau b(z)\,dz,$$

equations (11) reduce to equations (12). This enables us to express the required solution of equations (11) in the form

$$f(t, x;\ \tau, y) = \frac{1}{\sigma \sqrt{2\pi}} e^{-\frac{(y-x-A)^2}{2\sigma^2}},$$

where

$$A = \int_t^\tau a(z)\, dz, \qquad \sigma^2 = \int_t^\tau b(z)\, dz.$$

§ 55. Purely Discontinuous Stochastic Processes. The Kolmogorov-Feller Equations

In present-day physics, an important part is played by processes in which the changes in a system take place not continuously, but by jumps. Examples of problems of this sort were given in § 50.

We shall say that a stochastic process $\xi(t)$ is *purely discontinuous* if during any interval of time $(t, t + \Delta t)$ the value of the variable $\xi(t)$ remains equal to x with probability $1 - p(t, x)\Delta t + o(\Delta t)$ and may undergo a change only with probability $p(t, x)\Delta t + o(\Delta t)$. (We assume here that the probability of more than one change in $\xi(t)$ in the interval Δt is $o(\Delta t)$.) Of course, inasmuch as we are restricting ourselves to processes without aftereffect, the distribution function for the changes in $\xi(t)$ that occur after a jump does not depend on what values $\xi(t)$ had at times prior to the jump.

Let $P(t, x, y)$ denote the conditional distribution function of $\xi(t)$ under the condition that a jump has occurred at time t and that immediately before the jump $\xi(t)$ was equal to x, i.e., $\xi(t-0) = x$.

The distribution function $F(t, x;\ \tau, y)$ is readily expressible, as follows, in terms of the functions $p(t, x)$ and $P(t, x, y)$:

$F(t, x;\ \tau, y) =$

$$= [1 - p(t, x)(\tau - t)] E(x, y) + (\tau - t) p(t, x) P(t, x, y) + o(\tau - t). \quad (1)$$

According to definition, the functions $p(t, x)$ and $P(t, x, y)$ are non-negative and $P(t, x, y)$, like any distribution function, satisfies the conditions

$$P(t, x, -\infty) = 0,\ P(t, x, +\infty) = 1.$$

In addition, we assume that $p(t, x)$ is bounded and that both $p(t, x)$ and $P(t, x, y)$ are continuous in t and x (actually, it suffices to assume that they are Borel measurable in x).

X. THE THEORY OF STOCHASTIC PROCESSES

We make no assumptions concerning the function $F(t, x; \tau, y)$ except to retain its definition at $t = \tau$:

$$\lim_{\tau \to t+0} F(t, x; \tau, y) = \lim_{t \to \tau-0} F(t, x; \tau, y) = E(x, y) = \begin{cases} 0 & \text{for } y \leq x, \\ 1 & \text{for } y > x. \end{cases}$$

One of the aims of the present section is to prove the following theorem.

THEOREM: *The distribution function $F(t, x; \tau, y)$ for a purely discontinuous process without aftereffect satisfies the following two integro-differential equations*:

$$\frac{\partial F(t, x; \tau, y)}{\partial t} = p(t, x) [F(t, x; \tau, y) - \int F(t, z; \tau, y) d_z P(t, x, z)], \quad (2)$$

$$\frac{\partial F(t, x; \tau, y)}{\partial \tau} =$$

$$= - \int_{-\infty}^{y} p(t, z) d_z F(t, x; \tau, y) + \int p(\tau, z) P(\tau, z, y) d_z F(t, x; \tau, z). \quad (3)$$

Equation (2) was obtained by Kolmogorov in 1931; both equations (2) and (3) were derived by Feller in 1937 under our assumptions. These facts make it natural to call equations (2) and (3) the *Kolmogorov-Feller* equations.

Proof: By virtue of the generalized Markov equation,

$$F(t, x; \tau, y) = \int F(t + \Delta t, z; \tau, y) d_z F(t, x; t + \Delta t, z).$$

Substituting the value of $F(t, x; t + \Delta t, z)$ given by formula (1), we find:

$$F(t, x; \tau, y) = \int F(t + \Delta t, z; \tau, y) d_z[1 - p(t, x) \Delta(t) + o(\Delta t)] E(x, z) +$$
$$+ \int F(t + \Delta t, z; \tau, y) d_z [p(t, x) \Delta t + o(\Delta t)] P(t, x, z).$$

Since

$$\int F(t + \Delta t, z; \tau, y) d_z E(x, z) = F(t + \Delta t, x; \tau, y),$$

we have

$$F(t, x; \tau, y) = [1 - p(t, x) \Delta t] F(t + \Delta t, x; \tau, y) +$$
$$+ \Delta t\, p(t, x) \int F(t + \Delta t, z; \tau, y) d_z P(t, x, z) + o(\Delta t).$$

Hence

$$\frac{F(t + \Delta t, x; \tau, y) - F(t, x; \tau, y)}{\Delta t} = p(t, x) F(t + \Delta t, x; \tau, y) -$$
$$- p(t, x) \int F(t + \Delta t, z; \tau, y) d_z P(t, x, z) + o(1).$$

Passing to the limit, we arrive at (2).

§ 55. Purely Discontinuous Stochastic Processes

The Markov equation, equation (1), and also the definition of the function $E(x, z)$ enable us to write the following chain of equalities:

$$F(t, x; \tau + \Delta\tau, y) = \int F(\tau, z; \tau + \Delta\tau, y) \, d_z F(t, x; \tau, z) =$$

$$= \int \{[1 - p(\tau, z) \Delta\tau] E(z, y) + \Delta\tau \, p(\tau, z) P(\tau, z, y) +$$

$$+ o(\Delta\tau)\} d_z F(t, x; \tau, z) =$$

$$= \int_{-\infty}^{y} d_z F(t, x; \tau, z) - \Delta\tau \int_{-\infty}^{y} p(\tau, z) \, d_z F(t, x; \tau, z) +$$

$$+ \Delta t \int p(\tau, z) P(\tau, z, y) \, d_z F(t, x; \tau, z) + o(\Delta\tau).$$

By standard reasoning, it follows from this that the derivative $\partial F/\partial \tau$ exists and that equation (3) holds.

We shall solve one further problem important in applications: What is the probability of a system changing its state n times ($n = 0, 1, 2, \ldots$) in an interval of time from t to τ ($\tau > t$)?

Let $p_n(t, x, \tau)$ denote the probability that starting from the state x at time t, the system changes its state n times by the time τ. We begin the solution of the problem with the case $n = 0$.

To this end, we write down the following equation:

$$p_0(t, x, \tau) = p_0(t, x, \tau + \Delta\tau) + p_0(t, x, \tau)[1 - p_0(\tau, x, \tau + \Delta\tau)], \quad (4)$$

which expresses the fact that the absence of changes in state of the system in an interval of time (t, τ) can occur in the following two mutually exclusive ways: 1) The system has not changed its state in a greater interval of time $(t, \tau + \Delta\tau)$; 2) The system has not changed its state up to the time τ but in the interval of time $(\tau, \tau + \Delta\tau)$ it has done so. Since, by the definition of a purely discontinuous process,

$$p_0(\tau, x, \tau + \Delta\tau) = 1 - p(\tau, x) \Delta\tau + o(\Delta\tau),$$

equation (4) can also be expressed as:

$$\frac{p_0(t, x, \tau + \Delta\tau) - p_0(t, x, \tau)}{\Delta\tau} = -p_0(t, x, \tau) p(\tau, x) + o(1).$$

By letting $\Delta t \to 0$, we find from this that the derivative $\partial p_0(t, x, \tau)/\partial \tau$ exists and that

$$\frac{\partial p_0(t, x, \tau)}{\partial \tau} = - p_0(t, x, \tau)\, p(\tau, x).$$

Solving this differential equation, we obtain:

$$p_0(t, x, \tau) = C e^{-\int_t^\tau p(u, x)\, du}.$$

Since

$$p_0(\tau, x, \tau) = 1,$$

it follows that $C = 1$ and

$$p_0(t, x, \tau) = e^{-\int_t^\tau p(u, x)\, du}. \qquad (5)$$

We shall now see that a knowledge of $p_0(t, x, \tau)$ and the function $P(t, x, y)$ defined previously enables us to compute the probability $p_n(t, x, \tau)$ for any n. In fact, an n-fold change of state occurs in the following way:

1) Up to the instant s $(t < s < \tau)$, the system does not undergo a change of state (the probability of this event is $p_0(t, x, s)$);

2) In the interval $(s, s + \Delta s)$, the system changes state (the probability of this is $p_1(s, x, s + \Delta s) = p(s, x)\Delta s + o(\Delta s)$);

3) The probability that the new state of the system will be between y and $y + \Delta y$ is

$$P(s, x, y + \Delta y) - P(s, x, y) = \Delta_y P(s, x, y).$$

4) Finally, in the interval of time $(s + \Delta s, \tau)$, the system changes its state $n - 1$ times (the probability of this event is $p_{n-1}(s + \Delta s, y, \tau)$).

By the Multiplication Theorem, the probability that all four of these events will occur is

$$p_0(t, x, s)\,[p(s, x) + o(1)]\,\Delta s \cdot \Delta_y P(s, x, y) \cdot p_{n-1}(s + \Delta s, y, \tau).$$

Since s and y are arbitrary $(t < s < \tau$ and $-\infty < y < \infty)$, from the formula on total probability it follows that

$$p_n(t, x, \tau) = \int_t^\tau \int p_0(t, x, s)\, p(s, x)\, p_{n-1}(s, y, \tau)\, d_y P(s, x, y)\, ds$$

$$= \int_t^\tau p_0(t, x, s)\, p(s, x) \int p_{n-1}(s, y, \tau)\, d_y P(s, x, y)\, ds. \qquad (6)$$

§ 55. Purely Discontinuous Stochastic Processes

Hence, in particular,

$$p_1(t, x, \tau) = \int_t^\tau p_0(t, x, s)\, p(s, x) \int p_0(s, y, \tau)\, d_y P(s, x, y)\, ds. \tag{7}$$

The method for determining $p_n(t, x, \tau)$ is obvious: by formula (5) we find $p_0(t, x, \tau)$; by formula (7) we compute $p_1(t, x, \tau)$; and then, successively, $p_2(t, x, \tau)$, $p_3(t, x, \tau)$, ..., and finally, $p_n(t, x, \tau)$.

Example 1. Let the number of changes of state in the time from 0 up to τ be our variable $\xi(t)$. Under the assumption that $p(t, x) = a$, where a is a positive constant, find $p_n(t, x, \tau)$.

In this case, the possible states of the system will be precisely the non-negative integers ($x = 0, 1, 2, ...$). Since with each change of state, the variable $\xi(t)$ is increased by exactly 1, we have

$$P(t, x, y) = \begin{cases} 0 & \text{for } y \leq x+1, \\ 1 & \text{for } y > x+1. \end{cases}$$

By equation (5), we have:

$$p_0(t, x, \tau) = e^{-a(\tau-t)}.$$

According to (7),

$$p_1(t, x, \tau) = \int_t^\tau p_0(t, x, s)\, p(s, x)\, p_0(s, x+1, \tau)\, ds$$

$$= a \int_t^\tau e^{-(s-t)a} e^{-(\tau-s)a}\, ds = a(\tau-t)\, e^{-a(\tau-t)}.$$

By equation (6),

$$p_2(t, x, \tau) = \int_t^\tau p_0(t, x, s)\, p(s, x)\, p_1(s, x+1, \tau)\, ds$$

$$= \frac{[a(\tau-t)]^2}{2!} e^{-a(\tau-t)}.$$

Let us now suppose that

$$p_{n-1}(t, x, \tau) = \frac{[a(\tau-t)]^{n-1}}{(n-1)!} e^{-a(\tau-t)}.$$

Then by equation (6),

$$p_n(t, x, \tau) = \int_t^\tau p_0(t, x, s)\, p(s, x)\, p_{n-1}(s, x+1, \tau)\, ds$$

$$= \int_t^\tau \frac{a[a(\tau-s)]^{n-1}}{(n-1)!} e^{-a(\tau-t)}\, ds = \frac{[a(\tau-t)]^n}{n!} e^{-a(\tau-t)}.$$

This shows that for any integer $n \geqq 0$

$$p_n(t, x, \tau) = \frac{[a(\tau-t)]^n}{n!} e^{-a(\tau-t)}.$$

The solution of our problem is thus the Poisson law. In particular,

$$p_n(0, 0, \tau) = \frac{(a\tau)^n}{n!} e^{-a\tau}.$$

It is easy to show that the function

$$F(t, x; \tau, y) = \begin{cases} 0 & \text{for } y \leqq 0, \\ \sum_{n<y} \frac{[a(\tau-t)]^n}{n!} e^{-a(\tau-t)} & \text{for } y > 0 \end{cases}$$

is the solution of the integro-differential equations (2) and (3).

Example 2. At the time $t = 0$, there are N radioactive atoms. The probability of decay of an atom in the interval of time $(t, t + \Delta t)$ is $aN(t)\Delta t + o(\Delta t)$, where a is a positive constant and $N(t)$ is the number of atoms which have not decayed up to the time t. Find the probability that n atoms will decay in the interval of time t to τ.[6]

This is a typical purely discontinuous stochastic process. The quantity n can of course only assume the values $0, 1, 2, \ldots, N(t)$.

By the condition of the problem,

$$p(t, x) = \begin{cases} 0 & \text{for } x \leqq 0 \text{ and } x \geqq N, \\ a(N-x) & \text{for } 0 < x \leqq N \end{cases}$$

and

$$P(t, x, y) = \begin{cases} 0 & \text{for } y \leqq x+1, \\ 1 & \text{for } y > x+1. \end{cases}$$

---

[6] We assume here that the products of atomic disintegration do not decay any further and, at any rate, that they have no effect on the atoms which are still decaying.

§ 55. Purely Discontinuous Stochastic Processes

To begin with, we evaluate the probability of n disintegrations during the interval of time from 0 to τ.

By equation (5),

$$p_0(0, 0, \tau) = e^{-\int_0^\tau p(t, 0)\, dt} = e^{-aN\tau}.$$

Similarly,

$$p_0(t, k, \tau) = e^{-a(N-k)(\tau-t)}.$$

Further, by equation (7),

$$p_1(0, 0, \tau) = \int_0^\tau p_0(0, 0, s)\, p(s, 0)\, p_0(s, 1, \tau)\, ds$$

$$= \int_0^\tau e^{-aNs}\, a\, N\, e^{-a(N-1)(\tau-s)}\, ds =$$

$$= N e^{-aN\tau} \int_0^\tau a\, e^{a(\tau-s)}\, ds = N e^{-aN\tau} [e^{a\tau} - 1]. \tag{8}$$

By equation (6), it is easy to determine in succession $p_2(0, 0, \tau)$, $p_3(0, 0, \tau)$, etc., and to prove that

$$p_n(0, 0, \tau) = \binom{N}{n} e^{-aN\tau} [e^{a\tau} - 1]^n. \tag{9}$$

We leave this to the reader.

The equation

$$p_n(t, k, \tau) = \binom{N-k}{n} e^{-a(N-k)(\tau-t)} [e^{a(\tau-t)} - 1]^n. \tag{9'}$$

is clearly valid for $0 \leq n \leq N - k$.

We can now proceed to determine the probability we are seeking, which we shall denote by $p_n(t, \tau)$. By the formula on total probability and by use of (9) and (9'), we find:

$$p_n(t, \tau) = \sum_{k=0}^{N-n} p_k(0, 0, t) \cdot p_n(t, k, \tau)$$

$$= \sum_{k=0}^{N-n} \binom{N}{k} e^{-aNt} [e^{at} - 1]^k \binom{N-k}{n} e^{-a(N-k)(\tau-t)} [e^{a(\tau-t)} - 1]^n$$

$$= e^{-aN\tau} [e^{a(\tau-t)} - 1]^n \sum_{k=0}^{N-n} \binom{N}{k} \binom{N-k}{n} e^{ak(\tau-t)} [e^{at} - 1]^k.$$

Since
$$\binom{N}{k}\binom{N-k}{n} = \binom{N}{n}\binom{N-n}{k}$$
and
$$\sum_{k=0}^{N-n}\binom{N-n}{k}[e^{a(\tau-t)}(e^{at}-1)]^k = [1 + e^{a\tau} - e^{a(\tau-t)}]^{N-n},$$
we finally have
$$p_n(t, \tau) = \binom{N}{n}[e^{-at} - e^{-a\tau}]^n [e^{-a\tau} + 1 - e^{-at}]^{N-n}.$$

It is easy to show that the function
$$F(t, x;\, \tau, y) = \begin{cases} 0 & \text{for } y \leqq x, \\ \sum_{n<y} p_n(t, x, \tau) & \text{for } y < N - x, \\ 1 & \text{for } y > N - x \end{cases}$$
is the solution of the integro-differential equations (2) and (3).

§ 56. Homogeneous Stochastic Processes with Independent Increments

We now consider an important class of stochastic processes a complete characterization of which can be given in terms of characteristic functions.

By a *homogeneous stochastic process with independent increments* is meant the aggregate of random variables $\xi(t)$ depending on a single real parameter t and satisfying the following two conditions:

1) The distribution function of the quantity $\xi(t + t_0) - \xi(t_0)$ is independent of t_0 (the process is homogeneous in time);

2) For any finite number of non-overlapping intervals (a, b) of the parameter t the increments of the variable $\xi(t)$, i.e., the differences $\xi(b) - \xi(a)$, are mutually independent (the increments are independent).

Before proceeding to obtain specific results, we shall discuss a few examples. In these examples, the above conditions may be taken as a

§ 56. HOMOGENEOUS PROCESSES WITH INDEPENDENT INCREMENTS

working hypothesis. Of course, their admissibility can only be justified by the agreement that exists between the theoretical deductions and experiment.

Example 1. *Diffusion of a gas.* Consider a molecule of a certain gas moving among other molecules of the same gas under conditions of constant temperature and density. Let us introduce into space a rectangular coordinate system, and let us observe how one of the coordinates of the position of the molecule in question, say the x-coordinate, changes with time.

In consequence of the random collisions that the given molecule has with the other molecules, this coordinate will change with time by undergoing random increments. The requirement that the gas be under constant conditions itself clearly implies that the process under study is homogeneous in time. In view of the large number of molecules in motion and of the weak dependence of their motion, the process is one with independent increments.

Example 2. *Velocity of a molecule.* Again consider a molecule of a certain gas which is moving in a volume filled with the molecules of this or another gas under conditions of constant density and temperature. Let us again introduce a rectangular coordinate system into the space, and let us follow how one of the components of velocity with respect to these axes changes with time. As it moves, the molecules will undergo random collisions with the other molecules. In consequence of these collisions, the velocity component will acquire random increments. We again have a homogeneous stochastic process with independent increments.

Example 3. *Radioactive decay.* It is well known that radioactivity is the phenomenon in which the atoms of a substance are converted into those of some other substance with the concomitant liberation of a considerable amount of energy. Observations of comparatively large masses of a radioactive substance show that the various atoms decay independently of one another, so that the number of atomic disintegrations in nonoverlapping intervals of time are mutually independent. Moreover, the probability that a certain number of disintegrations will occur in an interval of time of given length depends on the length of this interval but is practically independent of where the interval is located in time. In reality, of course, the radioactivity of a substance gradually decreases as its mass decreases. However, for comparatively small intervals of time (and not too great an amount of substance) this change is so insignificant that it is possible to neglect it entirely.

One can readily cite many other examples in which a natural phenomenon or technical process can be considered to be a homogeneous process with independent increments. We indicate a few more examples: cosmic radiation (the number of cosmic particles that fall on a given area in a given interval of time), the breaking of yarn on a ring spinning frame, the work load of a telephone operator (the number of calls in a given interval of time),[7] and so on.

We now proceed to examine the characteristic property of homogeneous stochastic processes with independent increments.

Let the distribution function of the increment in the variable $\xi(t)$ for an interval of time of length τ be denoted by $F(x, \tau)$. Then, if the intervals of time of length τ_1 and τ_2 do not overlap, it follows that

$$F(x;\ \tau_1 + \tau_2) = \int F(x - y;\ \tau_1)\, d_y F(y, \tau_2). \tag{1}$$

If $f(z, \tau)$ is the characteristic function, i.e., if

$$f(z, \tau) = \int e^{izx}\, d_x F(x;\ \tau),$$

then, in terms of characteristic functions, equation (1) takes on the following form:

$$f(z;\ \tau_1 + \tau_2) = f(z, \tau_1) \cdot f(z, \tau_2). \tag{1'}$$

In general, if the intervals $\tau_1, \tau_2, \ldots, \tau_n$ are non-overlapping, then

$$f(z;\ \sum_{k=1}^{n} \tau_k) = \prod_{k=1}^{n} f(z;\ \tau_k).$$

In particular, if $\tau_1 = \tau_2 = \ldots = \tau_n$ and

$$\sum_{k=1}^{n} \tau_k = \tau,$$

then

$$f(z, \tau) = \left[f\left(z;\ \frac{\tau}{n}\right) \right]^n.$$

---

[7] See Chapter X of *Probability and its Engineering Uses*, by C. T. Fry (New York, 1928), one of the pioneering works in the field. [*Trans.*]

§ 56. Homogeneous Processes with Independent Increments

Thus, *the distribution function of any homogeneous stochastic process with independent increments is infinitely divisible.*

It should be noted that infinitely divisible distribution laws have come to be considered in probability theory owing to the study of homogeneous processes with independent increments. We have seen that the theory of infinitely divisible distribution laws has had a decisive effect on the solution and development of the classical problems of probability theory concerning sums of random variables. Whereas the earlier efforts of mathematicians, as we have indicated, were concentrated on determining the broadest conditions under which the law of large numbers was true and under which normalized sums would converge to the normal law, the general questions which were considered in the preceding chapter arose, of course, once Kolmogorov had completely described the laws governing homogeneous stochastic processes without aftereffect. The fundamental distribution laws which were obtained earlier as asymptotic laws are here found to play the part of exact solutions of the corresponding functional equations in the theory of stochastic processes. Furthermore, this new point of view has enabled us to explain the reason why only two limit distributions were considered in the classical theory of probability, namely, the normal law and the Poisson law.

Since for arbitrary $\tau > 0$

$$f(z, \tau) = [f(z, 1)]^\tau$$

for homogeneous processes with independent increments, it follows that such processes are completely determined by the assignment of the characteristic function of the quantity $\xi(1) - \xi(0)$. In § 45, we saw that the equation

$$\log f(z, 1) = i\gamma z + \int \{e^{izu} - 1 - izu\} \frac{1}{u^2} dG(u) \qquad (2)$$

holds for any infinitely divisible law with a finite variance, where γ is a real constant and $G(u)$ is a non-decreasing function of bounded variation. We shall confine ourselves to the consideration of this special case of a homogeneous process.

Let us introduce the following notation:

$$M(u) = \int_{-\infty}^{u} \frac{1}{x^2} dG(x) \quad \text{for} \quad u < 0,$$

$$N(u) = -\int_{u}^{\infty} \frac{1}{x^2} dG(x) \quad \text{for} \quad u > 0,$$

$$\sigma^2 = G(+0) - G(-0);$$

formula (2) then takes the form:

$$\log f(z, 1) = i\gamma z - \frac{\sigma^2 z^2}{2} + \int_{-\infty}^{0} \{e^{izu} - 1 - izu\} dM(u) +$$

$$+ \int_{0}^{\infty} \{e^{izu} - 1 - izu\} dN(u). \quad (2')$$

We shall now make clear what the probability-theoretic meaning is of the functions $M(u)$ and $N(u)$.

In deriving the canonical representation of infinitely divisible laws in § 45, we introduced the function

$$G_n(u) = n \int_{-\infty}^{u} x^2 \, d\Phi_n(x).$$

Let us set

$$M_n(u) = \int_{-\infty}^{u} \frac{1}{x^2} dG_n(x) = n\Phi_n(u) \quad \text{for} \quad u < 0$$

and

$$N_n(u) = -\int_{u}^{\infty} \frac{1}{x^2} dG_n(x) = -n[1 - \Phi_n(u)] \quad \text{for} \quad u > 0.$$

Since

$$G_n(u) \to G(u)$$

at the points of continuity of the function $G(u)$ as $n \to \infty$, we conclude from the Second Helly Theorem that

$$M_n(u) = n\Phi_n(u) \to M(u)$$

at the points of continuity of the function $M(u)$.

§ 56. HOMOGENEOUS PROCESSES WITH INDEPENDENT INCREMENTS

From the point of view of stochastic processes, $\Phi_n(x)$ ($x < 0$) is the probability that the random variable $\xi(\tau)$ will acquire a decrement larger in absolute value than x in the interval $k/n \leq \tau < (k+1)/n$. Thus, $M_n(x)$ is the sum over all k from 0 to $n-1$ of the probabilities that the variable $\xi(\tau)$ will acquire a decrement by means of jumps each larger than x in absolute value as the parameter τ varies over the intervals $(k/n, (k+1)/n)$. Since $M(u)$ and $N(u)$ are the limits of the respective functions $M_n(u)$ and $N_n(u)$ as $n \to \infty$, they are referred to as *jump functions*.

If $M(u) \equiv 0$ for $u < 0$ and $N(u) \equiv 0$ for $u > 0$, i.e., if there are no jump functions, then it is obvious from equation (2') that, in this case, the stochastic process is governed by the normal law. We thus see that a stochastic process governed by the normal law is continuous in the sense of probability theory. We shall now prove a stronger result.

THEOREM: *A homogeneous stochastic process with independent increments and a finite variance*[8] *is governed by the normal law*[9] *if and only if, for arbitrary $\varepsilon > 0$, the probability of the maximum of the absolute value of the increments in $\xi(\tau)$ for the intervals $((k-1)/n, k/n)$ ($k = 1, 2, 3, \ldots, n$) exceeding ε tends to zero with $1/n$.*[10]

Proof: We have just seen that a homogeneous stochastic process with independent increments is governed by the normal law if and only if for $x > 0$

$$M(-x) \equiv N(x) \equiv 0. \tag{3}$$

Since

$$M(u) = \lim_{n \to \infty} M_n(u) \text{ and } N(u) = \lim_{n \to \infty} N_n(u),$$

condition (3) is equivalent to the following:

$$\lim_{n \to \infty} n\, \Phi_n(-u) \equiv \lim_{n \to \infty} n\, [1 - \Phi_n(u)] \equiv 0. \tag{4}$$

Let ξ_{nk} denote the increment in $\xi(\tau)$ in the interval $((k-1)/n, k/n)$; then

$$p_{nk} = \Phi_n(-x) + 1 - \Phi_n(x+0) = P\{|\xi_{nk}| > x\}.$$

Clearly, relations (4) are equivalent to the following:

---

[8] The theorem is true even without the assumption that the variance exists.
[9] In particular, by the normal law with zero variance, i.e., by a distribution of the form $F(x) = 0$ for $x \leq a$ and $F(x) = 1$ for $x > a$.
[10] Thus, processes which are governed by the normal law, and such processes only are "uniformly continuous" in a probability-theoretic sense.

From the inequality

$$\lim_{n\to\infty} \sum_{k=1}^{n} p_{nk} = 0.$$

From the inequality

$$1 - \sum_{k=1}^{n} p_{nk} \leq \prod_{k=1}^{n}(1 - p_{nk}) \leq e^{-\sum_{k=1}^{n} p_{nk}} \leq 1,$$

we see that relations (4) are equivalent to the statement

$$\lim_{n\to\infty} \prod_{k=1}^{n}(1 - p_{nk}) = 1,$$

which signifies that the probability of the inequalities $|\xi_{nk}| < \varepsilon$ being realized for all k ($1 \leq k \leq n$) approaches one as $n \to \infty$. In other words, we have shown that relations (3) hold if and only if

$$\mathsf{P}\{\max_{1 \leq k \leq n} |\xi_{kn}| \geq \varepsilon\} \to 0$$

as $n \to \infty$, Q.E.D.

§ 57. The Concept of a Stationary Stochastic Process. Khintchine's Theorem on the Correlation Coefficient

The Markov processes, or processes without aftereffect, which we examined in the preceding sections by no means exhaust all of the questions science poses to the theory of probability. In fact, in many situations the past states of a system have a very influential effect on the probability of its future states, and the influences of the past cannot be neglected even in an approximate treatment of the matter. In principle, such a situation could be corrected by changing our conception of the state of the system by the introduction of new parameters. Thus, for example, if we were to consider the change in position of a particle in diffusion phenomena or Brownian motion as a process without aftereffect, this would then mean that we would not be taking the inertia of the particle into account; and the inertia, of course, plays an essential role in these phenomena. By including in our conception of the state of the system the velocity of the particle in addition to the coordinates of its position, we would correct the situation in this example. However, there are cases where such a correction does not facilitate the solution of

§ 57. Stationary Stochastic Process. Khintchine's Theorem

the problems that have been posed. In the first place, one could point to statistical mechanics, in which specifying that a particle is in a certain cell of phase space merely yields a probabilistic judgment concerning its future state. Here, a knowledge of the previous positions of the particle changes our judgments essentially concerning its future state. Khintchine has singled out, in this connection, an important class of stochastic processes with aftereffect, the so-called *stationary processes*, which proceed homogeneously with time.

A stochastic process $\xi(t)$ is called *stationary* if the n-dimensional distribution functions for the two finite groups of variables $\xi(t_1)$, $\xi(t_2)$, $\xi(t_3)$, ..., $\xi(t_n)$ and $\xi(t_1 + u)$, $\xi(t_2 + u)$, ..., $\xi(t_n + u)$ coincide, which means that they do not depend on u. Here, the numbers n and u as well as the moments of time $t_1, t_2, \ldots, t_n$ may be chosen quite arbitrarily.

If we introduce the notation

$$F(x_1, x_2, \ldots, x_n; t_1, t_2, \ldots, t_n) = \mathsf{P}\{\xi(t_1) < x_1, \xi(t_2) < x_2, \ldots, \xi(t_n) < x_n\}, \quad (1)$$

then by the definition just given, the equation

$$F(x_1, x_2, \ldots, x_n; t_1 + u, t_2 + u, \ldots, t_n + u) = F(x_1, x_2, \ldots, x_n; t_1, t_2, \ldots, t_n) \quad (1')$$

holds for any values of u and n.

Clearly, the distribution function of any stochastic process, i.e., $F(x_1, x_2, \ldots, x_n; t_1, t_2, \ldots, t_n)$, has to satisfy the following two conditions:

1) *The symmetry condition*: For any permutation $i_1, i_2, \ldots, i_n$ of the numbers $1, 2, \ldots, n$ the equation

$$F(x_{i_1}, x_{i_2}, \ldots, x_{i_n}; t_{i_1}, t_{i_2}, \ldots, t_{i_n}) = F(x_1, x_2, \ldots, x_n; t_1, t_2, \ldots, t_n)$$

holds;

2) *The compatibility condition*: If $m < n$, then for any $t_{m+1}, t_{m+2}, t_{m+3}, \ldots, t_n$

$$F(x_1, x_2, \ldots, x_m; t_1, t_2, \ldots, t_m)$$
$$= F(x_1, x_2, \ldots, x_m; \infty, \ldots, \infty; t_1, t_2, \ldots, t_m, t_{m+1}, \ldots, t_n). \quad (2)$$

In recent years, the theory of stationary processes has found considerable application in physics and engineering.

For example, the study of a number of acoustical phenomena, including those encountered in radio engineering (e.g. random noise), as well as the search for hidden periodicities of interest to astronomers, geophysicists, and meteorologists lead to stationary processes.

Often, in a steady technological process, one can readily observe occurrences which proceed according to the scheme of a stationary process. As an example, we discuss the process of spinning. Considerable inhomogeneity in the properties of the materials being spun (the length of the fibres, their strength, the transverse cross-sectional size, and so on), fluctuations in the rate and uniformity of feeding the material onto the machines at various stages of the spinning process, and many other reasons lead to variations in the properties of the yarn from one cross-section to the next. It turns out here that a knowledge of the various properties of the yarn in some particular portion of a skein does not give us a complete knowledge of its properties in any other portion of the skein. But since the spinning process may be regarded as steady, the probabilistic characterization of the quality of the yarn is a stationary process.

Clearly, any numerical characteristic of a stationary process $\xi(t)$ is independent of t and if, for example, $\xi(t)$ possesses a finite variance, then it is obvious that the following equations are valid:

$$\mathsf{E}\xi(t+u) = \mathsf{E}\xi(t) = \mathsf{E}\xi(0) = a,$$
$$\mathsf{D}\xi(t+u) = \mathsf{D}\xi(t) = \mathsf{D}\xi(0) = \sigma^2,$$
$$\mathsf{E}\{\xi(t+u)\,\xi(t)\} = \mathsf{E}\{\xi(u)\,\xi(0)\}.$$

This fact allows us to take $a = 0$ and $\sigma = 1$ without limiting the generality of the subsequent results (to see this, it obviously suffices to consider the ratio $(\xi(t) - a)/\sigma$ instead of $\xi(t)$).

We now consider the important example of a normal stationary process. For any n ($n = 1, 2, \ldots$), let the vector $\Xi_n = \{\xi(t_1), \xi(t_2), \ldots, \xi(t_n)\}$ be distributed normally. We assume that

$$\mathsf{E}\xi(t_j) = 0, \quad \mathsf{D}\xi(t_j) = 1 \quad (-\infty < t_j < \infty),$$

and suppose that

$$\mathsf{E}\xi(t_i)\xi(t_j) = R(t_i - t_j),$$

where $R(0) = 1$ and $R(t)$ is an even function of t. The function $R(t)$ is such that the quadratic form

$$\sum_{j=1}^{n} \sum_{i=1}^{n} R(t_i - t_j)\, x_i\, x_j$$

is positive-semidefinite.

§ 57. STATIONARY STOCHASTIC PROCESS. KHINTCHINE'S THEOREM

Since the characteristic function of the vector Ξ_n is

$$f_n(u_1, \ldots, u_n; t_1, \ldots, t_n) = \exp\{-\sum_{i=1}^{n}\sum_{j=1}^{n} R(t_i - t_j)\, u_i\, u_j\}$$

and that of the vector $\Xi_k = \{\xi(t_1), \ldots, \xi(t_k)\}$ for any $k < n$ is

$$f_k(u_1, \ldots, u_k; t_1, \ldots, t_k) = \exp\{-\sum_{i=1}^{k}\sum_{j=1}^{k} R(t_i - t_j)\, u_i\, u_j\} =$$
$$= f_k(u_1, \ldots, u_k, 0, 0, \ldots, 0;\ t_1, \ldots, t_k, t_{k+1}, \ldots, t_n),$$

we conclude that the stochastic process defined above satisfies a self-compatibility[11] condition. Moreover, it is immediately clear that this process is stationary.

A homogeneous Markov process, i.e., a Markov process for which the distribution of the random variable $\xi(t)$ is the same for every value of t and the transition probability $F(t, x; \tau, y)$ is a function of the three arguments x, y, and $\tau - t$ only, is also stationary.

In many theoretical questions and in practical applications, multi-dimensional distributions such as (1) are not used to study a stationary process and what is made use of is only the constancy of its expectation and variance and the fact that the correlation coefficient merely depends on the difference of values of the parameter t. It is therefore natural to generalize the definition of stationary process and to say that a *stochastic process is stationary in the wide sense* if the expectation and variance of $\xi(t)$ are independent of t and the correlation coefficient of $\xi(t)$ and $\xi(t+u)$ is a function of u alone.

Clearly, a process $\xi(t)$ cannot be determined solely from the knowledge of its second moments, and therefore this knowledge cannot replace the theory of stochastic processes which is based on the consideration of distribution functions. Nevertheless, the treatment of many theoretical questions on the basis of second moments (or, as it is called, *correlation theory*) turns out to be adequate and yields satisfactory results.

In the present section, we confine ourselves to an investigation of the *correlation function*, i.e., the correlation coefficient of $\xi(t)$ and $\xi(t+u)$:

$$R(u) = \frac{\mathsf{E}\{[\xi(t+u) - \mathsf{E}\,\xi(t+u)]\,[\xi(t) - \mathsf{E}\,\xi(t)]\}}{\sqrt{\mathsf{D}\,\xi(t)\,\mathsf{D}\,\xi(t+u)}}.$$

---

[11] By this, we mean the compatibility of all the distribution functions of the process.

By virtue of our assumption that $a = 0$ and $\sigma = 1$, the expression for $R(u)$ takes on the simpler form:

$$R(u) = \mathsf{E}\{\xi(u)\xi(0)\}.$$

In correlation theory, it is customary to call a stationary stochastic process *continuous* if

$$\mathsf{E}[\xi(t+u) - \xi(t)]^2 \to 0$$

as $u \to 0$. From Tchebychev's inequality, it follows that, for any $\varepsilon > 0$ and arbitrary t, the particular relation

$$\mathsf{P}\{|\xi(t+u) - \xi(t)| \geq \varepsilon\} \to 0 \qquad (u \to 0)$$

is valid for a continuous process. Also, the equality

$$\mathsf{E}\{\xi(t+u) - \xi(t)\}^2 = 2(1 - R(u))$$

implies that the relation

$$\lim_{u \to 0} R(u) = 1$$

holds for a continuous stationary process.

In the case of a continuous stationary process, $R(u)$ is a continuous function of u. In fact,

$$|R(u + \Delta u) - R(u)| = |\mathsf{E}\,\xi(u + \Delta u)\,\xi(0) - \mathsf{E}\,\xi(u)\,\xi(0)|$$
$$= |\mathsf{E}\{\xi(0)\,[\xi(u + \Delta u) - \xi(u)]\}|.$$

But by the Schwarz inequality,

$$|\mathsf{E}\{\xi(0)\,[\xi(u + \Delta u) - \xi(u)]\}| \leq \sqrt{\mathsf{E}\,\xi^2(0)\,\mathsf{E}\,[\xi(u + \Delta u) - \xi(u)]^2}.$$

And since

$$\mathsf{E}\xi^2(0) = 1,$$

and for a continuous process

$$\mathsf{E}\,[\xi(u + \Delta u) - \xi(u)]^2 \to 0$$

as $\Delta u \to 0$, it follows that

$$|R(u + \Delta u) - R(u)| \to 0$$

also as $\Delta u \to 0$. This proves our assertion.

§ 57. Stationary Stochastic Process. Khintchine's Theorem

In the theorem which will now be proved, the stationary property may be taken both in the wide sense and in the strict sense.

KHINTCHINE'S THEOREM: *The function $R(u)$ is the correlation function for a continuous stationary process if and only if it is representable in the form*

$$R(u) = \int \cos ux \, dF(x), \qquad (3)$$

where $F(x)$ is some distribution function.

Proof: (i) the condition of the theorem is necessary. In fact, if $R(u)$ is the correlation function for a continuous stationary process, then it is continuous and bounded. Let us show, moreover, that it is positive semi-definite. Indeed, for any real numbers $u_1, u_2, \ldots, u_n$, complex numbers $\eta_1, \eta_2, \ldots, \eta_n$, and integer n, the following relation is valid:

$$0 \leq \mathsf{E} \left| \sum_{k=1}^{n} \eta_k \xi(u_k) \right|^2 = \mathsf{E} \left\{ \sum_{i=1}^{n} \sum_{j=1}^{n} \eta_i \bar{\eta}_j \xi(u_i) \xi(u_j) \right\} = \sum_{j=1}^{n} \sum_{i=1}^{n} R(u_i - u_j) \eta_i \bar{\eta}_j.$$

Now, by virtue of the Bochner-Khintchine Theorem (§ 39), it follows from this that $R(u)$ is representable in the form

$$R(u) = \int e^{iux} dF(x),$$

where $F(x)$ is a non-decreasing function of bounded variation. Since the function $R(u)$ is real,[12] we obtain from this that

$$R(u) = \int \cos ux \, dF(x).$$

Finally, taking into account the assumption that the process is continuous and hence that

$$R(+0) = 1,$$

we find that $F(\infty) - F(-\infty) = 1$, i.e., $F(x)$ is a distribution function.

(ii) The condition is sufficient. We are given that $R(u)$ is a function of the form (3). It is required to prove that a stationary process $\xi(t)$ exists which has as its correlation function the function $R(u)$. To this end, we consider a normally distributed n-dimensional vector $\xi(t_1), \xi(t_2), \xi(t_3), \ldots, \xi(t_n)$, for any integer n and any set of real numbers $t_1, t_2, \ldots, t_n$ having the properties:

---

[12] In consequence of the result of Example 3 of § 36, the function $F(x)$ is symmetric, i.e.,
$$F(x+0) = 1 - F(-x).$$

$$E\xi(t_1) = E\xi(t_2) = \ldots = E\xi(t_n) = 0,$$
$$D\xi(t_1) = D\xi(t_2) = \ldots = D\xi(t_n) = 1;$$

for arbitrary i and j, the correlation coefficient of $\xi(t_i)$ and $\xi(t_j)$ is $R(t_i - t_j)$, i.e.,

$$E\xi(t_i)\,\xi(t_j) = R(t_i - t_j).$$

The form of the function $R(u)$ ensures the positive-semidefiniteness of the quadratic form appearing in the exponent of the n-dimensional normal law. The *normal* stochastic process thus defined is stationary both in the wide and in the strict sense.

This theorem plays a basic role in the theory of stationary processes and its physical applications. For details of this, we refer the reader to the specialized literature in the bibliography given at the end of the book.

Example 1. Let

$$\xi(t) = \xi \cos \lambda t + \eta \sin \lambda t,$$

where ξ and η are uncorrelated[13] random variables for which $E\xi = E\eta = 0$, $D\xi = D\eta = 1$, and λ is a constant.

Since

$$\begin{aligned}R(u) &= E\,\xi(t+u)\,\xi(t) \\ &= E\,[\xi \cos \lambda t + \eta \sin \lambda t]\,[\xi \cos \lambda (t+u) + \eta \sin \lambda (t+u)] \\ &= E\,[\xi^2 \cos \lambda t \cos \lambda (t+u) + \\ &\quad + \xi\eta\,(\sin \lambda t \cos \lambda (t+u) + \cos \lambda t \sin \lambda (t+u)) + \\ &\quad\quad + \eta^2 \sin \lambda t \sin \lambda (t+u)] \\ &= \cos \lambda t \cos \lambda (t+u) + \sin \lambda t \sin \lambda (t+u) = \cos \lambda u\end{aligned}$$

the process $\xi(t)$ is stationary in the wide sense. For this case, we must set

$$F(x) = \begin{cases} 0 & \text{for } x \leq -\lambda, \\ 1/2 & \text{for } -\lambda < x \leq \lambda, \\ 1 & \text{for } x > \lambda \end{cases}$$

in equation (3) (see footnote 12 above).

---

[13] The random variables ξ and η are said to be *uncorrelated* if

$$E\xi\eta = E\xi \cdot E\eta.$$

§ 57. Stationary Stochastic Process. Khintchine's Theorem

Example 2. Let

$$\xi(t) = \sum_{k=1}^{n} b_k \xi_k(t),$$

where $\xi_k(t) = \xi_k \cos \lambda_k t + \eta_k \sin \lambda_k t$, λ_k is a constant, $\sum_{k=1}^{n} b_k^2 = 1$, and the random variables ξ_k and η_k satisfy the following conditions:

$$\mathsf{E}\xi_k = \mathsf{E}\eta_k = 0, \ \mathsf{D}\xi_k = \mathsf{D}\eta_k = 1 \ (k = 1, 2, \ldots, n),$$
$$\mathsf{E}\xi_i \xi_j = \mathsf{E}\eta_i \eta_j \text{ for } i \neq j, \quad \mathsf{E}\xi_i \xi_j = 0 \ (i, j = 1, 2, \ldots, n).$$

It is easy to show that the correlation function for the process $\xi(t)$ is

$$R(u) = \sum_{k=1}^{n} b_k^2 \cos \lambda_k u$$

and therefore that the process $\xi(t)$ is stationary in the wide sense. The function $F(\dot{x})$ in formula (3) increases only at the points $\pm \lambda_k$ and has jumps of magnitude $b_k^2/2$ at these points.

Stochastic processes for which the function $F(x)$ increases only by jumps are called *processes with a discrete spectrum*.

It is easy to see that any process of the form

$$\xi(t) = \sum_{k=1}^{\infty} b_k \xi_k(t), \tag{4}$$

where the series $\sum b_k^2$ converges and $\xi_k(t)$ has the same meaning as in Example 2, is stationary in the wide sense and has a discrete spectrum. It is worth noting that Slutsky proved a deeper converse statement, as follows: *Any stationary process with a discrete spectrum is representable in the form* (4). The generalization of this theorem of Slutsky to the case of an arbitrary spectrum will be formulated in the next section.

In parallel with the theory of stationary processes there has been developed a theory of stationary sequences. A sequence of random variables

$$\ldots, \xi_{-2}, \xi_{-1}, \xi_0, \xi_1, \xi_2, \ldots \tag{5}$$

is called *stationary* if for any integers n, u, and t_j ($1 \leq j \leq n$) the condition (1) is satisfied. In exactly the same way, the sequence (5) is called *stationary in the wide sense* if every term in the sequence has the same expectation and variance

$$\ldots = \mathsf{E}\,\xi_{-2} = \mathsf{E}\,\xi_{-1} = \mathsf{E}\,\xi_{0} = \mathsf{E}\,\xi_{1} = \mathsf{E}\,\xi_{2} = \ldots = a,$$

$$\ldots = \mathsf{D}\,\xi_{-2} = \mathsf{D}\,\xi_{-1} = \mathsf{D}\,\xi_{0} = \mathsf{D}\,\xi_{1} = \mathsf{D}\,\xi_{2} = \ldots = \sigma^2$$

and the correlation coefficient of ξ_i and ξ_j is a function of $|i-j|$ only.

As an exercise, we suggest that the reader prove the following theorem: If, for a stationary sequence,

$$\lim_{s \to \infty} R(s) = 0$$

where $R(s)$ is the correlation coefficient of ξ_i and ξ_{i+s}, then the law of large numbers holds for this sequence, i.e., as $n \to \infty$,

$$\mathsf{P}\left\{\left|\frac{1}{n}\sum_{k=1}^{n} \xi_k - a\right| < \varepsilon\right\} \to 1,$$

for any positive constant ε.

§ 58. The Notion of a Stochastic Integral. Spectral Decomposition of Stationary Processes

In what follows, we shall need the notion of a stochastic integral. Let the stochastic process $\xi(t)$ and the point function $f(t)$ be given in the interval $a \leq t \leq b$. Subdivide the interval (a, b) by means of the points $a = t_0 < t_1 < \ldots < t_n = b$ and consider the sum

$$I_n = \sum_{i=1}^{n} f(t_i)\,\xi(t_i)\,(t_i - t_{i-1}).$$

If, as $\max_{1 \leq i \leq n}(t_i - t_{i-1}) \to 0$, this sum approaches a limit (which is, in general, a random variable), then this limit is called the *integral of the stochastic process* $\xi(t)$ and is denoted by the symbol

$$I = \int_a^b f(t)\,\xi(t)\,dt.$$

The improper integral (for $a = -\infty$, $b = \infty$) is defined, in the usual way, as the limit of the proper integral as $a \to -\infty$ and $b \to \infty$ (independently).

The convergence of the integral sums I_n is to be taken in the following sense: There exists a random variable I such that

$$E(I_n - I)^2 \to 0 \tag{1}$$

as $n \to \infty$.

By making use of familiar theorems of the theory of functions of a real variable, we can easily prove that the sequence of random variables I_n converges to a limit I in the sense of (1) if and only if

$$E(I_m - I_n)^2 \to 0 \tag{2}$$

as $\min(m, n) \to \infty$. We shall not stop to give a proof of this fact. Instead, we prove the following theorem.

THEOREM 1: *In order for the integral*

$$I = \int_a^b f(t)\, \xi(t)\, dt$$

to exist it is sufficient that the integral

$$A = \int_a^b \int_a^b R(t - s)\, f(t)\, f(s)\, ds\, dt$$

exist. Moreover,

$$A = E\, [\int_a^b f(t)\, \xi(t)\, dt]^2.$$

Proof: For the first half of the theorem, it suffices to show that if the integral A exists, then relation (2) holds. We have:

$$E(I_n - I_m)^2 = E\, [\sum_{i=1}^n f(t_i)\, \xi(t_i)\, \Delta t_i]^2 -$$

$$- 2 E \sum_{i=1}^n \sum_{j=1}^m f(t_i)\, f(s_j)\, \xi(t_i)\, \xi(s_j)\, \Delta t_i\, \Delta s_j + E\, [\sum_{j=1}^m f(s_j)\, \xi(s_j)\, \Delta s_j]^2 =$$

$$= \sum_{i=1}^n \sum_{k=1}^n f(t_i)\, f(\tau_k)\, R(t_i - \tau_k)\, \Delta t_i\, \Delta \tau_k -$$

$$- 2 \sum_{i=1}^n \sum_{j=1}^m f(t_i)\, f(s_j)\, R(t_i - s_j)\, \Delta t_i\, \Delta s_j + \sum_{i=1}^m \sum_{k=1}^m f(s_j)\, f(\sigma_k)\, R(s_j - \sigma_k)\, \Delta s_j\, \Delta \sigma_k.$$

Here, the numerical values of t_k and τ_k coincide, as do the values of s_k and σ_k.

By virtue of the assumption that the integral A exists,

$$A = \lim \sum_i \sum_k f(t_i) f(\tau_k) R(t_i - \tau_k) \Delta t_i \Delta \tau_k$$

$$= \lim \sum_i \sum_j f(t_i) f(s_j) R(t_i - s_j) \Delta t_i \Delta s_j$$

$$= \lim \sum_j \sum_k f(s_j) f(\sigma_k) R(s_j - \sigma_k) \Delta s_j \Delta \sigma_k ,$$

provided $\max (\Delta t_i, \Delta s_j) \to 0$. Thus, as $\min (m, n) \to \infty$,

$$\mathsf{E}(I_m - I_n)^2 \to 0.$$

For the proof of the second part of the theorem, we observe that

$$\mathsf{E}\,[\sum_i f(t_i)\,\xi(t_i)\,\Delta t_i]^2 = \mathsf{E} \sum_i \sum_j f(t_i) f(\tau_j) \xi(t_i) \xi(\tau_j) \Delta t_i \Delta \tau_j$$

$$= \sum_i \sum_j f(t_i) f(\tau_j) R(t_i - \tau_j) \Delta t_i \Delta \sigma_j ,$$

and as $\max_{1 \leq i \leq n} \Delta t_i \to 0$ the last sum approaches the integral A.

Together with the notion of stochastic integral just introduced, it is also possible to consider a *stochastic Stieltjes integral*, which is defined as the limit of the sum

$$\sum_{k=1}^{n} f(t_k) [\xi(t_k) - \xi(t_{k-1})] \tag{3}$$

as $\max (t_i - t_{i-1}) \to 0$. Here, as usual, $a = t_0 < \ldots < t_n = b$, and the limit is taken in the sense of (1). If the limit of the sums (3) exists, then we denote it by the symbol

$$\int_a^b f(t)\,d\xi(t).$$

At the end of § 57, we formulated a theorem of Slutsky which characterizes the relation existing between stationary processes with a discrete spectrum and Fourier series with uncorrelated random coefficients. It can be shown that the following property holds for any process stationary

§ 58. STOCHASTIC INTEGRAL

in the wide sense: *For any $\varepsilon > 0$ and for T as large as we wish, there exist pairwise uncorrelated random variables $\xi_1, \xi_2, \ldots, \xi_n, \eta_1, \eta_2, \ldots, \eta_n$ and real numbers*[14] *$\lambda_1, \lambda_2, \ldots, \lambda_n$ such that for any value of t in the interval $-T \leq t \leq T$, the inequality*

$$\mathsf{E}\,[\xi(t) - \sum_{k=1}^{n}(\xi_k \cos \lambda_k t + \eta_k \sin \lambda_k t)]^2 < \varepsilon$$

holds.

In particular, it follows from this that under the above conditions

$$\mathsf{P}\{|\xi(t) - \sum_{k=1}^{n}(\xi_k \cos \lambda_k t + \eta_k \sin \lambda_k t)| > \eta\} \leq \frac{\varepsilon}{\eta^2},$$

where η is some positive number specified in advance.

We next cite without proof the following important theorem.

THEOREM 2: *Any process which is stationary in the wide sense is representable in the form*

$$\xi(t) = \int_0^\infty \cos \lambda t\, dZ_1(\lambda) + \int_0^\infty \sin \lambda t\, dZ_2(\lambda), \qquad (4)$$

where the stochastic processes $Z_1(\lambda)$ and $Z_2(\lambda)$ ($\lambda \geq 0$) have the following properties:

a) $\mathsf{E}\,[Z_i(\lambda_1 + \Delta\lambda_1) - Z_i(\lambda_1)]\,[Z_j(\lambda_2 + \Delta\lambda_2) - Z_j(\lambda_2)] = 0 \qquad (i, j = 1, 2)$

if $i \neq j$, but if the intervals $(\lambda_1, \lambda_1 + \Delta\lambda_1)$ and $(\lambda_2, \lambda_2 + \Delta\lambda_2)$ do not overlap then i and j can be equal;

b) $\mathsf{E}\,[Z_1(\lambda + \Delta\lambda) - Z_1(\lambda)]^2 = \mathsf{E}\,[Z_2(\lambda + \Delta\lambda) - Z_2(\lambda)]^2.$

It is natural to refer to formula (4) as the *spectral decomposition of the process $\xi(t)$*.

The stochastic processes $Z_1(\lambda)$ and $Z_2(\lambda)$ figuring in formula (4) are determined by the equations

$$Z_1(\lambda) = \lim_{T \to \infty} \frac{1}{2\pi} \int_{-T}^{T} \frac{\sin \lambda t}{t}\, \xi(t)\, dt$$

and

---

[14] The numbers n and $\lambda_1, \lambda_2, \ldots, \lambda_n$, as well as the variables ξ_i and η_i, depend on ε and T.

$$Z_2(\lambda) = \lim_{T\to\infty} \frac{1}{2\pi} \int_{-T}^{T} \frac{1-\cos \lambda t}{t} \xi(t)\, dt.$$

It is easy to show that both of the above integrals exist (this is done by using the Khintchine formula derived in § 57). It is also possible to show that

$$F(\lambda + \Delta\lambda) - F(\lambda) = \mathsf{E}\,[Z_1(\lambda + \Delta\lambda) - Z_1(\lambda)]^2,$$

where $F(\lambda)$ is the function determined by the Khintchine Theorem.

The possibility of decomposing an arbitrary stochastic process which is stationary in the wide sense into the form (4) was indicated by Kolmogorov in 1940. He formulated this result in the terminology of the Geometry of Hilbert space and proved it by making use of the Spectral Theory of Operators. Many authors, Cramér, Karhunen, Loève, Blanc-Lapierre, etc., have since devoted many papers to the probability-theoretic interpretation of this decomposition and its probabilistic derivation.

We shall not indicate here how the spectral decomposition can be applied to questions in the theory of vibrations and geophysics; instead, we refer the reader to the articles of Yaglom, Blanc-Lapierre, and Fortet indicated in the bibliography at the end of the book.

§ 59. The Birkhoff-Khintchine Ergodic Theorem

In 1931, the American mathematician George Birkhoff proved a general theorem[15] in statistical mechanics which, as was shown by Khintchine three years later, admits a broad probability-theoretic generalization. This theorem is as follows: *If a continuous stationary process $\xi(t)$ has a finite expectation, then with probability one the limit*

$$\lim_{T\to\infty} \frac{1}{T} \int_0^T \xi(t)\, dt$$

exists. The process is understood here to be stationary in the strict sense.

Since this theorem is a special variant of the strong law of large numbers, we shall prove it for stationary sequences rather than stationary processes for the purpose of continuing directly the results given in Chapter VI.

---

[15] For an historical survey, see *Ergodentheorie*, by E. Hopf; for an introductory treatment, *Ergodic Theory*, by Paul R. Halmos. See also *Mathematical Foundations of Statistical Mechanics*, etc. [*Trans.*]

§ 59. BIRKHOFF-KHINTCHINE ERGODIC THEOREM

Theorem: If

$$\ldots, \xi_{-1}, \xi_0, \xi_1, \ldots$$

is a stationary sequence of random variables for which $\mathsf{E}\xi_j$ exists, then with probability one the sequence of arithmetic means

$$\frac{1}{n}\sum_{k=1}^{n}\xi_k$$

converges to a limit.

Proof:[16] Let us set

$$h_{ab} = \frac{\xi_a + \xi_{a+1} + \cdots \xi_{b-1}}{b-a}.$$

We have to prove that with probability one the quantity h_{0b} approaches a limit as $b \to \infty$. Let $\bar{K}$ denote the random event consisting in the existence of this limit. Hence, we are to show that $\mathsf{P}(\bar{K}) = 1$, or what is the same, $\mathsf{P}(K) = 0$.

Let us suppose the converse is true and that the event K has a positive probability (i.e., that the quantity h_{0b} does not approach a limit as $b \to \infty$); we shall show that this assumption leads to a contradiction.

To this end, consider all intervals (α_n, β_n) with rational endpoints such that $\alpha_n < \beta_n$. The set of all such intervals is denumerable. If the $\lim_{b\to\infty} h_{0b}$ does not exist, we can then find a first interval (α_n, β_n) such that $\limsup_{b\to\infty} h_{0b} > \beta_n$ and $\liminf_{b\to\infty} h_{0b} < \alpha_n$ (the event K_n). Thus, the event K decomposes into a denumerable number of mutually exclusive events K_n. Since by assumption $\mathsf{P}(K) > 0$, we can find an n for which $\mathsf{P}(K_n) > 0$.

Thus, it has been shown that if $\mathsf{P}(K) > 0$, then two numbers α and β $(\alpha < \beta)$ exist such that simultaneously

$$\limsup h_{0b} > \beta,$$
$$\liminf h_{0b} < \alpha. \tag{1}$$

Let us now suppose that all ξ_j have taken on some definite values, and let a and b be integers. Now, if (a, b) is an interval such that $h_{ab} > \beta$ but $h_{ab'} \leq \beta$ for all b' for which $a < b' < b$, this interval will be called *singular* (relative to β).

---

[16] This proof is due to Gnedenko. [*Trans.*]

It is easy to show that two singular intervals cannot overlap partially. In fact, if (a, b) and (a_1, b_1) were two singular intervals such that $a < a_1 < b < b_1$, then from the equality

$$h_{ab} = \frac{(a_1 - a) h_{aa_1} + (b - a_1) h_{a_1 b}}{b - a}$$

and the inequality $h_{ab} > \beta$, it would follow that at least $h_{aa_1} > \beta$ or $h_{a_1 b} < \beta$. However, the first of these two inequalities is impossible, because the interval (a, b) is singular; and the second is impossible, because the interval (a_1, b_1) is singular.

If the interval (a, b) is singular, possesses a length not exceeding s, and is not contained in any other singular interval of length not exceeding s, then such an interval will be called *s-singular*.

It is easy to see that every singular interval of length not exceeding s is contained in only one s-singular interval (or is itself such an interval). For, among all the singular intervals of length not exceeding s and containing the given interval, there will be one of maximum length. This interval will thus be the unique s-singular interval. If two such intervals existed, they would have to overlap, which was shown above to be impossible, or one of them would be contained in the other and therefore could not be an s-singular interval. It follows from the definition that two s-singular intervals must be completely disjoint.

Let K_s denote the event that the inequalities (1) are satisfied and that, moreover, there exists at least one $t \leq s$ such that $h_{0t} > \beta$. Then, $K_s \subset K_{s+1}$ and $K = \sum_{s=1}^{\infty} K_s$. Hence,

$$P(K) = \lim_{s \to \infty} P(K_s).$$

(This is a consequence of the Corollary to Lemma 1 of § 62.) It follows from this that $P(K_s) > 0$ for s sufficiently large. Hereafter, we shall confine ourselves to such values of s only.

Suppose the event K_s takes place. Then among those values of $t \leq s$ for which $h_{0t} > \beta$ there exists a smallest, t'. The interval $(0, t')$ is singular. Consequently, it lies in some s-singular interval (a, b) (or is itself such an interval), for which $a \leq 0 < b$. The converse is also true: If there exists an s-singular interval (a, b) such that $a \leq 0 < b$, then there is a $t \leq s$ for which $h_{0t} > \beta$. In fact, since the interval (a, b) is s-singular, we have $b \leq b - a \leq s$. For $a = 0$, it obviously suffices to set $t = b$. However, if $a < 0$, then from the identity

§ 59. Birkhoff-Khintchine Ergodic Theorem

$$h_a = \frac{-a\, h_{a0} + b\, h_{0b}}{b - a}$$

and the inequalities $h_{ab} > \beta$ and $h_{a0} \leq \beta$, it follows that $h_{0b} > \beta$. Thus, in this case, we can also set $t = b$.

We now let $p = -a$ and $q = b - a$, so that $1 \leq q \leq s$ and $0 \leq p \leq q-1$. Since, by the above, there can be only one s-singular interval $(-p, -p+q)$, the event K_s can be decomposed into the sum of mutually exclusive events K_{pq} corresponding to the s-singular intervals $(-p, -p+q)$:

$$K_s = \sum_{p,q} K_{pq} \quad (q = 1, \ldots, s,\; p = 0, 1, \ldots, q-1).$$

Now, a shift of index $j = j' + p$ in the sequence ξ_j carries the event K_{0q} into the event K_{pq}. Therefore, since the sequence is stationary,[17] $\mathsf{P}(K_{pq}) = \mathsf{P}(K_{0q})$ and $\mathsf{E}(\xi_0/K_{pq}) = \mathsf{E}(\xi_p/K_{0q})$. Now

$$\mathsf{P}(K_s)\, \mathsf{E}(\xi_0/K_s) = \sum_{p,q} \mathsf{P}(K_{pq})\, \mathsf{E}(\xi_0/K_{pq}) = \sum_q \mathsf{P}(K_{0q}) \sum_p \mathsf{E}(\xi_p/K_{0q})$$
$$= \sum_q \mathsf{P}(K_{0q})\, \mathsf{E}(q\, h_{0q}/K_{0q}),$$

and by taking into account that the inequality $h_{0q} > \beta$ holds in the event K_{0q}, we find:

$$\mathsf{P}(K_s)\, \mathsf{E}(\xi_0/K_s) > \sum_q \mathsf{P}(K_{0q})\, q\, \beta = \beta \sum_{p,q} \mathsf{P}(K_{pq}) = \beta\, \mathsf{P}(K_s).$$

By assumption $\mathsf{P}(K_s) \neq 0$, so that

$$\mathsf{E}(\xi_0/K_s) > \beta.$$

Since $\mathsf{P}(K_s) \to \mathsf{P}(K)$, we finally obtain

$$\mathsf{E}(\xi_0/K) \geq \beta.$$

In the same way (if singular intervals relative to a are considered), it can be shown that

$$\mathsf{E}(\xi_0/K) \leq a.$$

We have thus arrived at a contradiction. Hence, it follows that $\mathsf{P}(K) = 0$, Q.E.D.

---

[17] Notice that we have used the stationarity assumption only at this point.

To investigate what limit the quantity h_{0n} approaches as $n \to \infty$ requires subsidiary considerations. We confine ourselves here to a proof of the following theorem only.

THEOREM: *If the random variables ξ_j are stationary and have finite variances and if the correlation function $R(k) \to 0$ as $k \to \infty$, then*

$$\mathsf{P}\left\{h_{0n} \underset{n \to \infty}{\to} a\right\} = 1 \qquad (a = \mathsf{E}\xi_j).$$

Proof: Consider the variance of the quantity h_{0n}. Since the sequence ξ_j is stationary, we have

$$\mathsf{D}\,h_{0n} = \mathsf{E}\left[\frac{1}{n}\sum_{k=1}^{n}(\xi_k - a)\right]^2 = \frac{\mathsf{D}\,\xi_k}{n^2}\left[n + 2\sum_{1 \leq i < j \leq n} R(j-i)\right].$$

Clearly,

$$\sum_{1 \leq i < j \leq n} R(j-i) = \sum_{k=1}^{n-1}(n-k)\,R(k).$$

Consider a value of m sufficiently large that the inequality

$$|R(k)| \leq \varepsilon \quad (\varepsilon > 0)$$

holds for $k > m$. From this it follows that

$$\mathsf{D}\,h_{0n} \leq \frac{\mathsf{D}\,\xi_k}{n^2}\left[n + 2\sum_{k=1}^{m}(n-k)\,R(k) + 2\varepsilon \sum_{k=m+1}^{n-1}(n-k)\right].$$

Using the fact that $|R(k)| \leq 1$, we can obviously strengthen the inequality, as follows:

$$\mathsf{D}\,h_{0n} \leq \frac{\mathsf{D}\,\xi_k}{n^2}[n + 2m(n-1) + \varepsilon(n-m-1)(2n-m-1)].$$

Hence, it is clear that the right-hand side of this inequality can be made smaller than 3ε by taking n sufficiently large. Thus, as $n \to \infty$, the sequence h_{0n} converges to a in probability. But since h_{0n} converges with probability one as $n \to \infty$, the theorem follows.

The above theorem is not only of considerable theoretical interest but also finds wide application in statistical physics and engineering practice. The reason for this is that when a process is stationary such important characteristics as $\mathsf{E}\xi(t)$, $\mathsf{D}\xi(t)$, and $R(u)$ can be determined without our knowing the probability distribution of the possible values, which of course is needed to compute these characteristics from their respective

§ 59. BIRKHOFF-KHINTCHINE ERGODIC THEOREM

formulas. The determination of these *spatial averages*, as one says in physics, requires the research worker to have information which is frequently not at his disposal. Anyhow, the experimental estimation of these quantities necessitates the repeated realization of trials for the process (i.e., many realizations of the function $\xi(t)$ must be obtained from experiment). The Birkhoff-Khintchine Ergodic Theorem shows that with probability one we can confine ourselves (under certain conditions) to a single realization of the process $\xi(t)$.

CHAPTER XI

ELEMENTS OF STATISTICS

§ 60. Some Problems of Mathematical Statistics

The term *statistics* comes from the Latin word *status*, meaning state (of affairs). When statistics originally began to take shape as a scientific discipline in the 18th century, the term *statistics* was associated with the system of collecting facts characterizing the state of a country. It was not yet assumed that only phenomena of a mass nature fell within the province of statistics. At present, statistics has a larger and, at the same time, a more definite content. Specifically, one can say that statistics comprises the following three subdivisions:

1) *The collection of statistical data*, i.e., data characterizing the individual units of mass aggregates;

2) *The statistical investigation of the data obtained*, consisting in discovering what regularities (laws) can be established on the basis of the data of mass observation;

3) *The development of statistical control methods and the analysis of statistical data.*

This last subdivision is what constitutes the subject of *mathematical statistics* proper.

The collecting of statistical data—mostly having to do with populations—was begun long ago: data exists which indicates that the emperor Yao had taken a census of the population in China in the year 2238 B.C.; a census of the population had also been conducted in ancient Egypt, in Persia, and in the Roman empire; a count of the population is known to have been taken in Russia in the years 1245, 1259, 1273, 1287, and even earlier. It must be noted, it is true, that these censuses were extremely primitive; in China, for example, the population was accounted for, for 200 years, by copying the lists of the previous censuses. However, even such imperfect and incomplete censuses permitted the planning of important state measures.

§ 60. SOME PROBLEMS OF MATHEMATICAL STATISTICS

In the preceding chapters, we have introduced a number of important concepts for the numerical characterization of random variables and random events. Included among these concepts are, first of all, probability, distribution function, expectation, variance, etc. However, in only a limited number of cases are we able not only to state that a given random event has a probability or that a given variable that takes on chance values has a distribution function but also to indicate what that probability or distribution function is. The usual state of affairs that one encounters in practice is such that all of the numerical characteristics we have introduced have to be estimated on the basis of experimental data. The development of methods for making such estimates constitutes one of the most important problems of mathematical statistics.

But even when sufficiently strong grounds exist for the apriori estimation of these numerical characteristics, the following problem of mathematical statistics arises all the same: By means of appropriately conducted experiments, to determine whether or not all of the conditions on which the apriori estimation is based actually hold. For example, let a tetrahedron be given; on the assumption that the tetrahedron is geometrically regular and its mass uniformly distributed, we have drawn the conclusion that the probability is 1/4 that the tetrahedron, when tossed, will land on any particular one of its faces. But do our assumptions correspond to reality? This question cannot be answered by reasoning alone: one must resort to the experimental facts.

In the present chapter, we shall merely give a general introduction to the problems of mathematical statistics, referring the reader to the specialized literature for a systematic presentation and for details. We confine ourselves here to the consideration of the following problems only:

1. *Estimation of an unknown distribution function.*

This problem may be posed as follows: As the result of n independent trials, the values

$$x_1, x_2, \ldots, x_n$$

are obtained for the random variable ξ.

It is required to determine, at least approximately, the unknown distribution function for the variable ξ.

2. *Estimation of the unknown parameters of a distribution.*

General theoretical considerations often permit one to make sufficiently definite inferences concerning the type of distribution function of the random variable under consideration. Thus, for example, in certain cases Liapounov's Theorem enables us to say that a distribution is normal. As a result, the determination of the unknown distribution function is reduced to estimating, on the basis of experiment, the unknown parameters a and σ.

The general problem can be formulated thus: The random variable ξ has a distribution function of a certain form, depending on k parameters whose values are unknown. A series of observations are made of the variable ξ; on the basis of these observations it is required to estimate the values of these parameters.

Clearly, the estimation of the unknown probability p of an event A is a special case of the problem just formulated, since we could consider a random variable ξ which takes on the value 1 if the event A occurs and the value 0 if A does not occur. The distribution function for ξ would depend on the single parameter p.

The solution to the problem just posed will only be given for the normal distribution

$$p(x/a, \sigma) = \frac{1}{\sigma\sqrt{2\pi}} e^{-\frac{(x-a)^2}{2\sigma^2}}.$$

In this case, our second problem can obviously be broken up into the three particular problems:

1) The quantity σ is assumed to be known; it is required to estimate the unknown value of a.

2) The quantity a is assumed to be known; it is required to estimate the unknown value of σ.

3) Both of the parameters a and σ are unknown; it is required to estimate their values.

These questions can be stated more precisely in the following way: In n independent trials, the variable ξ has taken on the values

$$x_1, x_2, \ldots, x_n.$$

It is required to find functions $\bar{a} = a(x_1, \ldots, x_n)$ and $\bar{\sigma} = \sigma(x_1, \ldots, x_n)$ that would be reasonable approximations to the values of the quantities a and σ being estimated (in the first problem, a may also be a function of σ; and in the second, σ may be a function of a). Moreover, it is required to estimate the average accuracy of these approximations.

Rather than seek approximations for the unknown parameters a and σ as the functions $\bar{a}$ and $\bar{\sigma}$, it is sometimes preferable to look for functions of the observed values and of the known quantities, a' and a'' (σ' and σ''), such that it may be asserted with sufficient practical assurance that

$$a' < a < a''$$

and, correspondingly, $\sigma' < \sigma < \sigma''$.

The functions a' and a'' are called *confidence limits for a*; and σ' and σ'', *confidence limits for σ*. In what follows, we shall offer two approaches to the solution of these problems.

3. *The testing of statistical hypotheses.*

The problem which we shall consider here can be posed as follows: There are certain reasons for considering the distribution function of the random variable ξ to be $F(x)$; the question is: Are the observed values consistent with the hypothesis that ξ has the distribution $F(x)$?

In particular, if the form of the distribution function is not called into question and only the values of certain parameters characterizing the distribution require verification, then one may ask the question: Do the results of our observations confirm the hypothesis that the distribution parameters have the presumed values? This is the problem of *testing a simple hypothesis*. If the hypothesis in question is that the parameters assume values from certain specified sets rather than certain exact values (for example, the hypothesis $p < p_0$, in the case of the binomial distribution), then the hypothesis is called *composite*.

It should be noted that the above three problems far from exhaust the basic problems of mathematical statistics. Industry and science are constantly posing new problems to mathematical statistics. It suffices to mention, for example, that the sorting out of defective articles in mass production and quality control of manufactured articles necessitate not only the determination of unknown parameters and distribution functions but also a rational management of the process of production. Moreover, the working out of rational methods of utilizing observed data as well as the planning of tests are among the basic problems of mathematical statistics.

§ 61. Variational Series and Empirical Distribution Functions

The starting point for the statistical analysis of any random variable ξ is a series of n observations, in which the variable ξ takes on the values

$$x_1, x_2, \ldots, x_n. \tag{1}$$

We shall suppose in what follows that the trials are always mutually independent and produced under the same conditions. Of course, in certain problems these assumptions are excessively restrictive. Situations often arise in which one has to consider trials under variable conditions (for example, firing at a moving target) or even dependent sequences of trials (for example, the testing of specimens of yarn of a given length taken from one spool or skein). Although some general results exist for problems involving variable testing conditions, the statistics of dependent trials can be said to be not fully developed, and its development is one of the urgent problems of mathematical statistics.

Let us now arrange the set of outcomes of the trials in increasing order of magnitude. Denoting the observed value which is k-th in magnitude by x_k^*, we can write down the sequence (1) as follows:

$$x_1^* \leq x_2^* \leq \ldots \leq x_n^*.$$

This sequence, i.e., the sequence of observed values of a random variable arranged in increasing order of magnitude is called a *variational series*.

We shall call the function $F_n(x)$ defined by the following conditions

$$F_n(x) = \begin{cases} 0 \text{ for } x \leq x_1^*, \\ k/n \text{ for } x_k^* < x \leq x_{k+1}^*, \\ 1 \text{ for } x > x_n^*. \end{cases}$$

the *empirical distribution function*.

It is clear that the empirical distribution function is monotonic, continuous from the left, and has points of discontinuity at the values of its argument which are the terms of the variational series. The size of the jump at a point of discontinuity is an integral multiple of $1/n$. For what follows, we draw attention to the fact that for each value of x the ordinate $F_n(x)$ is a random variable whose possible values are $0, 1/n,$ $\ldots, (n-1)/n, n/n = 1$. The probability of the equality $F_n(x) = k/n$, as is easy to see, is

$$\mathsf{P}\left\{F_n(x) = \frac{k}{n}\right\} = \binom{n}{k} \{F(x)\}^k \{1 - F(x)\}^{n-k}.$$

In the simplest special case, where the random variable ξ can only take on a finite number of values $a_1, a_2, \ldots, a_s$, the terms of the variational series are necessarily the numbers in this sequence. If $m_1, m_2, \ldots, m_s$

$(m_1 + m_2 + \ldots + m_s = n)$ denote the respective number of trials in which $\xi = a_1, \xi = a_2, \ldots, \xi = a_s$, then, by the law of large numbers, for n sufficiently large, the relative frequencies m_i/n will approximate the unknown probabilities

$$p_1 = \mathsf{P}\{\xi = a_1\},\ p_2 = \mathsf{P}\{\xi = a_2\},\ \ldots,\ p_s = \mathsf{P}\{\xi = a_s\}.$$

Furthermore, the strong law of large numbers also holds for our present case. In § 62, we shall prove a general theorem concerning the convergence of the empirical distribution function to the theoretical distribution as the number of trials is increased.

We note that the summary numerical characteristics usually made use of in statistics: the sample mean of the observed values

$$\bar{x} = \frac{x_1 + x_2 + \ldots + x_n}{n},$$

and the sample standard deviation

$$s = \sqrt{\frac{1}{n} \sum_{k=1}^{n} (x_k - \bar{x})^2},$$

and so on, are certain functionals of the empirical distribution function. Thus, for example,

$$\bar{x} = \int x\, dF_n(x),$$
$$s^2 = \int (x - \bar{x})^2\, dF_n(x).$$

§ 62. Glivenko's Theorem and Kolmogorov's Compatibility Criterion

Before proceeding to formulate and prove an important theorem which Glivenko obtained in 1933, we have to establish two auxiliary theorems.

LEMMA 1: *If an event E is equivalent to the combined occurrence of the infinite collection of events $E_1, E_2, \ldots,$*

$$E = E_1 E_2 \ldots,$$

and if each event E_{n+1} implies the preceding event E_n, then

$$\mathsf{P}\{E\} = \lim_{n \to \infty} \mathsf{P}\{E_n\}.$$

XI. Elements of Statistics

Proof: The event E_1 can, in fact, be represented as a sum of mutually exclusive events in each of the two following ways:
$$E_1 = E_1\bar{E}_2 + E_2\bar{E}_3 + \ldots + E_{n-1}\bar{E}_n + E_n$$
and $E_1 = E_1\bar{E}_2 + E_2\bar{E}_3 + \ldots + E_{n-1}\bar{E}_n + E_n\bar{E}_{n+1} + \ldots + E.$

Hence, $\mathsf{P}\{E_1\} = \mathsf{P}\{E_1\bar{E}_2\} + \mathsf{P}\{E_2\bar{E}_3\} + \ldots + \mathsf{P}\{E_{n-1}\bar{E}_n\} + \mathsf{P}\{E_n\}$
and
$$\mathsf{P}\{E_1\} = \mathsf{P}\{E_1\bar{E}_2\} + \mathsf{P}\{E_2\bar{E}_3\} + \\ + \ldots + \mathsf{P}\{E_{n-1}\bar{E}_n\} + \mathsf{P}\{E_n\bar{E}_{n+1}\} + \ldots + \mathsf{P}\{E\}.$$

A comparison of the last two equations leads to the relation
$$\mathsf{P}\{E\} = \mathsf{P}\{E_n\} - \sum_{k=n}^{\infty}\mathsf{P}\{E_k\bar{E}_{k+1}\}.$$
Since the quantity subtracted on the right-hand side is the remainder of a convergent series, we have:
$$\mathsf{P}\{E\} = \lim_{n\to\infty}\mathsf{P}\{E_n\}.$$

Corollary: *If $E_1 \subset E_2 \subset \ldots$ and $E = E_1 + E_2 + \ldots$, then*
$$\mathsf{P}\{E\} = \lim_{n\to\infty}\mathsf{P}\{E_n\}.$$

Proof: Since $\bar{E}_1 \supset \bar{E}_2 \supset \ldots$, and $\bar{E} = \bar{E}_1\bar{E}_2\ldots$, it follows from Lemma 1 that
$$1 - \mathsf{P}\{E\} = \mathsf{P}\{\bar{E}\} = \lim_{n\to\infty}\mathsf{P}\{\bar{E}_n\} = \lim_{n\to\infty}(1 - \mathsf{P}\{E_n\}) = 1 - \lim_{n\to\infty}\mathsf{P}\{E_n\}.$$
Hence, $\qquad\qquad\qquad \mathsf{P}\{E\} = \lim_{n\to\infty}\mathsf{P}\{E_n\}.$

Lemma 2: *If each of the events in a finite or denumerable sequence $E_1, E_2, \ldots, E_n, \ldots$ has the probability one, then the probability of the combined occurrence of these events is also one.*

Proof: Let us first consider two events E_1 and E_2 for which
$$\mathsf{P}\{E_1\} = \mathsf{P}\{E_2\} = 1.$$
Since
$$\mathsf{P}\{E_1 + E_2\} = \mathsf{P}\{E_1\} + \mathsf{P}\{E_2\} - \mathsf{P}\{E_1E_2\}$$
and $\mathsf{P}\{E_1 + E_2\} = 1$, it follows that
$$\mathsf{P}\{E_1E_2\} = 1.$$

§ 62. GLIVENKO'S THEOREM, KOLMOGOROV'S COMPATIBILITY CRITERION

Hence, by mathematical induction, we conclude that for any n the events for which
$$P\{E_1\} = P\{E_2\} = \ldots = P\{E_n\} = 1$$
also satisfy the relation
$$P\{E_1 E_2 \ldots E_n\} = 1.$$

Let there now be given an infinite sequence of events $E_1, E_2, \ldots, E_n, \ldots$ for which
$$P\{E_1\} = P\{E_2\} = \ldots = P\{E_n\} = \ldots = 1.$$
Since it is clear that
$$E_1 E_2 E_3 \ldots = E_1(E_1 E_2)(E_1 E_2 E_3) \ldots$$

and each factor on the right-hand side of the equation implies the preceding one, we have, by Lemma 1,
$$P\{E_1 E_2 E_3 \ldots\} = \lim_{n \to \infty} P\{E_1 E_2 \ldots E_n\}.$$
This relation proves the lemma.

GLIVENKO'S THEOREM:* *Let $F(x)$ be the distribution function for the random variable ξ and $F_n(x)$ the empirical distribution function based on the outcomes of n independent observations for the variable ξ. Then as $n \to \infty$,*
$$P\{\sup_{-\infty < x < \infty} |F_n(x) - F(x)| \to 0\} = 1.$$

Proof: Let $x_{r,k}$ denote the smallest value of x satisfying the inequality
$$F(x-0) = F(x) \leq \frac{k}{r} \leq F(x+0) \qquad (k = 1, 2, \ldots, r).$$

Let A be the event that $\xi < x_{r,k}$ and let B denote the event $\xi \leq x_{r,k}$. Clearly,
$$P\{A\} = F(x_{r,k}) \text{ and } P\{B\} = F(x_{r,k} + 0).$$

Since the relative frequency of occurrence of the events A and B are $F_n(x_{r,k})$ and $F_n(x_{r,k} + 0)$ respectively, by Borel's Theorem (see § 34)
$$P\{F_n(x_{r,k}) \underset{n \to \infty}{\to} F(x_{r,k})\} = 1 \qquad (1)$$
and
$$P\{F_n(x_{r,k} + 0) \underset{n \to \infty}{\to} F(x_{r,k} + 0)\} = 1. \qquad (1')$$

Now let E_k^r and D_k^r denote the events that, as $n \to \infty$,

---
* Also known as the Glivenko-Cantelli Theorem. [*Trans.*]

$$F_n(x_{r,k}) \to F(x_{r,k})$$

and

$$F_n(x_{r,k}+0) \to F(x_{r,k}+0),$$

respectively ($k = 1, 2, \ldots, r$), and set

$$E^r = E_1^r E_2^r \ldots E_r^r D_1^r D_2^r \ldots D_r^r.$$

Clearly, the event E^r is equivalent to the fact that

$$\max_{\substack{1 \leq j \leq r \\ 1 \leq k \leq r}} \left\{ |F_n(x_{r,j}) - F(x_{r,j})|, |F_n(x_{r,k}+0) - F(x_{r,k}+0)| \right\} \to 0$$

as $n \to \infty$. Since by (1) and (1')

$$\mathsf{P}\{E_1^r\} = \ldots = \mathsf{P}\{E_r^r\} = \mathsf{P}\{D_1^r\} = \ldots = \mathsf{P}\{D_r^r\} = 1,$$

then by Lemma 2,

$$\mathsf{P}\{E^r\} = 1.$$

Further, let $E = E^1 E^2 E^3 \ldots$. By Lemma 2, $\mathsf{P}\{E\} = 1$. Finally, let S denote the event that

$$\sup_{-\infty < x < \infty} |F_n(x) - F(x)| \to 0$$

as $n \to \infty$. Now the inequalities

$$F_n(x_{r,k}+0) \leq F_n(x) \leq F_n(x_{r,k+1})$$

and

$$F(x_{r,k}+0) \leq F(x) \leq F(x_{r,k+1})$$

are satisfied for any x between $x_{r,k}$ and $x_{r,k+1}$, where

$$0 \leq F(x_{r,k+1}) - F(x_{r,k}+0) \leq \frac{1}{r}.$$

We conclude from this that

$$F_n(x) - F(x) \leq F_n(x_{r,k+1}) - F(x_{r,k}+0)$$

$$\leq F_n(x_{r,k+1}) - F(x_{r,k+1}) + \frac{1}{r}$$

and

$$F_n(x) - F(x) \geq F_n(x_{r,k}+0) - F(x_{r,k+1})$$

$$\geq F_n(x_{r,k}+0) - F(x_{r,k}+0) - \frac{1}{r},$$

that is,

$$\sup_{-\infty < x < \infty} |F_n(x) - F(x)| \leq$$

$$\leq \max_{\substack{1 \leq j \leq r \\ 1 \leq k \leq r}} \left\{ |F_n(x_{r,j}) - F(x_{r,j})|, |F_n(x_{r,k}+0) - F(x_{r,k}+0)| \right\} + \frac{1}{r}.$$

§ 62. Glivenko's Theorem, Kolmogorov's Compatibility Criterion

Since r is arbitrary, the last inequality implies that $E \subset S$. This obviously proves that

$$P\{\sup_{-\infty < x < \infty} |F_n(x) - F(x)| \to 0\} = 1.$$

Just as, in the theorems on the law of large numbers, where no precise estimates were given for the probabilities of the possible deviations of the arithmetic means of the random variables from their respective expectations, so Glivenko's Theorem, while it establishes the important fact that the empirical distribution function approaches the distribution of the observed random variable, does not indicate with what probabilities the possible deviations may occur. We thus have the problem of determining the distribution function for the variable

$$D_n = \sup_{-\infty < x < \infty} |F_n(x) - F(x)|.$$

The solution to this problem was given in 1933 by Kolmogorov, in the following limit theorem.

Kolmogorov's Theorem: *If the function $F(x)$ is continuous, then as $n \to \infty$,*

$$P\{\sqrt{n}\, D_n < z\} \to K(z) = \begin{cases} 0 & \text{for } z \leq 0, \\ \sum_{k=-\infty}^{\infty} (-1)^k e^{-2k^2 z^2} & \text{for } z > 0. \end{cases}$$

We shall not offer any proof of Kolmogorov's Theorem, since no proof exists which we consider to be sufficiently elementary. We pause merely to indicate how it may be used for practical purposes. This is all the more appropriate as the following argument is constantly encountered in mathematical statistics.

Let it be required to test the hypothesis that a random variable ξ has a continuous function $F(x)$ as its distribution function. For testing purposes, n independent trials are carried out, and on the basis of the results an empirical distribution function $F_n(x)$ is constructed. According to Glivenko's Theorem, the function $F_n(x)$ should serve as an approximation to the distribution function $F(x)$. The quantity D_n then gives a measure of the deviation of $F_n(x)$ from $F(x)$. Let ε be a sufficiently small positive number and z_n^0 a quantity such that

$$P\{D_n \geq z_n^0\} = \varepsilon.$$

If the conditions of the problem allow us to assume that it is practically impossible for the event with probability ε to occur in a single experiment, we arrive at the following criterion for testing our hypothesis. Suppose that n independent trials result in a certain value of D_n. If it should turn out that $D_n \geq z_n^0$, then this would mean that with probability ε this event has been realized in a single experiment (in the examination of the deviation of the empirical distribution function from $F(x)$, the n trials for determining the values of the random variable constitute a single experiment). According to our assumption, such an event is practically impossible; we must therefore conclude that the deviation that has resulted must be essential. Thus, in this case, we have to reject the hypothesis that our random variable has the distribution $F(x)$. At the same time, if it were found that $D_n < z_n^0$, then there would be no justification for giving up the hypothesis, and it should be regarded as being compatible with experiment at least for as long as it is shown to agree satisfactorily with experience.

We note that we have a state of affairs which is peculiar to every mathematical discipline: in order to disprove some theoretical premise, it suffices to give just one counter-example, but to prove its correctness, examples are no longer adequate. Corroborating examples merely make this premise more probable (we are employing the expression *more probable* in the everyday sense rather than in the sense in which it is used in this book).

We must make one further remark. If the number of trials is very large, then we can say that

$$\mathsf{P}\{\sqrt{n}D_n < z\} \sim \mathrm{K}(z).$$

The compatibility criterion based on Kolmogorov's Theorem can now be described as follows.

Let $D_n^{(0)}$ denote the maximum value of $|F_n(x) - F(x)|$ actually found, and set $\lambda_0 = \sqrt{n}D_n^{(0)}$. If the difference

$$1 - \mathrm{K}(\lambda_0) \sim \mathsf{P}\{\sqrt{n}\,D_n \geq \lambda_0\}$$

is small, then a very unlikely event has occurred, and the difference between $F_n(x)$ and $F(x)$ must be regarded as significant and no longer explained by the randomness of the observed values. However, if $1 - \mathrm{K}(\lambda_0)$ is large, then the difference between $F_n(x)$ and $F(x)$ must be considered insignificant, and our hypothesis may be regarded as being compatible with experiment.

§ 63. Comparison of Two Distribution Functions

One of the most essential problems arising in mathematical statistics is that of answering the question: Is it possible to establish with reasonable certainty that two series of independent measurements,

$$x_1, x_2, \ldots, x_m \tag{1}$$

and
$$y_1, y_2, \ldots, y_n, \tag{2}$$

have been obtained from observations on random variables with the same probability distribution? This problem is encountered in the solution of many practically and theoretically important problems. For illustrative purposes, we discuss two examples of this kind.

Consider a method of production in some industrial enterprise. To establish whether the quality of the items in different lots (and also within each of large lots) is uniform, a random sample is taken from each lot and a measurement of some particular characteristic of each of the items is made. Suppose that the results of measuring one batch are given by the numbers (1) and those of another by the numbers (2). Can the difference between these two systems of values be explained by the unavoidable random fluctuations in measurement, or is the difference so great that one must conclude that there is an actual difference in quality of the articles in the two lots in question?

In any physical experiment in which measurements are made, it is of importance to establish that the conditions under which the experiment is being performed have not varied during the process of experimentation. It is especially important to ascertain this invariability for those experiments which are carried out over a long period of time, or during the course of which one has to replace certain portions of the apparatus. This raises the following question: Can we establish on the basis of the measurements (1) taken at the start of the experiment and those in (2) taken at some other of its stages that the conditions under which the experiment is being performed have essentially not varied?

Of course, in many problems of practical importance, one is interested in establishing that the empirical data disagrees with the hypothesis that the distribution is invariable. One has to treat this sort of formulation of the problem, for example, in comparing the quality of different varieties of seed with regard to productivity, early ripening, and resistance to disease. However, it is clear that both of the problems just posed differ from each other in form only, and the solution of one of them implies the solution of the other.

Let us suppose that the two series of measurements (1) and (2) are independent, both within each of the series and with respect to each other, and that they have a common continuous distribution function $F(x)$. This assumption is the statistical hypothesis which is now subject to verification. Let $S_m(x)$ and $T_n(x)$ denote the empirical distribution functions for the first and second samples, respectively, and set

$$D^+_{m,n} = \sup_{-\infty < x < \infty} [S_m(x) - T_n(x)],$$

$$D_{m,n} = \sup_{-\infty < x < \infty} |S_m(x) - T_n(x)|.$$

Smirnov has established that under the above assumptions and the fulfillment of the condition $\lim_{n \to \infty} m/n = \tau$ $(0 < \tau < \infty)$[1] the following limit relations hold:

$$\lim_{n \to \infty} \mathsf{P}\left\{\sqrt{\frac{mn}{m+n}} D^+_{m,n} < z\right\} = 1 - e^{-2z^2} \quad \text{for } z > 0; \quad (3)$$

$$\lim_{n \to \infty} \mathsf{P}\left\{\sqrt{\frac{mn}{m+n}} D_{m,n} < z\right\} = \mathsf{K}(z) = \sum_{k=-\infty}^{\infty} (-1)^k e^{-2k^2 z^2} \quad \text{for } z > 0. \quad (4)$$

These theorems are used in the way described at the end of the preceding section for testing that the distribution function is compatible with the observed data.

In the special case where $m = n$, it is not difficult to obtain not only the above results of Smirnov but also the exact distributions of the variables $D^+_{n,n}$ and $D_{n,n}$. For brevity, we introduce the notation:

$$D^+_n = D^+_{n,n}, \quad D_n = D_{n,n}, \quad c = [z\sqrt{2n}].$$

THEOREM: *The following equalities hold under the assumptions stated above*:

$$\Phi_n(z) = 0 \text{ for } z \leq \frac{1}{\sqrt{2n}}, \qquad \Phi^+_n(z) = 0 \text{ for } z < \frac{1}{\sqrt{2n}}$$

$$\Phi^+_n(z) = \mathsf{P}\left\{\sqrt{\frac{n}{2}} D^+_n < z\right\} = 1 - \frac{\binom{2n}{n-c}}{\binom{2n}{n}} \text{ for } \frac{1}{\sqrt{2n}} \leq z \leq \sqrt{\frac{n}{2}}, \quad (5)$$

---

[1] This restriction is unessential, and it is possible to drop it.

§ 63. Comparison of Two Distribution Functions

$$\Phi_n(z) = \mathsf{P}\left\{\sqrt{\frac{n}{2}}\,D_n < z\right\} = \sum_{k=-\left[\frac{n}{c}\right]}^{\left[\frac{n}{c}\right]} (-1)^k \frac{\binom{2n}{n-kc}}{\binom{2n}{n}} \quad \text{for} \quad \frac{1}{\sqrt{2n}} < z \leq \sqrt{\frac{n}{2}}, \quad (6)$$

$$\Phi_n^+(z) = \Phi_n(z) = 1 \qquad \text{for} \quad z > \sqrt{\frac{n}{2}}.$$

Proof: Let us arrange the results of both series of observations (1) and (2) in increasing order of magnitude $z_1 \leq z_2 \leq \ldots \leq z_{2n}$ and consider the sequence of auxiliary random variables $\xi_1, \xi_2, \ldots, \xi_{2n}$ defined by the rule:

$$\xi_k = \begin{cases} +1 \text{ if } z_k \text{ belongs to the first series of observations;} \\ -1 \text{ if } z_k \text{ belongs to the second series of observations.} \end{cases}$$

By setting
$$S_0 = 0, \; S_k = \xi_1 + \xi_2 + \ldots + \xi_k,$$
we find:
$$n\,D_n^+ = \sup_{0 \leq k \leq 2n} S_k, \quad n\,D_n = \sup_{0 \leq k \leq 2n} |S_k|.$$

Let us plot the points (k, S_k) for $k = 0, 1, \ldots, 2n$ in the tx-plane, and let us connect these points by straight-line segments. The x-value on the polygonal line thus obtained will increase by one unit at n of the points and decrease by one unit at the remaining points. Thus, our curve (which we shall call a *trajectory*) starts at the point $(0, 0)$ and terminates at the point $(2n, 0)$. Since the number of increases in x is equal to n and the number of decreases is also n, the overall number of different trajectories is $\binom{2n}{n}$. Furthermore, since all of the observations, by assumption, have the same distribution function, all of the trajectories are equally likely. The probability of each trajectory is thus equal to $1 \big/ \binom{2n}{n}$. In our geometric interpretation, the required probabilities

1. $\mathsf{P}\{n\,D_n^+ < c\}$ and 2. $\mathsf{P}\{n\,D_n < c\}$

are the probabilities that:

1) The entire trajectory will be below the line $x = c$; and
2) The entire trajectory will stay between the lines $x = c$ and $x = -c$.

We shall further confine ourselves to integral values of the parameter c, because the quantities nD_n^+ and nD_n can only take on integral values.

We determine the number of trajectories favorable to the event $nD_n^+ < c$ in the following way. We set each trajectory that reaches the line $x = c$ in correspondence with a new trajectory by the following rule: From the point $(0, 0)$ up to the first point where a trajectory meets the line $x = c$, the new curve coincides with the trajectory. Starting from this point of intersection, the new curve is the mirror image of the remaining portion of the trajectory with respect to the line $x = c$. The new curve thus starts at the point $(0, 0)$ and terminates at the point $(2n, 2c)$. It is easy to see that the total number of distinct new curves (and hence, also, the number of trajectories reaching the line $x = c$) is equal to $\binom{2n}{n-c}$. In fact, the number of increases in x for a new curve is $n + c$ and the number of decreases is $n - c$. The number of trajectories which do not reach the line $x = c$ is thus equal to $\binom{2n}{n} - \binom{2n}{n-c}$. This clearly implies equation (5).

To prove equation (6), a slightly more complicated argument is required. We decompose the set $\mathfrak{M}$ of all possible trajectories into the following disjoint subsets: $\mathfrak{A}_0$, the trajectories that reach neither the line $x = c$ (the line a) nor the line $x = -c$ (the line β); $\mathfrak{A}_1$, the trajectories that reach[2] a but not β; $\mathfrak{B}_1$, the trajectories that reach β but do not reach a; $\mathfrak{A}_2$, the trajectories that first reach a, then β, and then never reach a again; $\mathfrak{B}_2$, the trajectories that first reach β and then a and then never reach β again; $\mathfrak{A}_3$, the trajectories that first reach a, then β, then a again, and then never reach β again, and so on. It is evident that beginning from a certain point, the sets of trajectories so generated will be empty. Furthermore, it is clear that

$$\mathfrak{M} = \mathfrak{A}_0 + \sum_{i=1} \{\mathfrak{A}_i + \mathfrak{B}_i\}.$$

Together with the disjoint sets just constructed, we form the following subsets of trajectories: A_1, those reaching a at least once; B_1, those reaching β at least once; A_2, those reaching a at least once and then β; B_2, those reaching β at least once and then a; A_3, those that reach the lines a and β at least once in the order $a\beta a$. We continue this process to its natural conclusion. Since we have the relations

---

[2] Or intersect.

§ 63. Comparison of Two Distribution Functions

$$A_1 = \mathfrak{A}_1 + \sum_{i=2} \{\mathfrak{A}_i + \mathfrak{B}_i\}, \quad B_1 = \mathfrak{B}_1 + \sum_{i=2} \{\mathfrak{A}_i + \mathfrak{B}_i\},$$

$$A_2 = \mathfrak{A}_2 + \sum_{i=3} \{\mathfrak{A}_i + \mathfrak{B}_i\}, \quad B_2 = \mathfrak{B}_2 + \sum_{i=3} \{\mathfrak{A}_i + \mathfrak{B}_i\}$$

$$\dots\dots\dots\dots\dots\dots\dots\dots\dots\dots\dots\dots\dots\dots,$$

we have, for any $i \geq 1$,

$$[A_{2i-1} - A_{2i}] + [B_{2i-1} - B_{2i}] = \mathfrak{A}_{2i-1} + \mathfrak{A}_{2i} + \mathfrak{B}_{2i-1} + \mathfrak{B}_{2i},$$

and therefore

$$\mathfrak{A}_0 = \mathfrak{M} - \sum_{i=1} \{[A_{2i-1} - A_{2i}] + [B_{2i-1} - B_{2i}]\}. \tag{7}$$

In order to solve our problem, we still have to compute the number of paths in each of the sets $A_{2i-1}, A_{2i}, B_{2i-1}, B_{2i}$ for $i = 1, 2, \dots$. We shall carry out this computation for the sets A_1 and A_2. With each trajectory starting at the point $(0, 0)$ and reaching the line a, we set in correspondence a path which starts from the point $(0, 0)$, coincides with the original trajectory up to the point where it first reaches the line a, and then is the mirror image of the original trajectory with respect to the line a. This path terminates at the point $(2n, 2c)$. We have already shown that there exist $\binom{2n}{n-c}$ such paths. If the trajectory first reaches the line a and then the line β, then the path just constructed reaches the line $x = 3c$. To compute the number of trajectories in the set A_2, we construct a new path as follows: From the point $(0, 0)$ up to the line a it coincides with the original trajectory; then it proceeds as the mirror image of the original trajectory with respect to the line a until it reaches the line $x = 3c$; and, finally, in the last stage, it is the reflection of the first reflection with respect to the line $x = 3c$. The new path thus terminates at the point $(2n, 4c)$. The number of new paths (which also means the number of original trajectories) is equal to $\binom{2n}{n-2c}$. By analogous reasoning (performing the requisite number of reflections), we find that the number of trajectories in each of the sets A_i and B_i is equal to $\binom{2n}{n-ic}$. Thus, according to equation (7), the number of trajectories in the set $\mathfrak{A}_0$ is

$$\binom{2n}{n} - 2 \sum_{i=1} \left\{ \binom{2n}{n-(2i-1)c} - \binom{2n}{n-2ic} \right\}.$$

This proves equation (6).

XI. ELEMENTS OF STATISTICS

The first and last equations of the theorem are obvious.

We can now prove the limit relations (3) and (4) of Smirnov for the special case in question. For, consider the ratio

$$I_k = \binom{2n}{n-kc} \bigg/ \binom{2n}{n} = \frac{(n!)^2}{(n-kc)!\,(n+kc)!}$$

for fixed k. By Stirling's formula, we find:

$$I_k = \left(1 - \frac{kc}{n}\right)^{-n+kc} \left(1 + \frac{kc}{n}\right)^{-n-kc} (1 + o(1)).$$

By then making use of the MacLaurin expansion for the logarithm and recalling that $c = [z\sqrt{2n}]$, we obtain:

$$\log I_k = -\frac{k^2 c^2}{n} + o(1) = -2\,k^2\,z^2 + o(1).$$

Thus,

$$I_k = e^{-2k^2 z^2}(1 + o(1)). \tag{8}$$

For $k = 1$, this equation implies (3).

To prove relation (4), consider $z > 0$, $\varepsilon > 0$, and an integer N such that

$$e^{-2N^2 z^2} < \frac{\varepsilon}{16}.$$

Under these assumptions,

$$\left|\sum_{|k|>N} (-1)^k e^{-2k^2 z^2}\right| < \frac{\varepsilon}{8}.$$

Since

$$\binom{2n}{n-kc} > \binom{2n}{n-(k+1)c},$$

it follows that

$$\left|\sum_{N<|k|\leq[\frac{n}{c}]} (-1)^k \frac{\binom{2n}{n-ck}}{\binom{2n}{n}}\right| < 4\,\frac{\binom{2n}{n-Nc}}{\binom{2n}{n}},$$

and therefore, for sufficiently large values of n,

$$\left|\sum_{N<|k|\leq[\frac{n}{c}]} (-1)^k \frac{\binom{2n}{n-kc}}{\binom{2n}{n}}\right| < 4\,e^{-2N^2 z^2}(1 + o(1)) < \frac{\varepsilon}{3}.$$

Thus,

§ 64. COMPARISON OF TWO STATISTICAL HYPOTHESES 401

$$\left| \sum_{N < |k| \leq \left[\frac{n}{c}\right]} (-1)^k \frac{\binom{2n}{n-kc}}{\binom{2n}{n}} - \sum_{|k| > N} (-1)^k e^{-2k^2 z^2} \right| < \frac{\varepsilon}{2}.$$

But since, by (8),

$$\left| \sum_{|k| \leq N} (-1)^k e^{-2k^2 z^2} - \sum_{|k| \leq N} (-1)^k \frac{\binom{2n}{n-kc}}{\binom{2n}{n}} \right| < \frac{\varepsilon}{2}$$

for arbitrarily large n, it follows that for each $z > 0$ and any $\varepsilon > 0$

$$\left| P\left\{ \sqrt{\frac{n}{2}} D_n < z \right\} - K(z) \right| < \varepsilon$$

for sufficiently large values of n, Q.E.D.

We must make one further remark, as follows, in regard to the results presented in this section: The rule, suggested by our theorems, for testing the invariability of a distribution function does not depend on what the distribution function is. This fact is very important in practical applications, since the true distribution of a random variable is, as a rule, unknown, and it becomes of interest to ascertain just one thing: Is it possible to say that the unknown distribution has remained unchanged? The sole condition which has been imposed on the distribution function in deducing the theorems of this section is its continuity. This requirement often does not have to be verified, since it is a consequence of the physical character of the phenomenon studied.

§ 64. The Concept of Critical Region. Type I and Type II Errors. Comparison of Two Statistical Hypotheses

In the preceding sections, we presented two problems concerning the testing of statistical hypotheses. Both results did not depend on the distribution function of the random variable under consideration. Here, we shall consider a problem of a somewhat different nature.

Suppose that we know the functional form of the distribution of the random variable ξ but that we do not know the values of the parameters $\theta_1, \theta_2, \ldots, \theta_k$ upon which it depends. There exists some basis for presuming that the parameters have certain specific values $\theta_1 = \theta_1{}^0$, $\theta_2 = \theta_2{}^0$, ..., $\theta_k = \theta_k{}^0$ (simple hypothesis) or else belong to a certain set (composite hypothesis). It is required to ascertain whether or not the observed values of the variable ξ confirm this hypothesis.

In order to emphasize the practical importance of this problem, we discuss some examples.

Example 1. Consider a large lot of the products of a certain manufacture. Each unit of this product belongs to one of two categories: serviceable or defective. Any lot is regarded as suitable for shipment if the relative number p of defective units of the product is not too great— say, not larger than a certain number p_0 ($0 < p_0 < 1$). The quantity p is not known; its value is to be determined by investigating a comparatively small number of articles (small, that is, in comparison with the entire lot). Consider the random variable ξ, equal to zero if a randomly sampled item proves to be serviceable and equal to one if it is defective. The distribution function for ξ is

$$F(x) = \begin{cases} 0 & \text{for } x \leq 0, \\ 1-p & \text{for } 0 < x \leq 1, \\ 1 & \text{for } x > 1. \end{cases}$$

The parameter p on which the distribution depends is unknown. Our problem is to test the hypothesis that $p \leq p_0$.

Example 2. The random variable ξ is distributed normally:

$$p(x) = \frac{1}{\sigma \sqrt{2\pi}} e^{-\frac{(x-a)^2}{2\sigma^2}},$$

the parameters a and σ being unknown. It is required to test the hypothesis that

$$|a - a_0| \leq \alpha \text{ and } \sigma \leq \sigma_0,$$

where a_0, σ_0 and α are certain given quantities.

This, and analogous problems, are constantly arising in the theory of measurement as well as in the natural sciences and in production problems.

Let n be the number of observations on the basis of which it must be decided whether a given hypothesis is to be accepted or rejected. Let

$$x_1, x_2, \ldots, x_n \tag{1}$$

§ 64. COMPARISON OF TWO STATISTICAL HYPOTHESES

be the observed values. The testing process leading to the acceptance or rejection of the hypothesis is then a certain rule according to which the set of all possible outcomes of the n observations can be divided into two disjoint sets R_{n1} and R_{n2}. If the collection of numbers (1) is in the set R_{n1}, then this will correspond to the acceptance of the hypothesis in question, and if it is in R_{n2}, to the rejection of the hypothesis. If we represent the numbers in (1) as coordinates of a point in n-dimensional euclidean space R_n, then clearly every testing procedure implies a partitioning of the space R_n into two parts, R_{n1} and R_{n2}. If the point $(x_1, x_2, x_3, \ldots, x_n)$ is in the part R_{n1}, then the hypothesis is accepted, and if $(x_1, \ldots, x_n)$ is in R_{n2}, then the hypothesis is rejected. The set R_{n2} is referred to as the *critical region*. The selection of a rule for testing an hypothesis is thus equivalent to the selection of a critical region.

To illustrate this, let us return to Example 1. The set R_n in this case consists of all possible aggregates of n quantities each of which may take on the values 0 or 1 only. The critical region R_{n2} comprises those elements of R_n for which

$$(x_1 + x_2 + \ldots + x_n)/n > p_0.$$

We now turn to the following special problem of testing an hypothesis for which there is a complete solution: There are given two simple hypotheses H_1 and H_2. Hypothesis H_1 is that $\theta_i = \theta_i'$ $(i = 1, 2, \ldots, k)$, and hypothesis H_2 is that $\theta_i' = \theta_i''$ $(i = 1, 2, \ldots, k)$. These hypotheses are the alternative hypotheses between which we must choose, and it is required to make our choice on the basis of the observations.

Now, two kinds of error may be committed in accepting or rejecting the hypothesis H_1. The first type of error is committed when H_1 is rejected when it is actually true. In other words, an error of the first type occurs when the point $(x_1, x_2, \ldots, x_n)$ falls in the region R_{n2} when hypothesis H_1 is true. An error of the second type is made if we accept H_1 when it is false. These two possible types of error will be called errors of Type I and errors of Type II, respectively. If the critical region has been selected, we can then compute the probabilities of the errors of Type I and II; for given n and R_{n2}, let these probabilities be denoted by a_1 and a_2, respectively.

Clearly, the smaller the values of a_1 and a_2 for a given critical region, the better has been the choice of the critical region. However, for a given number of trials n, it is impossible, no matter what the choice of critical region, to make the quantities a_1 and a_2 simultaneously arbitrarily small.

However, by changing the critical region, we can achieve arbitrarily small errors of Type I and of Type II separately. Thus, if we set $R_{n2} = R_n$, it it is clear that $a_2 = 0$ in this case; however, if we set $R_{n1} = R_n$, then $a_1 = 0$. From this we derive the following rational principle for the selection of the critical region: For given values of a_1 and n, one should choose that region R_{n2} for which a_2 attains a minimum. Of course, when we do this, the smaller the value of a_1 that we choose, the larger is the minimum value attained by a_2. One cannot say in advance what value of a_1 should be chosen so that the method of testing the hypothesis be the most advantageous, because the practical side of the matter plays an integral part in the matter.

For example, suppose that the acceptance or rejection of hypothesis H_1 is concerned with the expenditure of money. If the acceptance of hypothesis H_1 when it is actually false results in a large expenditure (let us say, in having to adjust manually certain details of an assembly) whereas the rejection of hypothesis H_1 when true results in comparatively little expense, then clearly it is necessary to select a_2 as small as possible and in so doing we may have to reconcile ourselves to comparatively large values of a_1.

Let us suppose that practical considerations have been taken into account and that the quantity a_1 has been chosen; then the following theorem of Neyman and E. Pearson holds. We prove it, however, only for the case where the variable ξ has a finite density function both for the hypotheses H_1 and H_2.

THEOREM: *Of all the possible critical regions for which the probability of the error of Type* I *is* a_1, *the probability of the error of Type* II *assumes its least value for that critical region* R_{n2}^* *which consists of all those points* $(x_1, x_2, \ldots, x_n)$ *for which*[3]

$$\prod_{k=1}^{n} f(x_k/H_2) \geqq c \prod_{k=1}^{n} f(x_k/H_1). \tag{2}$$

The number c is determined by the condition

$$\psi(c) = \mathsf{P}\{(x_1, x_2, \ldots, x_n) \subset R_{n2}^*/H_1\} = a_1. \tag{3}$$

Proof: Since (for independent trials) the probability of the point $(x_1, x_2, \ldots, x_n)$ being in some region S is

---

[3] R_{n2}^* is thus the optimum critical region.

§ 64. Comparison of Two Statistical Hypotheses

$$P\{S/H_1\} = \int \ldots \int_S \prod_{k=1}^{n} f(x_k/H_1)\, dx_1\, dx_2 \ldots dx_n,$$

given that the hypothesis H_1 is true, and

$$P\{S/H_2\} = \int \ldots \int_S \prod_{k=1}^{n} f(x_k/H_2)\, dx_1\, dx_2 \ldots dx_n,$$

given that the hypothesis H_2 is true, it follows by assumption that

$$P\{R_{n2}^*/H_1\} = \alpha_1$$

and, for any other of the regions R_{n2} in question, that

$$P\{R_{n2}/H_1\} = \alpha_1.$$

By the axiom on the addition of probabilities,

$$P\{R_{n2} - R_{n2}\, R_{n2}^*/H_1\} = P\{R_{n2}/H_1\} - P\{R_{n2}\, R_{n2}^*/H_1\} =$$
$$= \alpha_1 - P\{R_{n2}\, R_{n2}^*/H_1\}$$

and

$$P\{R_{n2}^* - R_{n2}\, R_{n2}^*/H_1\} = \alpha_1 - P\{R_{n2}\, R_{n2}^*/H_1\},$$

i.e.,

$$P\{R_{n2} - R_{n2}\, R_{n2}^*/H_1\} = P\{R_{n2}^* - R_{n2}\, R_{n2}^*/H_1\}.$$

From the definition of R_{n2}^* and the last equality,

$$P\{R_{n2}^* - R_{n2}\, R_{n2}^*/H_2\} \geq c\, P\{R_{n2}^* - R_{n2}\, R_{n2}^*/H_1\} =$$
$$= c\, P\{R_{n2} - R_{n2}\, R_{n2}^*/H_1\}. \quad (4)$$

But for any point $(x_1, x_2, \ldots, x_n)$ not belonging to R_{n2}^*,

$$\prod_{k=1}^{n} f(x_k/H_2) < c \prod_{k=1}^{n} f(x_k/H_1),$$

and therefore, since the region $R_{n2} - R_{n2}\, R_{n2}^*$ lies entirely outside of R_{n2}^*, we must have

$$c\, P\{R_{n2} - R_{n2}\, R_{n2}^*/H_1\} > P\{R_{n2} - R_{n2}\, R_{n2}^*/H_2\}.$$

This inequality, together with (4), leads to the inequality

$$P\{R_{n2}^* - R_{n2}\, R_{n2}^*/H_2\} > P\{R_{n2} - R_{n2}\, R_{n2}^*/H_2\}.$$

Adding $\mathsf{P}\{R_{n2}R_{n2}^*/H_2\}$ to both sides of this inequality, we find:

$$\mathsf{P}\{R_{n2}^*/H_2\} > \mathsf{P}\{R_{n2}/H_2\}.$$

But since
$$\mathsf{P}\{R_n/H_2\} = 1$$
and
$$R_{n2}^* = R_n - R_{n1}^*, \quad R_{n2} = R_n - R_{n1},$$
it follows that
$$\mathsf{P}\{R_{n1}^*/H_2\} < \mathsf{P}\{R_{n1}/H_2\}.$$

This proves the theorem, since $\mathsf{P}\{R_{n1}/H_2\}$ and $\mathsf{P}\{R_{n1}^*/H_2\}$ are the probabilities of the errors of Type II for the critical regions R_{n2} and R_{n2}^*, respectively.

We must still confirm that the selection of the constant c can actually be effected by the rule (3). For this purpose, we note that the function

$$\psi(c) = \mathsf{P}\{R_{n2}^*/H_1\}$$

can only decrease as c increases (since the inequality (2) will be satisfied by an ever-shrinking set of points $(x_1, x_2, \ldots, x_n)$). Moreover, $\psi(0) = 1$ (since, for each point $(x_1, x_2, \ldots, x_n)$,

$$\prod_{k=1}^{n} f(x_k/H_2) \geq 0).$$

From (2) it follows, further, that

$$\mathsf{P}\{R_{n2}^*/H_2\} \geq c\,\mathsf{P}\{R_{n2}^*/H_1\}.$$

By substituting one on the left-hand side of this inequality and recalling the definition of $\psi(c)$, we obtain the inequality

$$1 \geq c\psi(c).$$

Thus
$$0 \leq \psi(c) \leq 1/c.$$

Therefore, $\psi(c) \to 0$ as $c \to \infty$. Since the function $\psi(c)$ does not increase, we can find a value of c such that for any α_1 ($0 < \alpha_1 < 1$)

$$\psi(c-0) \geq \alpha_1 \geq \psi(c+0).$$

This justifies the selection of the constant c according to equation (3) if the function $\psi(c)$ is continuous at the point c; however, if the function

§ 64. COMPARISON OF TWO STATISTICAL HYPOTHESES

$\psi(c)$ has a discontinuity at the point c, the situation becomes slightly more complicated and requires a slight change in the definition of R_{n2}^*, namely, the removal from it of a set of points $(x_1, x_2, \ldots, x_n)$ for which

$$\prod_{k=1}^{n} f(x_k/H_2) = c \prod_{k=1}^{n} f(x_k/H_1)$$

and their addition to the set R_{n1}^*, so that the probability of an error of Type I is equal to a_1.

Let us consider an example. Suppose given that ξ is normally distributed and that the value of the variance is σ^2. There exist two hypotheses concerning the expectation a, namely, that $a = a_1$ (hypothesis H_1) and $a = a_2$ (hypothesis H_2). It is required to determine the optimum critical region.

In this example, relation (2) is expressible in the following form:

$$e^{-\frac{1}{2\sigma^2} \sum_{k=1}^{n} [(x_k - a_2)^2 - (x_k - a_1)^2]} \geq c.$$

It is easy to show that this inequality is equivalent to the following (under the assumption that $a_2 > a_1$):

$$\sum_{k=1}^{n} x_k \geq \frac{\sigma^2 \log c}{a_2 - a_1} + \frac{n}{2}(a_2 + a_1),$$

or, what is the same, the inequality

$$\frac{1}{\sigma \sqrt{n}} \sum_{k=1}^{n} (x_k - a_1) \geq \frac{\sigma \log c}{(a_2 - a_1) \sqrt{n}} + \frac{\sqrt{n}}{2\sigma}(a_2 - a_1) = k_1.$$

The inequality obtained defines the optimum critical region R_{n2}^*.

Since the quantity

$$\frac{1}{\sigma \sqrt{n}} \sum_{k=1}^{n} (x_k - a_1)$$

is normally distributed with expectation 0 and variance 1, provided the hypothesis H_1 holds, we can readily determine k_1 (and hence c) for given a_1 by use of normal distribution tables. Suppose, for definiteness, that $a_1 = 0.05$. Then $k_1 = 1.645$ and, consequently, the optimum critical region for $a_1 = 0.05$ is given by the inequality

$$\sum_{k=1}^{n}(x_k - a_1) \geq 1.645\, \sigma \sqrt{n}.$$

It is interesting to note that the critical region in our example is independent of the alternative value $a_2 > a_1$.

The region R_{n1}^* is defined by the inequality

$$\sum_{k=1}^{n} x_k < \frac{\sigma^2 \log c}{a_2 - a_1} + \frac{n}{2}(a_2 + a_1),$$

which is clearly expressible as follows:

$$\frac{1}{\sigma \sqrt{n}} \sum_{k=1}^{n}(x_k - a_2) < k_1 - \frac{\sqrt{n}}{\sigma}(a_2 - a_1).$$

Under the assumption that the hypothesis H_2 holds, the quantity on the left-hand side of the inequality is normally distributed with expectation 0 and variance 1. It follows from this that the probability of an error of Type II is

$$\Phi\left(k_1 - \frac{\sqrt{n}}{\sigma}(a_2 - a_1)\right) = \frac{1}{\sqrt{2\pi}} \int_{-\infty}^{k_1 - \frac{\sqrt{n}}{\sigma}(a_2 - a_1)} e^{-\frac{z^2}{2}} dz.$$

If the quantities α_1 and α_2 are assigned, then there arises the problem of determining the minimal number of trials $n = n(\alpha_1, \alpha_2)$ that are necessary for the probabilities of drawing erroneous conclusions to be not larger than α_1 and α_2.

As n increases, the quantity $\alpha_{2n} = \alpha_2(\alpha_1, n)$ is non-increasing and, in general, approaches zero. It is evident that $n(\alpha_1, \alpha_2)$ is the smallest of those values of n for which $\alpha_2(\alpha_1, n) \leq \alpha_2$.

In the example just considered, the number n can be found quite simply for given values of α_1 and α_2. In fact, from

$$1 - \Phi(k_1) = \alpha_1$$

and

$$\Phi\left(k_1 - \frac{\sqrt{n}}{\sigma}(a_2 - a_1)\right) = \alpha_2,$$

we obtain the two equations:

§ 65. CLASSICAL PROCEDURE FOR ESTIMATING THE PARAMETERS

and
$$k_1 = \psi(1 - \alpha_1)$$

$$k_1 - \frac{\sqrt{n}}{\sigma}(a_2 - a_1) = \psi(\alpha_2),$$

where ψ is the function inverse to $\Phi(x)$. Hence,

$$n = \frac{\sigma^2}{(a_2 - a_1)^2}[\psi(1 - \alpha_1) - \psi(\alpha_2)]^2.$$

We shall now give a short numerical example. Let

$$a_1 = 135, \; a_2 = 150, \; \sigma = 25, \; \alpha_1 = 0.01, \; \alpha_2 = 0.03.$$

Since
$$\psi(0.99) = 2.33, \; \psi(0.03) = -1.88,$$

we find that

$$n = \frac{25^2}{15^2}[2.33 + 1.88]^2 = \frac{25}{9} \cdot 4.21^2 \approx 49.$$

Thus, the minimum number of observations that are needed to make a selection between the hypotheses H_1 and H_2 in the case of a normal distribution with the data just given is 49. Only for this number of trials can we be certain that if the hypothesis H_1 is true, then the probability that we will reject it is not larger than 0.01 and if the hypothesis H_2 is true, then the probability that we will reject it is not larger than 0.03.

§ 65. The Classical Procedure for Estimating the Distribution Parameters

The classical procedure for estimating the unknown parameters in the distribution function of a random variable ξ is based on the following. Before any observations are made, the quantities to be estimated are considered to be random variables obeying some *apriori* (pre-experimental) distribution law. By assuming this apriori distribution law to be known and by applying Bayes' Theorem, we are able to compute the *aposteriori* (post-experimental) distribution law for the parameters under the condition that the observed values of ξ have been found to be $x_1, x_2, \ldots, x_n$.

XI. Elements of Statistics

As already stated earlier, the entire discussion that follows will be concerned with determining the unknown parameters a and σ appearing in the normal distribution law

$$p(x/a, \sigma) = \frac{1}{\sigma\sqrt{2\pi}} e^{-\frac{(x-a)^2}{2\sigma^2}},$$

which our random variable ξ obeys.

The density function

$$f(x_1, x_2, \ldots, x_n/a, \sigma) = \frac{1}{(\sigma\sqrt{2\pi})^n} e^{-\frac{s^2}{2\sigma^2}},$$

where

$$s^2 = \sum_{k=1}^{n} (x_k - a)^2$$

corresponds to the probability that $x_1, x_2, \ldots, x_n$ will be obtained as the result of n independent observations for the values of the variable ξ, subject to the condition that the unknown parameters have the values a and σ.

If we introduce the notation

$$\bar{x} = \frac{1}{n} \sum_{k=1}^{n} x_k,$$

$$s_1^2 = \frac{1}{n} \sum_{k=1}^{n} (x_k - \bar{x})^2,$$

then a simple computation shows that

$$f(x_1, \ldots, x_n/a, \sigma) = \frac{1}{(\sigma\sqrt{2\pi})^n} e^{-\frac{n}{2\sigma^2}[s_1^2 + (\bar{x}-a)^2]}. \tag{1}$$

Let us recall the following three problems that were formulated in § 60:

1) σ is known; it is required to estimate a;
2) a is known; it is required to estimate σ;
3) a and σ are unknown and are to be estimated.

Let us suppose that σ is known, and let $\varphi_1(a)$ denote the apriori density function of the quantity a; then for a given value of σ and the observed values $x_1, x_2, \ldots, x_n$ we obtain as the conditional density function of the variable a the following expression:

§ 65. Classical Procedure for Estimating the Parameters

$$\varphi_1(a/x_1, x_2, \ldots, x_n; \sigma) = \frac{f(x_1, x_2, \ldots, x_n/a, \sigma)\,\varphi_1(a)}{\int f(x_1, x_2, \ldots, x_n/a, \sigma)\,\varphi_1(a)\,da}.$$

By substituting in this the value of the function f given by equation (1) and making obvious simplifications, we find:

$$\varphi_1(a/x_1, x_2, \ldots, x_n; \sigma) = \frac{e^{-\frac{n(a-\bar{x})^2}{2\sigma^2}} \cdot \varphi_1(a)}{\int e^{-\frac{n(a-\bar{x})^2}{2\sigma^2}} \cdot \varphi_1(a)\,da}. \tag{2}$$

The corresponding expressions in the second and third problems are

$$\varphi_2(\sigma/x_1, x_2, \ldots, x_n; a) = \frac{f(x_1, x_2, \ldots, x_n/a, \sigma)\,\varphi_2(\sigma)}{\int f(x_1, x_2, \ldots, x_n/a, \sigma)\,\varphi_2(\sigma)\,d\sigma}$$

and

$$\varphi_3(a, \sigma/x_1, x_2, \ldots, x_n) = \frac{f(x_1, x_2, \ldots, x_n/a, \sigma)\,\varphi_3(a, \sigma)}{\int\int f(x_1, x_2, \ldots, x_n/a, \sigma)\,\varphi_3(a, \sigma)\,da\,d\sigma},$$

where the functions $\varphi_2(\sigma)$ and $\varphi_3(a, \sigma)$ denote the apriori density functions of the variable σ and the pair (a, σ).

By substituting in these formulas the value of f given by equation (1) and by making simple reductions, we find:

$$\varphi_2(\sigma/x_1, x_2, \ldots, x_n; a) = \frac{\sigma^{-n} e^{-\frac{s^2}{2\sigma^2}} \varphi_2(\sigma)}{\int_0^\infty \sigma^{-n} e^{-\frac{s^2}{2\sigma^2}} \varphi_2(\sigma)\,d\sigma} \tag{3}$$

and

$$\varphi_3(a, \sigma/x_1, x_2, \ldots, x_n) = \frac{\sigma^{-n} e^{-\frac{n}{2\sigma^2}(s_1^2+(a-\bar{x})^2)} \varphi_3(a, \sigma)}{\int\int_0^\infty \sigma^{-n} e^{-\frac{n}{2\sigma^2}(s_1^2+(a-\bar{x})^2)} \varphi_3(a, \sigma)\,da\,d\sigma}. \tag{4}$$

The expressions obtained are not suitable for practical application not only because of their complexity but also, and mainly, because the apriori probabilities appearing in them are, as a rule, unknown. Since these apriori densities are often not known, assumptions about them are made which are more or less arbitrary, and on the basis of these, synoptical-type formulas are obtained for practical application. We shall take a different path; we shall make perfectly general assumptions about the character of the apriori distributions and from these assumptions we shall deduce limit laws (as $n \to \infty$) for the aposteriori probabilities.

THEOREM 1: *If the apriori density function $\varphi_1(a)$ has a bounded first derivative and*

$$\varphi_1(\bar{x}) \neq 0,$$

then

$$\psi_1(a/x_1, x_2, \ldots, x_n; \sigma) = \frac{1}{\sqrt{2\pi}} e^{-\frac{1}{2}\alpha^2} \left[1 + O\left(\frac{\sigma}{\sqrt{n}}\right)(1 + |\alpha|)\right] \quad (5)$$

uniformly in a, where

$$\alpha = \frac{\sqrt{n}}{\sigma}(a - \bar{x}) \quad (6)$$

and $\psi_1(a/x_1, x_2, \ldots, x_n; \sigma)$ denotes the aposteriori density function of the variable a.

Proof: For, from (6) we find:

$$a = \bar{x} + \frac{\alpha\sigma}{\sqrt{n}}. \quad (7)$$

and this means[4]

$$\frac{\sigma}{\sqrt{n}} \varphi_1(a/x_1, x_2, \ldots, x_n; \sigma) = \frac{e^{-\frac{\alpha^2}{2}} \varphi_1\left(\bar{x} + \frac{\alpha\sigma}{\sqrt{n}}\right)}{\int e^{-\frac{\alpha^2}{2}} \varphi_1\left(\bar{x} + \frac{\alpha\sigma}{\sqrt{n}}\right) d\alpha}.$$

By the mean-value theorem

$$\varphi_1\left(\bar{x} + \frac{\alpha\sigma}{\sqrt{n}}\right) = \varphi_1(\bar{x}) + \frac{\alpha\sigma}{\sqrt{n}} \varphi_1'(z),$$

where $z = \bar{x} + \theta a\sigma/\sqrt{n}$ and $0 < \theta < 1$.

By assumption,

$$|\varphi_1'(z)| \leq C < +\infty,$$

and therefore

$$\frac{\sigma}{\sqrt{n}} \varphi_1(a/x_1 x_2, \ldots, x_n; \sigma) = \frac{e^{-\frac{\alpha^2}{2}} \left[\varphi_1(\bar{x}) + \frac{\alpha\sigma}{\sqrt{n}} \varphi_1'(z)\right]}{\sqrt{2\pi} \left[\varphi_1(\bar{x}) + r_n\right]},$$

where

---

[4] Note that the density function of the variable a is given by

$$\psi_1(a/x_1, x_2, \ldots, x_n; \sigma) = \frac{\sigma}{\sqrt{n}} \varphi_1(a/x_1, x_2, \ldots, x_n; \sigma).$$

§ 65. Classical Procedure for Estimating the Parameters

$$r_n = \frac{\sigma}{\sqrt{2\pi n}} \int \alpha e^{-\frac{\alpha^2}{2}} \varphi'_1(z)\, d\alpha.$$

We can easily show that

$$|r_n| < \frac{2C\sigma}{\sqrt{2\pi n}}. \tag{8}$$

Simple transformations lead to the expression

$$\psi_1(\alpha/x_1, x_2, \ldots, x_n; \sigma) = \frac{1}{\sqrt{2\pi}} e^{-\frac{\alpha^2}{2}} \left[1 + \frac{\sigma}{\sqrt{n}} \frac{\alpha \varphi'_1(z) - \frac{\sqrt{n}}{\sigma} r_n}{\varphi_1(\bar{x}) + r_n} \right].$$

This expression, in conjunction with (8), proves the theorem.

THEOREM 2: *Under the conditions of the preceding theorem, we have*

$$\left. \begin{aligned} \mathsf{E}\,(a/x_1, x_2, \ldots, x_n;\, \sigma) &= \bar{x} + O\!\left(\frac{\sigma^2}{n}\right), \\ \mathsf{E}\,[(a-\bar{x})^2/x_1, x_2, \ldots, x_n;\, \sigma] &= \frac{\sigma^2}{n}\left[1 + O\!\left(\frac{\sigma}{\sqrt{n}}\right)\right]. \end{aligned} \right\} \tag{5'}$$

Proof: From (7), we find:

$$\mathsf{E}\,(a/B) = \bar{x} + \frac{\sigma}{\sqrt{n}}\,\mathsf{E}\,(\alpha/B)$$

and

$$\mathsf{E}\,[(a-\bar{x})^2/B] = \frac{\sigma^2}{n}\,\mathsf{E}\,(\alpha^2/B),$$

where B stands for some event. Consequently,

$$\mathsf{E}\,(a/x_1, x_2, \ldots, x_n;\, \sigma) = \bar{x} + \frac{\sigma}{\sqrt{n}} \int \alpha\, \psi_1(\alpha/x_1, \ldots, x_n;\, \sigma)\, d\alpha$$

and

$$\mathsf{E}\,[(a-\bar{x})^2/x_1, x_2, \ldots, x_n;\, \sigma] = \frac{\sigma^2}{n} \int \alpha^2\, \psi_1(\alpha/x_1, \ldots, x_n;\, \sigma)\, d\alpha.$$

The substitution in these expressions of the value of the function ψ_1 given in (5) followed by simple reductions then yields the theorem.

This theorem permits us to write the following approximation:

$$a \sim \bar{x},$$

the mean-square error of which is approximately σ^2/n.

Theorem 1 enables us to obtain the probability that a will lie within given limits, under the condition that the quantities $\sigma, x_1, x_2, \ldots, x_n$ have taken on specific values. In fact,

$$\mathsf{P}\left\{|a-\bar{x}|<\frac{\sigma z}{\sqrt{n}}\Big/x_1,\ldots,x_n;\ \sigma\right\}=\mathsf{P}\{|\alpha|<z/x_1,\ldots,x_n;\ \sigma\},$$

and therefore, in view of (5),

$$\mathsf{P}\left\{|a-\bar{x}|<\frac{\sigma z}{\sqrt{n}}\Big/x_1,\ldots,x_n;\ \sigma\right\}=\frac{2}{\sqrt{2\pi}}\int_0^z e^{-\frac{t^2}{2}}dt+O\left(\frac{\sigma}{\sqrt{n}}\right).$$

Neglecting the remainder $O(\sigma/\sqrt{n})$ (which, in general, may be done only when σ is small or when n is sufficiently large), we can state that

$$\mathsf{P}\left\{|a-\bar{x}|<\frac{\sigma}{\sqrt{n}}z/x_1, x_2,\ldots,x_n;\ \sigma\right\}=\frac{2}{\sqrt{2\pi}}\int_0^z e^{-\frac{t^2}{2}}dt.$$

THEOREM 3: *If the apriori density function $\varphi_2(\sigma)$ has a bounded first derivative and $\varphi_2(\bar{s}) \neq 0$, then*

$$\psi_2(z/x_1, x_2,\ldots,x_n;\ a)=\frac{1}{\sqrt{\pi}}e^{-z^2}\left[1+O\left(\frac{1}{\sqrt{n}}\right)(1+|z|)\right]$$

uniformly in z, where ψ_2 is the aposteriori density of the variable

$$\beta=\frac{\sigma-\bar{s}}{\bar{s}}\sqrt{n} \qquad (9)$$

and $\bar{s} = s/\sqrt{n}$.

Proof: From (9), we find:

$$\sigma=\bar{s}\left(1+\frac{\beta}{\sqrt{n}}\right),$$

and

$$\psi_2(z/x_1, x_2,\ldots,x_n;\ a)=\frac{\bar{s}}{\sqrt{n}}\varphi_2\left(\bar{s}\left(1+\frac{z}{\sqrt{n}}\right)\Big/x_1, x_2,\ldots,x_n;\ a\right)$$

$$=\frac{\left(1+\frac{z}{\sqrt{n}}\right)^{-n}e^{-\frac{n}{2\left(1+\frac{z}{\sqrt{n}}\right)^2}}\varphi_2\left(\bar{s}+\frac{z\bar{s}}{\sqrt{n}}\right)}{\int_{-\sqrt{n}}^{\infty}\left(1+\frac{\beta}{\sqrt{n}}\right)^{-n}e^{-\frac{n}{2\left(1+\frac{\beta}{\sqrt{n}}\right)^2}}\varphi_2\left(\bar{s}+\frac{\beta\bar{s}}{\sqrt{n}}\right)d\beta}.$$

§ 65. Classical Procedure for Estimating the Parameters

By the mean-value theorem

$$\varphi_2\left(\bar{s} + \frac{z\bar{s}}{\sqrt{n}}\right) = \varphi_2(\bar{s}) + \frac{z\bar{s}}{\sqrt{n}}\varphi_2'(u),$$

where $u = \bar{s} + \theta z\bar{s}/\sqrt{n}$ and $0 < \theta < 1$.

By assumption,

$$|\varphi_2'(u)| \leq C < +\infty,$$

and therefore

$\psi_2(z/x_1, x_2, \ldots, x_n; a) =$

$$= \frac{\left(1+\dfrac{z}{\sqrt{n}}\right)^{-n} e^{-\dfrac{n}{2\left(1+\dfrac{z}{\sqrt{n}}\right)^2}} \left[\varphi_2(\bar{s}) + z\bar{s}\, O\!\left(\dfrac{1}{\sqrt{n}}\right)\right]}{\varphi_2(\bar{s})\displaystyle\int_{-\sqrt{n}}^{\infty}\left(1+\dfrac{\beta}{\sqrt{n}}\right)^{-n} e^{-\dfrac{n}{2\left(1+\dfrac{\beta}{\sqrt{n}}\right)^2}} d\beta \left[1 + O\!\left(\dfrac{1}{\sqrt{n}}\right)\right]}$$

$$= \frac{\left(1+\dfrac{z}{\sqrt{n}}\right)^{-n} e^{-\dfrac{n}{2\left(1+\dfrac{z}{\sqrt{n}}\right)^2}} \left[1 + O\!\left(\dfrac{1}{\sqrt{n}}\right)(1+|z|)\right]}{\displaystyle\int_{-\sqrt{n}}^{\infty}\left(1+\dfrac{\beta}{\sqrt{n}}\right)^{-n} e^{-\dfrac{n}{2\left(1+\dfrac{\beta}{\sqrt{n}}\right)^2}} d\beta}. \tag{10}$$

But

$$-n\log\left(1+\frac{z}{\sqrt{n}}\right) - \frac{n}{2\left(1+\dfrac{z}{\sqrt{n}}\right)^2} = -n\left(\frac{z}{\sqrt{n}} - \frac{1}{2}\frac{z^2}{n} + \ldots\right) -$$

$$- \frac{n}{2}\left(1 - 2\frac{z}{\sqrt{n}} + 3\frac{z^2}{n} - \ldots\right) = -\frac{n}{2} - z^2 + O\!\left(\frac{1}{\sqrt{n}}\right) \tag{11}$$

and

$$J = \int_{-\sqrt{n}}^{\infty}\left(1+\frac{\beta}{\sqrt{n}}\right)^{-n} e^{-\dfrac{n}{2\left(1+\dfrac{\beta}{\sqrt{n}}\right)^2}} d\beta =$$

$$= n^{-\frac{n}{2}+1} 2^{\frac{n-3}{2}} \int_0^\infty z^{\frac{n-3}{2}} e^{-z}\, dz = n^{-\frac{n}{2}+1} 2^{\frac{n-3}{2}} \Gamma\!\left(\frac{n-1}{2}\right).$$

By Stirling's formula,

$$\Gamma\left(\frac{n-1}{2}\right) = \sqrt{2\pi\frac{n-3}{2}}\left(\frac{n-3}{2}\right)^{\frac{n-3}{2}} e^{-\frac{n-3}{2}} [1 + o(1)],$$

and therefore

$$J = \sqrt{\pi}\left(\frac{n-3}{2}\right)^{\frac{n}{2}-1} 2^{\frac{n-3}{2}} n^{-\frac{n}{2}+1} e^{-\frac{n}{2}+\frac{3}{2}} [1 + o(1)]$$

$$= \sqrt{\pi}\left(1 - \frac{3}{n}\right)^{\frac{n}{2}} e^{-\frac{n}{2}+\frac{3}{2}} [1 + o(1)] = \sqrt{\pi} e^{-\frac{n}{2}} [1 + o(1)]. \quad (12)$$

Equations (10), (11), and (12) prove the theorem.

Theorem 3 clearly implies the following result:

THEOREM 4: *Under the conditions of Theorem 3, the following relations hold:*

$$\mathsf{E}\,(\sigma/x_1,\,x_2,\,\ldots,\,x_n;\,a) = \bar{s} + O\left(\frac{1}{n}\right)$$

and

$$\mathsf{E}\,[(\sigma - \bar{s})^2/x_1\,x_2,\,\ldots,\,x_n;\,a] = \frac{\bar{s}^2}{2n}\left[1 + O\left(\frac{1}{\sqrt{n}}\right)\right].$$

This theorem allows us to conclude that the approximate relation

$$\sigma \sim \sqrt{\frac{1}{n}\sum_{k=1}^{n}(x_k - a)^2}$$

holds for large values of n and that the mean-square error of σ is approximately $\bar{s}^2/2n$.

Theorem 3 can also be used to determine the probability that σ lies within given limits. Thus, by neglecting quantities of the order $1/\sqrt{n}$, we can state that for given values of $x_1, x_2, \ldots, x_n$, and a, σ will lie in the interval

$$\bar{s}\left(1 - \frac{z}{\sqrt{n}}\right) < \sigma < \bar{s}\left(1 + \frac{z}{\sqrt{n}}\right)$$

with probability

$$\frac{2}{\sqrt{\pi}}\int_{0}^{+z} e^{-t^2}\,dt.$$

As to the third of our three problems, we shall confine ourselves to the statement of results only, since their derivation in no way differs from the proof of Theorem 3. Let us introduce the notation

§ 65. Classical Procedure for Estimating the Parameters

$$\alpha_1 = \frac{a-\bar{x}}{\bar{s}_1}\sqrt{n}, \quad \beta_1 = \frac{\sigma-\bar{s}_1}{\bar{s}_1}\sqrt{n},$$

where $\bar{s}_1 = s_1\sqrt{n}/(n-1)$.

THEOREM 5: *If the apriori density function $\varphi_3(a, \sigma)$ has bounded first derivatives with respect to a and σ and $\varphi_3(x, s_1) \neq 0$, then*

$$\psi_3(\alpha_1, \beta_1/x_1, x_2, \ldots, x_n) = \frac{1}{\pi\sqrt{2}} e^{-\frac{\alpha_1^2}{2} - \beta_1^2}\left(1 + O\left(\frac{1}{\sqrt{n}}\right)[1 + |\alpha_1| + |\beta_1|]\right)$$

uniformly in α_1 and β_1, where ψ_3 stands for the aposteriori density function of the pair (α_1, β_1).

From Theorem 5 there follows the result:

THEOREM 6: *Under the conditions of Theorem 5,*

$$E(a/x_1, x_2, \ldots, x_n) = \bar{x} + O\left(\frac{1}{\sqrt{n}}\right),$$

$$E[(a-\bar{x})^2/x_1, x_2, \ldots, x_n] = \frac{\bar{s}_1^2}{n}\left[1 + O\left(\frac{1}{\sqrt{n}}\right)\right],$$

$$E[\sigma/x_1, x_2, \ldots, x_n] = \bar{s}_1\left[1 + O\left(\frac{1}{n}\right)\right],$$

$$E[(\sigma-\bar{s}_1)^2/x_1, x_2, \ldots, x_n] = \frac{\bar{s}_1^2}{2n}\left[1 + O\left(\frac{1}{\sqrt{n}}\right)\right].$$

Just as in the case of Theorems 1 and 3, Theorem 5 can be used to determine the probability that a and σ will lie within given limits, subject to the condition that the observed values have been found to be $x_1, x_2, x_3, \ldots, x_n$.

The practical significance of Theorems 1, 3, and 5 is not the same. According to Theorem 1, the accuracy of the approximations (5) and (5') not only increases with increasing n but also with decreasing σ. Therefore, to determine a for given σ, there is justification for making use of formulas (5) and (5') even for small values of n, provided that σ is small. However, in the case of Theorems 3 and 5, the remainder terms in the corresponding formulas decrease only with increasing n, and therefore they are inapplicable for small values of n.

The theorems just proved are, in a certain sense, inversions of the following elementary statements. If the random variable ξ is normally distributed, if the parameters a and σ are known, and if $x_1, x_2, \ldots, x_n$ are the outcomes of independent observations for ξ, then:

XI. Elements of Statistics

1. The density function of the variable
$$\alpha = \frac{\sqrt{n}}{\sigma}(\bar{x}-a)$$
is
$$\psi(x/a,\sigma) = \frac{1}{\sqrt{2\pi}}e^{-\frac{x^2}{2}}.$$

2. $\mathsf{E}(\bar{x}/a,\sigma) = a$ and $\mathsf{D}(\bar{x}/a,\sigma) = \sigma^2/n$;

3. The density function of the variable
$$\beta = \frac{\bar{s}-\sigma}{\sigma}\sqrt{2n}$$
is asymptotically equal to
$$\psi_2(x/a,\sigma) = \frac{1}{\sqrt{2\pi}}e^{-\frac{x^2}{2}}.$$

4. $\mathsf{E}(\bar{s}/a,\sigma) = \sigma\left\{1+O\left(\frac{1}{n}\right)\right\}$; $\mathsf{D}(\bar{s}/a,\sigma) = \frac{\sigma^2}{2n}\left[1+O\left(\frac{1}{n}\right)\right]$.

5. The variables α and β are independent and the density function of the variable (α,β) is asymptotically equal to
$$\psi_3(x,y/a,\sigma) = \frac{1}{2\pi}e^{-\frac{x^2+y^2}{2}}.$$

6. $\mathsf{E}(\bar{x}/a,\sigma) = a$, $\mathsf{D}(\bar{x}/a,\sigma) = \frac{\sigma^2}{n}$;

$\mathsf{E}(\bar{s}/a,\sigma) \approx \sigma$, $\mathsf{D}(\bar{s}/a,\sigma) \approx \frac{\sigma^2}{2n}$.

Statements 1. and 2. require no proof.

Let us prove 3. In § 24, we found that the density function for the variable $\bar{s}$ is
$$\varphi(y) = \frac{\sqrt{2n}}{\sigma\Gamma\left(\frac{n}{2}\right)}\left(\frac{y\sqrt{n}}{\sigma\sqrt{2}}\right)^{n-1}e^{-\frac{ny^2}{2\sigma^2}}.$$

It is easy to verify that the density function for β is
$$\psi_2(x/a,\sigma) = \frac{\sigma}{\sqrt{2n}}\varphi\left(\frac{\sigma x}{\sqrt{2n}}+\sigma\right).$$

After simplification, we arrive at the expression

$$\psi_2(x/a, \sigma) = \frac{1}{\sqrt{2\pi}} e^{-\frac{x^2}{2}}.$$

To prove 4., we observe that elementary computations lead to the relations

$$\mathsf{E}\,\bar{s} = \sqrt{\frac{2}{n}} \frac{\Gamma\left(\frac{n+1}{2}\right)}{\Gamma\left(\frac{n}{2}\right)} \sigma \sim e^{-\frac{1}{4n}} \sigma, \quad \mathsf{E}\,\bar{s}^2 = \frac{2}{n} \frac{\Gamma\left(\frac{n}{2}+1\right)}{\Gamma\left(\frac{n}{2}\right)} \sigma^2 = \sigma^2.$$

Hence

$$\mathsf{E}\,\bar{s} \approx \sigma\left(1 - \frac{1}{4n}\right)$$

and

$$\mathsf{D}\,\bar{s} \approx \frac{\sigma^2}{2n}\left(1 - \frac{1}{8n}\right).$$

We shall prove the independence of $\bar{x}$ and $\bar{s}$ later. Once this is done, the remaining statements contained in 5. and 6. become obvious.

§ 66. Confidence Limits

The procedure given in § 65 for estimating the unknown parameters started from the assumption that any parameter subject to estimation was a random variable with a certain apriori probability distribution. Actually, such an assumption often makes no sense, simply because what we are estimating is an unknown constant. But even when the parameter is in fact a random variable, it turns out to be difficult to apply the exact formulas of the preceding section, since the apriori distribution of this random variable is not known.

Another approach to estimating unknown parameters was proposed by the English statistician, R. A. Fisher. His idea was developed and reformulated in the papers of the well-known contemporary statistician, J. Neyman.

In the classical formulation of the problem, it was required to establish certain fixed limits for an unknown parameter such that the probability of its lying between these limits would be comparable to (or no less than) a sufficiently large number ω. In the modern formulation of the problem, one is required to find functions of the observed values, θ' and θ'', where

$$\theta'(x_1, x_2, \ldots, x_n) < \theta''(x_1, x_2, \ldots, x_n)$$

(these functions are thus random variables) such that the value of the parameter a lies in the interval (θ', θ'') with a probability not less than ω.

The functions θ' and θ'' are called *confidence limits*.

We illustrate this approach by estimating the parameters in the normal distribution.

In the first of our three problems (p. 410), we set

$$a' = \bar{x} + \frac{z_1}{\sqrt{n}} \sigma, \quad a'' = \bar{x} + \frac{z_2}{\sqrt{n}} \sigma.$$

For any values of the parameters a and σ, we then obviously have

$$\mathsf{P}\{a' \leq a < a'' | a, \sigma\} = \mathsf{P}\left\{a - \frac{z_2 \sigma}{\sqrt{n}} \leq \bar{x} \leq a - \frac{z_1 \sigma}{\sqrt{n}} \Big| a, \sigma\right\}$$

$$= \frac{1}{\sqrt{2\pi}} \int_{-z_2}^{-z_1} e^{-\frac{t^2}{2}} dt = \frac{1}{\sqrt{2\pi}} \int_{z_1}^{z_2} e^{-\frac{t^2}{2}} dt.$$

From this, we conclude that the probability of the inequality

$$\bar{x} + \frac{z_1 \sigma}{\sqrt{n}} \leq a < \bar{x} + \frac{z_2 \sigma}{\sqrt{n}}$$

is

$$\omega = \frac{1}{\sqrt{2\pi}} \int_{z_1}^{z_2} e^{-\frac{t^2}{2}} dt.$$

Therefore we have, in particular,

$$\mathsf{P}\left\{|a - \bar{x}| \leq \frac{z}{\sqrt{n}} \sigma\right\} = \frac{2}{\sqrt{2\pi}} \int_0^z e^{-\frac{t^2}{2}} dt.$$

In the second problem, we regard the parameter a as known, whereas the parameter σ is to be estimated. We set

$$\sigma' = \frac{\bar{s}\sqrt{n}}{t_1}, \quad \sigma'' = \frac{\bar{s}\sqrt{n}}{t_2},$$

where

§ 66. Confidence Limits

$$\bar{s} = \sqrt{\frac{1}{n}\sum_{k=1}^{n}(x_k-a)^2}.$$

It is easy to see that

$$P\{\sigma' \leq \sigma < \sigma''/a, \sigma\} = P\left\{\frac{\sigma t_2}{\sqrt{n}} < \bar{s} \leq \frac{t_1\sigma}{\sqrt{n}} \bigg/ a, \sigma\right\}.$$

In § 24 (the χ^2-distribution), we found the density function of the variable $\bar{s}$ under the assumption that a and σ are given. Namely,

$$\varphi(y/a, \sigma) = \frac{\sqrt{2n}}{\sigma\,\Gamma\left(\frac{n}{2}\right)}\left(\frac{y\sqrt{n}}{\sigma\sqrt{2}}\right)^{n-1} e^{-\frac{y^2 n}{2\sigma^2}}.$$

From this, we find:

$$P\{\sigma' \leq \sigma < \sigma''/a, \sigma\} = \frac{\sqrt{2n}}{\sigma\,\Gamma\left(\frac{n}{2}\right)} \int_{\frac{\sigma t_2}{\sqrt{n}}}^{\frac{\sigma t_1}{\sqrt{n}}} \left(\frac{y\sqrt{n}}{\sigma\sqrt{2}}\right)^{n-1} e^{-\frac{y^2 n}{2\sigma^2}} dy$$

$$= \frac{1}{2^{\frac{n-2}{2}}\,\Gamma\left(\frac{n}{2}\right)} \int_{t_2}^{t_1} z^{n-1} e^{-\frac{z^2}{2}} dz.$$

We thus see that the conditional probability of the inequality

$$\sigma' \leq \sigma < \sigma''$$

under the assumption that the parameters a and σ are known, does not depend on the values of these parameters. Therefore, by the preceding discussion, the probability of the inequality

$$\frac{1}{t_1}\sqrt{\sum_{k=1}^{n}(x_k-a)^2} \leq \sigma < \frac{1}{t_2}\sqrt{\sum_{k=1}^{n}(x_k-a)^2}$$

is

$$\omega = \frac{1}{2^{\frac{n-2}{2}}\,\Gamma\left(\frac{n}{2}\right)} \int_{t_2}^{t_1} z^{n-1} e^{\frac{z^2}{2}} dz.$$

We finally pass to a consideration of the last problem, where both of the parameters, a and σ, are unknown. Set

$$a'_1 = \bar{x} + c_1 \sqrt{n}\, s_1 \quad \text{and} \quad a''_1 = \bar{x} + c_2 \sqrt{n}\, s_1,$$

$$\sigma'_1 = \frac{\sqrt{n}\, s_1}{t_1}, \quad \sigma''_1 = \frac{\sqrt{n}\, s_1}{t_2},$$

where

$$s_1 = \sqrt{\frac{1}{n} \sum_{k=1}^{n} (x_k - \bar{x})^2}.$$

Under the condition that a and σ are given, we have

$$\mathsf{P}\{a'_1 \leq a < a''_1 / a, \sigma\} = \mathsf{P}\left\{c_1 \leq \frac{a - \bar{x}}{s_1 \sqrt{n}} < c_2 / a, \sigma\right\}$$

and

$$\mathsf{P}\{\sigma'_1 \leq \sigma < \sigma''_1 / a, \sigma\} = \mathsf{P}\left\{t_2 < \frac{s_1 \sqrt{n}}{\sigma} \leq t_1 / a, \sigma\right\}.$$

We must now determine the conditional density functions for $(a - \bar{x})/s_1 \sqrt{n}$ and $s_1 \sqrt{n}/\sigma$ under the condition that a and σ are known. Now

$$\frac{\bar{x} - a}{s_1 \sqrt{n}} = \frac{\frac{1}{n} \sum_{k=1}^{n}(x_k - a)}{\sqrt{\sum_{k=1}^{n}[(x_k - a) - (\bar{x} - a)]^2}} = \frac{\bar{x}'}{\sqrt{\sum_{k=1}^{n}(x'_k - \bar{x}')^2}},$$

where $x'_k = x_k - a$ and $\bar{x}' = \frac{1}{n} \sum_{k=1}^{n} x'_k$ (the variables x_k' are independent and normally distributed with expectation 0 and variance σ^2).

We now introduce a new system of orthogonal coordinates $(y_1, y_2, y_3, \ldots, y_n)$ into the n-dimensional space $(x_1', x_2', \ldots, x_n')$, so that $y_1 = \sqrt{n}\bar{x}'$. Then

$$n s_1^2 = \sum_{k=1}^{n}(x'_k - \bar{x}')^2 = \sum_{k=1}^{n} x_k'^2 - n\bar{x}'^2 = \sum_{k=1}^{n} y_k^2 - y_1^2 = \sum_{k=2}^{n} y_k^2,$$

and therefore

§ 66. Confidence Limits

$$\frac{\bar{x}-a}{s_1/\sqrt{n}} = \frac{y_1}{\sqrt{n\sum_{k=2}^{n} y_k^2}}.$$

Since

$$y_k = \sum_{i=1}^{n} \alpha_{ki}\, x_i',$$

where the quantities α_{ki} satisfy the relations

$$\sum_{j=1}^{n} \alpha_{ij}\, \alpha_{jk} = \begin{cases} 1 & \text{for } i=k, \\ 0 & \text{for } i \neq k \end{cases}$$

and the variables x_i' are normally distributed, the variables y_k are also normally distributed. Further, $\mathsf{E} y_k = 0$ $(k = 1, 2, \ldots, n)$. Finally, from

$$\mathsf{E}\, y_i\, y_k = \sum_{j=1}^{n} \alpha_{ij}\, \alpha_{jk}\, \mathsf{E}\, x_j'^2 = \sigma^2 \sum_{j=1}^{n} \alpha_{ij}\, \alpha_{jk} = \begin{cases} \sigma^2 & \text{for } i = k, \\ 0 & \text{for } i \neq k, \end{cases}$$

we conclude that the variables y_k $(k = 1, 2, \ldots, n)$ are independent and that $\mathsf{D} y_k = \sigma^2$ $(k = 1, 2, \ldots, n)$.

Further, since

$$\frac{y_1}{\sqrt{\sum_{k=2}^{n} y_k^2}} = \frac{y_1/\sqrt{n-1}}{\sqrt{(\sum_{k=2}^{n} y_k^2)/(n-1)}},$$

and the numerator and denominator in this fraction are independent, and since, moreover, the density function of the numerator is

$$\frac{\sqrt{n-1}}{\sigma\sqrt{2\pi}}\, e^{-\frac{x^2}{2\sigma^2}(n-1)}$$

and that of the denominator (according to the χ^2-distribution in § 24) is

$$\frac{\sqrt{2(n-1)}}{\sigma\, \Gamma\!\left(\frac{n-1}{2}\right)} \left(\frac{y\sqrt{n-1}}{\sigma\sqrt{2}}\right)^{n-2} e^{-\frac{y^2}{2\sigma^2}(n-1)},$$

it follows from § 24 (the Student distribution) that the density function of the quotient $y_1 \Big/ \sqrt{\sum_{k=2}^{n} y_k^2}$ is

$$\frac{\Gamma\left(\frac{n}{2}\right)}{\sqrt{\pi}\,\Gamma\left(\frac{n-1}{2}\right)}(1+x^2)^{-\frac{n}{2}}.$$

It is easy to verify that the density function of the variable

$$y_1 \Big/ \sqrt{n \sum_{k=2}^{n} y_k^2}$$

is

$$\varphi(x/a, \sigma) = \frac{\sqrt{n}\,\Gamma\left(\frac{n}{2}\right)}{\sqrt{\pi}\,\Gamma\left(\frac{n-1}{2}\right)}(1+n x^2)^{-\frac{n}{2}},$$

and therefore

$$\mathsf{P}\left\{c_1 \leq \frac{a-\bar{x}}{s_1\sqrt{n}} < c_2 / a, \sigma\right\} = \frac{\sqrt{n}\,\Gamma\left(\frac{n}{2}\right)}{\sqrt{\pi}\,\Gamma\left(\frac{n-1}{2}\right)} \int_{c_1}^{c_2} (1+n x^2)^{-\frac{n}{2}} dx.$$

This probability does not depend on the values taken on by the parameters a and σ. We can therefore say that in the third problem the probability of the inequality

$$c_1 s_1 \sqrt{n} \leq a - \bar{x} < c_2 s_1 \sqrt{n}$$

is

$$\omega = \frac{\sqrt{n}\,\Gamma\left(\frac{n}{2}\right)}{\sqrt{\pi}\,\Gamma\left(\frac{n-1}{2}\right)} \int_{c_1}^{c_2} (1+n x^2)^{-\frac{n}{2}} dx.$$

It still remains to determine the probability of the inequality which fixes limits for σ. By making use of the above transformations, we find:

$$\mathsf{P}\{\sigma_1' \leq \sigma < \sigma_2'/a, \sigma\} = \mathsf{P}\left\{t_2 < \frac{s_1 \sqrt{n}}{\sigma} \leq t_1/a, \sigma\right\}$$

$$= \mathsf{P}\left\{t_2 \sigma < \sqrt{\sum_{k=2}^{n} y_k^2} \leq t_1 \sigma/a, \sigma\right\}.$$

In virtue of the results of § 24 (χ^2-distribution), this yields

§ 66. CONFIDENCE LIMITS

$$P\{\sigma_1' \leqq \sigma < \sigma_2' | a, \sigma\} = \frac{\sqrt{2(n-1)}}{\sigma \Gamma\left(\frac{n-1}{2}\right)} \int\limits_{\frac{t_2 \sigma}{\sqrt{n-1}}}^{\frac{t_1 \sigma}{\sqrt{n-1}}} \left(\frac{y\sqrt{n-1}}{\sigma\sqrt{2}}\right)^{n-2} e^{-\frac{y^2(n-1)}{2\sigma^2}} dy$$

$$= \frac{1}{2^{\frac{n-3}{2}} \Gamma\left(\frac{n-1}{2}\right)} \int\limits_{t_2}^{t_1} z^{n-2} e^{-\frac{z^2}{2}} dz.$$

Again, this probability does not depend on the values of the parameters a and σ. Consequently, the inequality

$$\frac{\sqrt{n}\, s_1}{t_1} \leqq \sigma < \frac{\sqrt{n}\, s_1}{t_2}$$

has the probability

$$\omega = \frac{1}{2^{\frac{n-3}{2}} \Gamma\left(\frac{n-1}{2}\right)} \int\limits_{t_2}^{t_1} t^{n-2} e^{-\frac{t^2}{2}} dt.$$

TABLES

Table of Values of the Function $\varphi(x) = \dfrac{1}{\sqrt{2\pi}} e^{-\frac{x^2}{2}}$

x	0	1	2	3	4	5	6	7	8	9
0.0	0.3989	3989	3989	3988	3986	3984	3982	3980	3977	3973
0.1	3970	3965	3961	3956	3951	3945	3939	3932	3925	3918
0.2	3910	3902	3894	3885	3876	3867	3857	3847	3836	3825
0.3	3814	3802	3790	3778	3765	3752	3739	3726	3712	3697
0.4	3683	3668	3653	3637	3621	3605	3589	3572	3555	3538
0.5	3521	3503	3485	3467	3448	3429	4410	3391	3372	3352
0.6	3332	3312	3292	3271	3251	3230	3209	3187	3166	3144
0.7	3123	3101	3079	3056	3034	3011	2989	2966	2943	2920
0.8	2897	2874	2850	2827	2803	2780	2756	2732	2709	2685
0.9	2661	2637	2613	2589	2565	2541	2516	2492	2468	2444
1.0	0.2420	2396	2371	2347	2323	2299	2275	2251	2227	2203
1.1	2179	2155	2131	2107	2083	2059	2036	2012	1989	1965
1.2	1942	1919	1895	1872	1849	1826	1804	1781	1758	1736
1.3	1714	1691	1669	1647	1626	1604	1582	1561	1539	1518
1.4	1497	1476	1456	1435	1415	1394	1374	1354	1334	1315
1.5	1295	1276	1257	1238	1219	1200	1182	1163	1145	1127
1.6	1109	1092	1074	1057	1040	1023	1006	0989	0973	0957
1.7	0940	0925	0909	0893	0878	0863	0848	0883	0818	0804
1.8	0790	0775	0761	0748	0734	0721	0707	0694	0681	0669
1.9	0656	0644	0632	0620	0608	0596	0584	0573	0562	0551
2.0	0.0540	0529	0519	0508	0498	0488	0478	0468	0459	0449
2.1	0440	0431	0422	0413	0404	0396	0387	0379	0371	0363
2.2	0355	0347	0339	0332	0325	0317	0310	0303	0297	0290
2.3	0283	0277	0270	0264	0258	0252	0246	0241	0235	0229
2.4	0224	0219	0213	0208	0203	0198	0194	0189	0184	0180
2.5	0175	0171	0167	0163	0158	0154	0151	0147	0143	0139
2.6	0136	0132	0129	0126	0122	0119	0116	0113	0110	0107
2.7	0104	0101	0099	0096	0093	0091	0088	0086	0084	0081
2.8	0079	0077	0075	0073	0071	0069	0067	0065	0063	0061
2.9	0060	0058	0056	0055	0053	0051	0050	0048	0047	0046
3.0	0.0044	0043	0042	0040	0039	0038	0037	0036	0035	0034
3.1	0033	0032	0031	0030	0039	0028	0027	0026	0025	0025
3.2	0024	0023	0022	0022	0021	0020	0020	0019	0018	0018
3.3	0017	0017	0016	0016	0015	0015	0014	0014	0013	0013
3.4	0012	0012	0012	0011	0011	0010	0010	0010	0009	0009
3.5	0009	0008	0008	0008	0008	0007	0007	0007	0007	0006
3.6	0006	0006	0006	0005	0005	0005	0005	0005	0005	0004
3.7	0004	0004	0004	0004	0004	0004	0003	0003	0003	0003
3.8	0003	0003	0003	0003	0003	0002	0002	0002	0002	0002
3.9	0002	0002	0002	0002	0002	0002	0002	0002	0001	0001

Table of Values of the Function $\Phi(x) = \dfrac{1}{\sqrt{2\pi}} \displaystyle\int_0^x e^{-\frac{z^2}{2}} dz$

x	0	1	2	3	4	5	6	7	8	9
0.0	0.00000	00399	00798	01197	01595	01994	02392	02790	03188	03586
0.1	03983	04380	04776	05172	05567	05962	06356	06749	07142	07535
0.2	07926	08317	08706	09095	09483	09871	10257	10642	11026	11409
0.3	11791	12172	12552	12930	13307	13683	14058	14431	14803	15173
0.4	15542	15910	16276	16640	17003	17364	17724	18082	18439	18793
0.5	19146	19497	19487	21094	20540	20884	21226	21566	21904	22240
0.6	22575	22907	23237	23565	23891	24215	24537	24857	25175	25490
0.7	25804	26115	26424	26730	27035	27337	27637	27935	28230	28524
0.8	28814	29103	29389	29673	29955	30234	30511	30785	31057	31327
0.9	31594	31859	23121	32381	32639	32894	33147	33398	33646	33891
1.0	34134	34375	34614	34850	35083	35314	35543	35769	35993	36214
1.1	36433	36650	36864	37076	37286	37493	37698	37900	38100	38298
1.2	38493	38686	38877	39065	39251	39435	39617	39796	39973	40147
1.3	40320	40490	40658	40824	40988	41149	41309	41466	41621	41774
1.4	41924	42073	42220	42364	42507	42647	42786	42922	43056	43189
1.5	43319	43448	43574	43699	43822	43943	44062	44179	44295	44408
1.6	44520	44630	44738	44845	44950	45053	45154	45254	45352	45449
1.7	45543	45637	45728	45818	45907	45994	46080	46164	46246	46327
1.8	46407	46485	46562	46638	46712	46784	46856	46926	46995	47062
1.9	47128	47193	47257	47320	47381	47441	47500	47558	47615	47670
2.0	47725	47778	47831	47882	47932	47982	48030	48077	48124	48169
2.1	48214	48257	48300	48341	48382	48422	48461	48500	48537	48574
2.2	48610	48645	48679	48713	48745	48778	48809	48840	48870	48899
2.3	48928	48956	48983	49010	49036	49061	49086	49111	49134	49158
2.4	49180	49202	49224	49245	49266	49286	49305	49324	49343	49361
2.5	49379	49396	49413	49430	49446	49461	49477	49492	49506	49520
2.6	49534	49547	49560	49573	49585	49598	49609	49621	49632	49643
2.7	49653	49664	49674	49683	49693	49702	49711	49720	49728	49736
2.8	49744	49752	49760	49767	49774	49781	49788	49795	49801	49807
2.9	49813	49819	49825	49831	49836	49841	49846	49851	49856	49861

3.0	0.49865	3.1	49903	3.2	49931	3.3	49952	3.4	49966
3.5	49977	3.6	49984	3.7	49989	3.8	49993	3.9	49995
4.0	499968								
4.5	499997								
5.0	49999997								

Table of Values of the Function $P_k(a) = \dfrac{a^k e^{-a}}{k!}$

k \ a	0.1	0.2	0.3	0.4	0.5	0.6
0	0.904837	0.818731	0.740818	0.670320	0.606531	0.548812
1	0.090484	0.163746	0.222245	0.268128	0.303265	0.329287
2	0.004524	0.016375	0.033337	0.053626	0.075816	0.098786
3	0.000151	0.001091	0.003334	0.007150	0.012636	0.019757
4	0.000004	0.000055	0.000250	0.000715	0.001580	0.002964
5		0.000002	0.000015	0.000057	0.000158	0.000356
6			0.000001	0.000004	0.000013	0.000035
7					0.000001	0.000003

k \ a	0.7	0.8	0.9	1.0	2.0	3.0
0	0.496585	0.449329	0.406570	0.367879	0.135335	0.049787
1	0.347610	0.359463	0.365913	0.367879	0.270671	0.149361
2	0.121663	0.143785	0.164661	0.183940	0.270671	0.224042
3	0.028388	0.038343	0.049398	0.061313	0.180447	0.224042
4	0.004968	0.007669	0.011115	0.015328	0.090224	0.168031
5	0.000695	0.001227	0.002001	0.003066	0.036089	0.100819
6	0.000081	0.000164	0.000300	0.000511	0.012030	0.050409
7	0.000008	0.000019	0.000039	0.000073	0.003437	0.021604
8		0.000002	0.000004	0.000009	0.000859	0.008101
9				0.000001	0.000191	0.002701
10					0.000038	0.000810
11					0.000007	0.000221
12					0.000001	0.000055
13						0.000013
14						0.000003
15						0.000001

(Cont'd)

k \ a	4.0	5.0	6.0	7.0	8.0	9.0
0	0.018316	0.006738	0.002479	0.000912	0.000335	0.000123
1	0.073263	0.033690	0.014873	0.006383	0.002684	0.001111
2	0.146525	0.084224	0.044618	0.022341	0.010735	0.004998
3	0.195367	0.140374	0.089235	0.052129	0.028626	0.014994
4	0.195367	0.175467	0.133853	0.091226	0.057252	0.033737
5	0.156293	0.175467	0.160623	0.127717	0.091604	0.060727
6	0.104194	0.146223	0.160623	0.149003	0.122138	0.091090
7	0.059540	0.104445	0.137677	0.149003	0.139587	0.117116
8	0.029770	0.065278	0.103258	0.130377	0.139587	0.131756
9	0.013231	0.036266	0.068838	0.101405	0.124077	0.131756
10	0.005292	0.018133	0.041303	0.070983	0.099262	0.118580
11	0.001925	0.008242	0.022529	0.045171	0.072190	0.097020
12	0.000642	0.003434	0.011262	0.026350	0.048127	0.072765
13	0.000197	0.001321	0.005199	0.014188	0.029616	0.050376
14	0.000056	0.000472	0.002228	0.007094	0.016924	0.032384
15	0.000015	0.000157	0.000891	0.003311	0.009026	0.019431
16	0.000004	0.000049	0.000334	0.001448	0.004513	0.010930
17	0.000001	0.000014	0.000118	0.000596	0.002124	0.005786
18		0.000004	0.000039	0.000232	0.000944	0.002893
19		0.000001	0.000012	0.000085	0.000397	0.001370
20			0.000004	0.000030	0.000159	0.000617
21			0.000001	0.000010	0.000061	0.000264
22				0.000003	0.000022	0.000108
23				0.000001	0.000008	0.000042
24					0.000003	0.000016
25					0.000001	0.000006
26						0.000002
27						0.000001

Table of Values of the Function $\sum P_m(a) = \sum_{m=0}^{k} \dfrac{a^m e^{-a}}{m!}$

k \ a	0.1	0.2	0.3	0.4	0.5	0.6
0	0.904837	0.818731	0.740818	0.670320	0.606531	0.548812
1	0.995321	0.982477	0.963063	0.938448	0.909796	0.878099
2	0.999845	0.998852	0.996390	0.992074	0.985612	0.977885
3	0.999996	0.999943	0.999724	0.999224	0.998248	0.997642
4	1.000000	0.999998	0.999974	0.999939	0.999828	0.999606
5	1.000000	1.000000	0.999999	0.999996	0.999986	0.999962
6	1.000000	1.000000	1.000000	1.000000	0.999999	0.999997
7	1.000000	1.000000	1.000000	1.000000	1.000000	1.000000

k \ a	0.7	0.8	0.9	1.0	2.0	3.0
0	0.496585	0.449329	0.406570	0.367879	0.135335	0.049787
1	0.844195	0.808792	0.772483	0.735759	0.406006	0.199148
2	0.965858	0.952577	0.937144	0.919699	0.676677	0.423190
3	0.994246	0.990920	0.988542	0.981012	0.857124	0.647232
4	0.999214	0.998589	0.997657	0.996340	0.947348	0.815263
5	0.999909	0.999816	0.999658	0.999406	0.983437	0.916082
6	0.999990	0.999980	0.999958	0.999917	0.995467	0.966491
7	0.999998	0.999999	0.999997	0.999990	0.998904	0.988095
8	0.000000	1.000000	1.000000	0.999999	0.999763	0.996196
9				1.000000	0.999954	0.998897
10					0.999992	0.999707
11					0.999999	0.999928
12					1.000000	0.999983
13						0.999996
14						0.999999
15						1.000000

TABLES

(Cont'd)

k \ a	4.0	5.0	6.0	7.0	8.0	9.0
0	0.018316	0.006738	0.002479	0.000912	0.000335	0.000123
1	0.091579	0.040428	0.017352	0.007295	0.003019	0.001234
2	0.238105	0.124652	0.061970	0.029636	0.013754	0.006232
3	0.433472	0.265026	0.151205	0.081765	0.042380	0.021228
4	0.628839	0.440493	0.285058	0.172991	0.099632	0.054963
5	0.785132	0.615960	0.445681	0.300708	0.191236	0.115690
6	0.889326	0.762183	0.606304	0.449711	0.313374	0.206780
7	0.948866	0.866628	0.743981	0.598714	0.452961	0.323896
8	0.978636	0.931806	0.847239	0.729091	0.592548	0.455652
9	0.991867	0.968172	0.916077	0.830496	0.716625	0.587408
10	0.997159	0.986205	0.957380	0.901479	0.815887	0.705988
11	0.999084	0.994547	0.979909	0.946650	0.888077	0.803008
12	0.999726	0.997981	0.991173	0.973000	0.936204	0.875773
13	0.999923	0.999202	0.996372	0.987188	0.965820	0.926149
14	0.999979	0.999774	0.998600	0.994282	0.982744	0.958533
15	0.999994	0.999931	0.999491	0.997593	0.991770	0.977964
16	0.999998	0.999980	0.999825	0.999041	0.996283	0.988894
17	0.999999	0.999994	0.999943	0.999637	0.998407	0.994680
18	0.999999	0.999998	0.999982	0.999869	0.999351	0.997575
19	0.999999	0.999999	0.999994	0.999955	0.999748	0.998943
20	1.000000	0.999999	0.999998	0.999985	0.999907	0.999560
21		1.000000	0.999999	0.999995	0.999997	0.999824
22			0.999999	0.999998	0.999989	0.999932
23			1.000000	0.999999	0.999997	0.999974
24				0.999999	0.999999	0.999990
25				1.000000	0.999999	0.999996
26					1.000000	0.999998
27						0.999999
28						1.000000

Table of Values of the Function $P(x) = \dfrac{1}{2^{\frac{k-2}{2}} \Gamma\left(\dfrac{k}{2}\right)} \int_x^\infty z^{k-1} e^{-\frac{z^2}{2}} dz$

x \ k	1	2	3	4	5	6	7	8
1	0.3173	0.6065	0.8013	0.9098	0.9626	0.9856	0.9948	0.9982
2	1574	3679	5724	7358	8491	9197	9598	9810
3	0833	2231	3916	5578	7000	8088	8850	9344
4	0455	1353	2615	4060	5494	6767	7798	8571
5	0254	0821	1718	2873	4159	5438	6600	7576
6	0143	0498	1116	1991	3062	4232	5398	6472
7	0081	0302	0719	1359	2206	3208	4289	5366
8	0047	0183	0460	0916	1562	2381	3326	4335
9	0027	0111	0293	0611	1091	1736	2527	3423
10	0016	0067	0186	0404	0752	1247	1886	2650
11	0009	0041	0117	0266	0514	0884	1386	2017
12	0005	0025	0074	0174	0348	0620	1006	1512
13	0003	0015	0046	0113	0234	0430	0721	1119
14	0002	0009	0029	0073	0146	0296	0512	0818
15	0001	0006	0018	0047	0104	0203	0360	0591
16	0001	0003	0011	0030	0068	0138	0251	0424
17	0000	0002	0007	0019	0045	0093	0174	0301
18		0001	0004	0012	0029	0062	0120	0212
19		0001	0003	0008	0019	0042	0082	0149
20		0000	0002	0005	0013	0028	0056	0103
21		0000	0001	0003	0008	0018	0038	0071
22		0000	0001	0002	0005	0012	0025	0049
23		0000	0000	0001	0003	0008	0017	0034
24		0000	0000	0001	0002	0005	0011	0023
25		0000	0000	0001	0001	0003	0008	0016
26		0000	0000	0000	0001	0002	0005	0010
27		0000	0000	0000	0001	0001	0003	0007
28		0000	0000	0000	0000	0001	0002	0005
29		0000	0000	0000	0000	0001	0001	0003
30		0000	0000	0000	0000	0000	0001	0002

(Cont'd)

k \ x	9	10	11	12	13	14	15
1	0.9994	0.9998	0.9999	1.0000	1.0000	1.0000	1.0000
2	9915	9963	9985	0.9994	0.9998	0.9999	1.0000
3	9643	9814	9907	9955	9979	9991	9996
4	9114	9473	9699	9834	9912	9955	9977
5	8343	8912	9312	9580	9752	9858	9921
6	7399	8153	8734	9161	9462	9665	9797
7	6371	7254	7991	8576	9022	9347	9576
8	5341	6288	7133	7851	8436	8893	9238
9	4373	5321	6219	7029	7729	8311	8775
10	3505	4405	5304	6160	6939	7622	8197
11	2757	3575	4433	5289	6108	6860	7526
12	2133	2851	3626	4457	5276	6063	6790
13	1626	2237	2933	3690	4478	5265	6023
14	1223	1730	2330	3007	3738	4497	5255
15	0909	1321	1825	2414	3074	3782	4514
16	0669	0996	1411	1912	2491	3134	3821
17	0487	0744	1079	1496	1993	2562	3189
18	0352	0550	0816	1157	1575	2068	2627
19	0252	0403	0611	0885	1231	1649	2137
20	0179	0293	0453	0671	0952	1301	1719
21	0126	0211	0334	0504	0729	1016	1368
22	0089	0151	0244	0375	0554	0786	1078
23	0062	0107	0177	0277	0417	0603	0841
24	0043	0076	0127	0203	0311	0458	0651
25	0030	0053	0091	0148	0231	0346	0499
26	0020	0037	0065	0107	0170	0259	0380
27	0014	0026	0046	0077	0124	0193	0287
28	0010	0018	0032	0055	0090	0142	0216
29	0006	0012	0023	0039	0065	0104	0161
30	0004	0009	0016	0028	0047	0076	0119

(Cont'd)

k\x	16	17	18	19	20	21	22
1	1.0000	1.0000	1.0000	1.0000	1.0000	1.0000	1.0000
2	1.0000	1.0000	1.0000	1.0000	1.0000	1.0000	1.0000
3	0.9998	0.9999	1.0000	1.0000	1.0000	1.0000	1.0000
4	9989	9995	0.9998	0.9999	1.0000	1.0000	1.0000
5	9958	9978	9989	9994	0.9997	0.9999	0.9999
6	9881	9932	9962	9979	9989	9994	9997
7	9783	9835	9901	9942	9967	9981	9990
8	9489	9665	9786	9867	9919	9951	9972
9	9134	9403	9597	9735	9829	9892	9933
10	8666	9036	9319	9539	9682	9789	9863
11	8095	8566	8944	9238	9462	9628	9747
12	7440	8001	8472	8856	9161	9396	9574
13	6728	7362	7916	8386	8774	9086	9332
14	5987	6671	7291	7837	8305	8696	9015
15	5246	5955	6620	7226	7764	8230	8622
16	4530	5238	5925	6573	7166	7696	8159
17	3856	4544	5231	5899	6530	7111	7634
18	3239	3888	4557	5224	5874	6490	7060
19	2687	3285	3918	4568	5218	5851	6453
20	2202	2742	3328	3946	4579	5213	5830
21	1785	2263	2794	3368	3971	4589	5207
22	1432	1847	2320	2843	3405	3995	4599
23	1137	1493	1906	2373	2888	3440	4017
24	0895	1194	1550	1962	2424	2931	3472
25	0698	0947	1249	1605	2014	2472	2971
26	0540	0745	0998	1302	1658	2064	2517
27	0415	0581	0790	1047	1353	1709	2112
28	0316	0449	0621	0834	1094	1402	1757
29	0239	0345	0484	0660	0878	1140	1449
30	0180	0263	0374	0518	0699	0920	1185

(Cont'd)

x \ k	23	24	25	26	27	28	29
1	1.0000	1.0000	1.0000	1.0000	1.0000	1.0000	1.0000
2	1.0000	1.0000	1.0000	1.0000	1.0000	1.0000	1.0000
3	1.0000	1.0000	1.0000	1.0000	1.0000	1.0000	1.0000
4	1.0000	1.0000	1.0000	1.0000	1.0000	1.0000	1.0000
5	1.0000	1.0000	1.0000	1.0000	1.0000	1.0000	1.0000
6	0.9999	0.9999	1.0000	1.0000	1.0000	1.0000	1.0000
7	9995	9997	0.9999	0.9999	1.0000	1.0000	1.0000
8	9984	9991	9995	9997	0.9999	0.9999	1.0000
9	9960	9976	9986	9992	9995	9997	0.9999
10	9913	9945	9967	9980	9988	9993	9996
11	9832	9890	9929	9955	9972	9983	9990
12	9705	9799	9866	9912	9943	9964	9977
13	9520	9661	9765	9840	9892	9929	9954
14	9269	9466	9617	9730	9813	9872	9914
15	8946	9208	9414	9573	9694	9784	9850
16	8553	8881	9148	9362	9529	9658	9755
17	8093	8487	8818	9091	9311	9486	9622
18	7575	8030	8429	8758	9035	9261	9443
19	7012	7520	7971	8364	8700	9981	9213
20	6419	6968	7468	7916	8308	8645	8926
21	5811	6387	6926	7420	7863	8253	8591
22	5203	5793	6357	6887	7374	7813	8202
23	4608	5198	5776	6329	6850	7330	7765
24	4038	4616	5194	5760	6303	6815	8289
25	3503	4058	4624	5190	5745	6278	6782
26	3009	3532	4076	4631	5186	5730	6255
27	2560	3045	3559	4093	4638	5182	5717
28	2158	2600	3079	3585	4110	4644	5179
29	1803	2201	2639	3111	3609	4125	4651
30	1494	1848	2243	2676	3142	3632	4140

Table of Values of the Function $S(x) = \dfrac{\Gamma\left(\dfrac{n}{2}\right)}{\sqrt{(n-1)\pi}\,\Gamma\left(\dfrac{n-1}{2}\right)} \int\limits_{-\infty}^{x} \left(1 + \dfrac{z^2}{n-1}\right)^{-\dfrac{n}{2}} dz$

x \ n	2	3	4	5	6	7	8	9	10	11
0.0	0.500	0.500	0.500	0.500	0.500	0.500	0.500	0.500	0.500	0.500
0.1	532	535	537	537	538	538	538	539	539	539
0.2	563	570	573	574	575	576	576	577	577	577
0.3	593	604	608	610	612	613	614	614	614	615
0.4	621	636	642	645	647	648	650	650	651	651
0.5	648	667	674	678	681	683	684	685	686	686
0.6	672	695	705	710	713	715	716	717	718	719
0.7	694	722	733	739	742	745	747	748	749	750
0.8	715	746	759	766	770	773	775	777	778	779
0.9	733	768	783	790	795	799	801	803	804	805
1.0	750	789	805	813	818	822	825	828	828	830
1.1	765	807	824	834	839	843	846	848	850	851
1.2	779	824	842	852	858	862	865	868	870	871
1.3	791	838	858	868	875	879	883	885	887	889
1.4	803	852	872	883	890	894	898	900	902	904
1.5	813	864	885	896	903	908	911	914	916	918
1.6	822	875	896	908	915	920	923	926	928	930
1.7	831	884	906	918	925	930	934	936	938	940
1.8	839	893	915	927	934	939	943	945	947	949
1.9	846	901	923	935	942	947	950	953	955	957
2.0	852	908	930	942	949	954	957	960	962	963
2.2	864	921	942	954	960	965	968	970	972	974
2.4	874	931	952	963	969	973	976	978	980	981
2.6	883	938	960	970	976	980	982	984	986	987
2.8	891	946	966	976	981	984	987	988	990	991
3.0	898	952	971	980	985	988	990	992	992	993
3.2	904	957	975	984	988	991	992	994	995	995
3.4	909	962	979	986	990	993	994	995	996	997
3.6	914	965	982	989	992	994	996	996	997	998
3.8	918	969	984	990	994	997	997	997	998	998
4.0	922	971	986	992	995	996	997	998	998	999
4.2	926	974	988	993	996	997	998	998	999	999
4.4	929	976	989	994	996	998	998	999	999	999
4.6	932	978	990	995	997	998	999	999	999	1.000
4.8	935	980	991	996	998	998	999	999	1.000	
5.0	937	981	992	996	998	999	999	1.000		
5.2	940	982	993	997	998	999	999			
5.4	942	984	994	997	998	999	1.000			
5.6	944	985	994	998	999	999				
5.8	946	986	995	998	999	999				
6.0	947	987	995	998	999	1.000				

(Cont'd)

x \ n	12	13	14	15	16	17	18	19	20	∞
0.0	0.500	0.500	0.500	0.500	0.500	0.500	0.500	0.500	0.500	0.50000
0.1	539	539	539	539	539	539	539	539	539	53983
0.2	577	578	578	578	578	578	578	578	578	57926
0.3	615	615	616	616	616	616	616	616	616	61791
0.4	652	652	652	652	653	653	653	653	653	65542
0.5	686	687	687	688	688	688	688	688	689	69146
0.6	720	720	721	721	721	722	722	722	722	72575
0.7	751	751	752	752	753	753	753	754	754	75804
0.8	780	780	781	782	782	782	783	783	783	78814
0.9	806	807	808	808	809	809	810	810	810	81594
1.0	831	832	832	833	833	834	834	835	835	84134
1.1	853	854	854	855	856	856	857	857	858	86433
1.2	872	873	874	875	876	876	877	877	878	88493
1.3	890	891	892	893	893	894	894	895	895	90320
1.4	906	907	908	908	909	910	910	911	911	91924
1.5	919	920	921	922	923	924	924	924	925	93319
1.6	931	932	933	934	935	935	936	936	937	94520
1.7	941	943	944	945	945	946	946	947	947	95543
1.8	950	952	952	953	954	955	955	956	956	96407
1.9	958	959	960	961	962	962	963	963	964	97128
2.0	965	967	967	967	968	969	969	970	970	97725
2.2	975	976	977	977	978	979	979	979	980	98610
2.4	982	983	984	985	985	986	986	986	987	99180
2.6	988	988	989	990	990	990	991	991	991	99534
2.8	991	992	992	993	993	994	994	994	994	99744
3.0	994	994	995	995	996	996	996	996	996	99865
3.2	996	996	996	997	997	997	997	998	998	99931
3.4	997	997	998	998	998	998	998	998	998	99966
3.6	998	998	998	999	999	999	999	999	999	99984
3.8	998	999	999	999	999	999	999	999	999	99993
4.0	999	999	999	999	999	1.000	1.000	1.000	1.000	99997
4.2	999	999	1.000	1.000	1.000					99999
4.4	1.000	1.000								99999

Table of Values of the Function $K(x) = \sum_{k=-\infty}^{\infty} (-1)^k e^{-2k^2 x^2}$

x	$K(x)$	x	$K(x)$	x	$K(x)$
0.28	0.000001	0.75	0.372833	1.22	0.898104
0.29	0.000004	0.76	0.389640	1.23	0.902972
0.30	0.000009	0.77	0.406372	1.24	0.907648
0.31	0.000021	0.78	0.423002	1.25	0.912132
0.32	0.000046	0.79	0.439505	1.26	0.916432
0.33	0.000091	0.80	0.455857	1.27	0.920556
0.34	0.000171	0.81	0.472041	1.28	0.924505
0.35	0.000303	0.82	0.488030	1.29	0.928288
0.36	0.000511	0.83	0.503808	1.30	0.931908
0.37	0.000826	0.84	0.519366	1.31	0.935370
0.38	0.001285	0.85	0.534682	1.32	0.938682
0.39	0.001929	0.86	0.549744	1.33	0.941848
0.40	0.002808	0.87	0.564546	1.34	0.944872
0.41	0.003972	0.88	0.579070	1.35	0.947756
0.42	0.005476	0.89	0.593316	1.36	0.950512
0.43	0.007377	0.90	0.607270	1.37	0.953142
0.44	0.009730	0.91	0.620928	1.38	0.955650
0.45	0.012590	0.92	0.634286	1.39	0.958040
0.46	0.016005	0.93	0.647338	1.40	0.960318
0.47	0.020022	0.94	0.660082	1.41	0.962486
0.48	0.024682	0.95	0.672516	1.42	0.964552
0.49	0.030017	0.96	0.684636	1.43	0.966516
0.50	0.036055	0.97	0.696444	1.44	0.968382
0.51	0.042814	0.98	0.707940	1.45	0.970158
0.52	0.050306	0.99	0.719126	1.46	0.971846
0.53	0.058534	1.00	0.730000	1.47	0.973448
0.54	0.067497	1.01	0.740566	1.48	0.974970
0.55	0.077183	1.02	0.750826	1.49	0.976412
0.56	0.087577	1.03	0.760780	1.50	0.977782
0.57	0.098656	1.04	0.770434	1.51	0.979080
0.58	0.110395	1.05	0.779794	1.52	0.980310
0.59	0.122760	1.06	0.788860	1.53	0.981476
0.60	0.135718	1.07	0.797636	1.54	0.982578
0.61	0.149229	1.08	0.806128	1.55	0.983622
0.62	0.163225	1.09	0.814342	1.56	0.684610
0.63	0.177753	1.10	0.822282	1.57	0.985544
0.64	0.192677	1.11	0.829950	1.58	0.986426
0.65	0.207987	1.12	0.837356	1.59	0.987260
0.66	0.223637	1.13	0.844502	1.60	0.988048
0.67	0.239582	1.14	0.851394	1.61	0.988791
0.68	0.255780	1.15	0.858038	1.62	0.989492
0.69	0.272189	1.16	0.864442	1.63	0.990154
0.70	0.288765	1.17	0.870612	1.64	0.990777
0.71	0.305471	1.18	0.876548	1.65	0.991364
0.72	0.322265	1.19	0.882258	1.66	0.991917
0.73	0.339113	1.20	0.887750	1.67	0.992438
0.74	0.355981	1.21	0.893030	1.68	0.992928

(Cont'd)

x	$K(x)$	x	$K(x)$	x	$K(x)$
1.69	0.993389	2.00	0.999329	2.31	0.999954
1.70	0.993828	2.01	0.999380	2.32	0.999958
1.71	0.994230	2.02	0.999428	2.33	0.999962
1.72	0.994612	2.03	0.999474	2.34	0.999965
1.73	0.994972	2.04	0.999516	2.35	0.999968
1.74	0.995309	2.05	0.999552	2.36	0.999970
1.75	0.995625	2.06	0.999588	2.37	0.999973
1.76	0.995922	2.07	0.999620	2.38	0.999976
1.77	0.996200	2.08	0.999650	2.39	0.999978
1.78	0.996460	2.09	0.999680	2.40	0.999980
1.79	0.996704	2.10	0.999705	2.41	0.999982
1.80	0.996932	2.11	0.999723	2.42	0.999984
1.81	0.997146	2.12	0.999750	2.43	0.999986
1.82	0.997346	2.13	0.999770	2.44	0.999987
1.83	0.997533	2.14	0.999790	2.45	0.999988
1.84	0.997707	2.15	0.999806	2.46	0.999989
1.85	0.997870	2.16	0.999822	2.47	0.999990
1.86	0.998023	2.17	0.999838	2.48	0.999991
1.87	0.998145	2.18	0.999852	2.49	0.999992
1.88	0.998297	2.19	0.999864	2.50	0.9999925
1.89	0.998421	2.20	0.999874	2.55	0.9999956
1.90	0.998536	2.21	0.999886	2.60	0.9999974
1.91	0.998644	2.22	0.999896	2.65	0.9999984
1.92	0.998744	2.23	0.999904	2.70	0.9999990
1.93	0.998837	2.24	0.999912	2.75	0.9999994
1.94	0.998924	2.25	0.999920	2.80	0.9999997
1.95	0.999004	2.26	0.999926	2.85	0.99999982
1.96	0.999079	2.27	0.999934	2.90	0.99999990
1.97	0.999149	2.28	0.999940	2.95	0.99999994
1.98	0.999213	2.29	0.999944	3.00	0.99999997
1.99	0.999273	2.30	0.999949		

BIBLIOGRAPHY

BIBLIOGRAPHY

Items in the Russian language are indicated by *

POPULAR

BOREL, E.: *Les hasard.* Paris, 2nd ed., 1948.
*GNEDENKO, B. V.: *How Mathematics Studies Random Phenomena.* Kiev: Izd. Akad. Nauk Ukr. SSR, 1947. [English trans. in prep.]
GNEDENKO, B. V. and A. YA. KHINTCHINE: *Introduction à la théorie des probabilités.* Paris: Dunod, 1960.
*YAGLOM, A. M. and I. M. YAGLOM: *Probability and Information.* GTTI, 1957. [English translation in prep.]

TEXTBOOKS AND MONOGRAPHS

BARTLETT, M. S.: *An Introduction to Stochastic Processes.* Cambridge: Cambridge Univ. Press, 1955.
*BERNSTEIN, S. N.: *Probability Theory.* Moscow: Gostekhizdat, 4th ed., 1946.
BHARUCHA-REID, A. T.: *Elements of the Theory of Markov Processes and Their Applications.* New York: McGraw-Hill, 1960.
BLACKWELL, D. and M. A. GIRSHICK: *Theory of Games and Statistical Decisions.* New York: Wiley, 1954.
BLANC-LAPIERRE, A. and R. FORTET: *Théorie des fonctions aléatoires.* Paris, 1953.
CHANDRASEKAR, S.: *Stochastic Problems in Physics and Astronomy.* Rev. Modern Phys., vol. 15, 1943.
CRAMÉR, H.: *Random Variables and Probability Distributions* (Cambridge Tracts in Math. no. 36). Cambridge: Cambridge Univ. Press, 1937.
——— *Mathematical Methods of Statistics.* Princeton: Princeton Univ. Press, 1946.
DOOB, J. L.: *Stochastic Processes.* New York: Wiley, 1953.
*DUNIN-BARKOVSKIĬ, I. V. and N. V. SMIRNOV: *Theory of Probability and Mathematical Statistics in Engineering (General Part).* Moscow: Gostekhizdat, 1955.
*EINSTEIN, A. and SMOLUKHOVSKIĬ: *Collected Articles on the Theory of Brownian Motion.* ONTI, 1936.
FELLER, W.: *An Introduction to Probability Theory and Its Applications.* New York: Wiley, 2nd ed., 1957.
FISZ, M.: *Wahrscheinlichkeitsrechnung und mathematische Statistik.* Berlin: Deutscher Verlag der Wissenschaften, 1958. [Translated from the 2nd Polish edition.]

FRECHET, M.: *Recherches théoriques modernes sur la théorie des probabilités* (2 volumes; In the collection: *Traité du calcul des probabilités*). Paris: Gauthiers-Villars, 1950 and 1937.

*GILENKO, N. D.: *Problems in the Theory of Probability*. Uchpedgiz, 1943.

*GLIVENKO, V. I.: *Course in the Theory of Probability*. GONTI, 1939.

*——— *The Stieltjes Integral*. ONTI, 1936.

GNEDENKO, B. V. and A. N. KOLMOGOROV: *Limit Distributions for Sums of Independent Random Variables*. Reading, Mass.: Addison-Wesley, 1954.

*GONCHAROV, V. L.: *The Theory of Probability*. Oborongiz, 1939.

GRENANDER, U. and M. ROSENBLATT: *Statistical Analysis of Stationary Time Series*. New York: Wiley, 1957.

HALD, A.: *Statistical Theory With Engineering Applications*. New York: Wiley, 1952.

KEMENY, J. G. and J. L. SNELL: *Finite Markov Chains*. Princeton: Van Nostrand, 1960.

*KHINTCHINE, A. YA.: *Fundamental Laws of Probability Theory*. GTTI, 1932.

——— *Asymptotische Gesetze der Wahrscheinlichkeitsrechnung*. Berlin, 1933; New York: Chelsea, 1949.

*——— *Limit Theorems for Sums of Independent Random Variables*. ONTI, 1938.

——— *Mathematical Foundations of Statistical Mechanics*. New York: Dover, 1949.

——— *Mathematical Foundations of Quantum Statistics*. Albany: Graylock, 1960.

KOLMOGOROV, A. N.: *Foundations of the Theory of Probability*. New York: Chelsea, 2nd ed., 1956.

LÉVY, P.: *Processus stochastiques et mouvement Brownien*. Paris: Gauthier-Villars, 1948.

——— *Théorie de l'addition des variables aléatoires* (Calcul des probabilités et ses applications, fasc. I). Paris: Gauthier-Villars, 1937.

LOÈVE, M.: *Probability Theory*. Princeton: Van Nostrand, 2nd ed., 1960.

*MARKOV, A. A.: *The Calculus of Probabilities*, 4th ed. GIZ, 1924.

MCSHANE, E. J.: *Integration*. Princeton: Princeton Univ. Press, 1957.

MISES, R. VON: *Probability, Statistics, and Truth*. New York: Macmillan, 1939.

——— *Wahrscheinlichkeitsrechnung*. Leipzig and Vienna, 1931.

ONICESCU, O., MIHOC, G., and C. T. IONESCU TULCEA: *The Calculus of Probabilities and Its Applications*. Bucharest, 1956. [In Rumanian.]

RÉNYI, A.: *The Theory of Probability*. Budapest, 1954. [In Hungarian.]

*ROMANOVSKIĬ, V. I.: *Mathematical Statistics*. ONTI, 1938.

*——— *Discrete Markov Chains*. Moscow: Gostekhizdat, 1949.

*——— *Fundamental Problems in the Theory of Errors*. Moscow: Gostekhizdat, 1947.

*SARIMSAKOV, T. A.: *Fundamentals of the Theory of Markov Processes*. Moscow: Gostekhizdat, 1954.

SCHMETTERER, L.: *Einführung in die mathematische Statistik*. Vienna, 1956.

*SIRAZHDINOV, S. KH.: *Limit Theorems for Homogeneous Markov Chains*. Izd. Akad. Nauk Uzbek. SSR, 1955.

TODHUNTER, I.: *A History of the Mathematical Theory of Probability*. Reprint. New York: Chelsea, 1949.

USPENSKY, I. V.: *Introduction to Mathematical Probability.* New York: McGraw-Hill, 1937.
WIENER, N.: *Extrapolation, Interpolation, and Smoothing of Stationary Time Series.* New York: Wiley, 1949.
WOLD, H.: *A Study in the Analysis of Stationary Time Series.* Stockholm, 1954.

JOURNALS

CHAPTER I

*BERNSTEIN, S. N.: *On the axiomatic foundation of probability theory,* Soob. Khar'k. ob-va, vol. 15 (1917).
Collection: *Théorie des probabilités. Exposés sur les fondemont et ses applications.* Paris, 1952.
*GNEDENKO, B. V.: *Development of the theory of probability in Russia,* Trudi In-ta Istorii Estestvoznaniya, Akad. Nauk SSSR, vol. 2 (1948).
*——— *Probability theory and the perception of the real world,* Uspekhi Mat. Nauk, vol. 5, no. 1 (1950).
*GNEDENKO, B. V. and A. N. KOLMOGOROV: *The theory of probability,* in the Collection ''Thirty years of mathematics in the SSSR.'' Moscow: Gostekhizdat, 1948.
KHINTCHINE, A. YA.: *Die Methode der willkürlichen Funktionen und der Kampf gegen den Idealismus in der Wahrscheinlichkeitsrechnung,* Sovietwissenschaft, naturwis. Abt. (2) (1954).
*——— *The studies of von Mises on probability and the principles of physical statistics,* Uspekhi Fiz. Nauk, vol. 9, no. 2 (1929).
*KOLMOGOROV, A. N.: *The role of Russian science in the development of probability theory,* Uchën. Zar. MGU, no. 91 (1947).
*SMIRNOV, N. V.: *Mathematical statistics,* in the Collection ''Thirty years of mathematics in the SSSR.'' Moscow: Gostekhizdat, 1948.

CHAPTER II

*BERNSTEIN, S. N.: *Return to the question of the accuracy of the Laplace limit formula,* Izv. Akad. Nauk SSSR, vol. 7 (1943).
FELLER, W.: *On the normal approximation to the binomial distribution,* Ann. of Math. Stat., vol. 16 (1945).
KHINTCHINE, A. YA.: *Über einen neuen Grenzwertsatz der Wahrscheinlichkeitsrechnung,* Math. Ann., vol. 101 (1929).
*PROKHOROV, YU. V.: *Asymptotic behavior of the binomial distribution,* Uspekhi Mat. Nauk, vol. 8, no. 3, pp. 136-142 (1953).
*SMIRNOV, N. V.: *Über Wahrscheinlichkeiten grosser Abweichungen,* Mat. Sbornik, vol. 40, no. 4 (1933) [German summary].

CHAPTER III

*DOBRUSHIN, R. L.: *Limit theorems for Markov chains of two states*, Izv. Akad. Nauk, Ser. Mat., 17 (1953).

*——— *The central limit theorem for non-homogeneous Markov chains*, Teor. Veroyatnost. i Primenen., vol. 1, nos. 1 and 4 (1956).

DOEBLIN, W.: *Exposé de la théorie des chaines simples constantes de Markoff à un nombre fini d'etats*, Rev. math. de l'Union Interbalkanique, II, 1 (1938).

*KOLMOGOROV, A. N.: *Markov chains with a denumerable number of possible states*, Bull. MGU, vol. 1, no. 3 (1937).

*MARKOV, A. A.: *Investigation of a noteworthy case of dependent trials*, Izv. Ros. Akad. Nauk, vol. 1 (1907).

PEPPER, E. D.: *Asymptotic expression for the probability of trials connected in a chain*, Ann. of Math., vol. 28, no. 3 (1927).

See also the corresponding chapters in the books of Bernstein and Feller named above in the list of textbooks and monographs as well as the book of Romanovskiĭ and the second volume of that of Fréchet. Doob's book contains an extensive bibliography on Markov chains.

CHAPTER IV

CRAMÉR, H.: *Über eine Eigenschaft der normalen Verteilungsfunktion*, Math. Zeit., vol. 41 (1936).

*GNEDENKO, B. V. and RVACHËVA, E. L.: *On a characteristic property of the normal law*, Trudi Mat. In-ta Akad. Nauk Ukr. SSR, no. 2 (1948).

*OBUKHOV, A. M.: *Theory of correlation of vectors*, Uchën. Zar. MFG, no. 45 (1940).

*RAIKOV, D. A.: *On the decomposition of the Gaussian and Poisson laws*, Izv. Akad. Nauk SSSR, Ser. Mat. (1938).

*SKITOVICH, V. P.: *Linear forms of independent random variables and the normal distribution law*, Izv. Akad. Nauk SSSR, 18 (952) (1954).

CHAPTER VI

*BERNSTEIN, S. N.: *On the law of large numbers*, Soob. Khar'k Mat. Ob-va, vol. 16 (1918).

HÁJEK, J. and A. RÉNYI: *Generalization of an inequality of Kolmogorov*, Acta Math. Acad. Sci. Hungar., vol. 6 (1955).

KHINTCHINE, A. YA.: *Sur la loi des grands nombres*, C. R. Acad. Sci. Paris, 188 (1929).

KOLMOGOROV, A. N.: *Sur la loi fort des grands nombres*, C. R. Acad. Sci. Paris, 191 (1930).

*PROKHOROV, YU. V.: *On the strong law of large numbers*, Dokl. Akad. Nauk SSSR, vol. 69, no. 5 (1949).

SLUTSKY, E. E.: *Über stochastische Asymptoten und Grenzwerte*, Metron, vol. 5, no. 3 (1925).

TCHEBYCHEV, P. L.: *Des valeurs moyennes*, Journal de math. pures et appliquées, 2 série, 12 (1867). Oeuvres, vol. 1, New York: Chelsea, 1962.

CHAPTER VII

*GNEDENKO, B. V.: *On characteristic functions,* Bull. MGU, vol. 1, no. 5 (1937).

*KHINTCHINE, A. YA.: *On a criterion for characteristic functions,* Bull. MGU, vol. 1, no. 5 (1937).

*KREĬN, M. G.: *On representing functions by means of Fourier-Stieltjes integrals,* Uchën. Zar Kuïbish. Red. In-ta, no. 7 (1943).

*RAIKOV, D. A.: *On positive-definite functions,* Dokl. Akad. Nauk SSSR, vol. 26, no. 9 (1940).

CHAPTER VIII

BERNSTEIN, S. N.: *Sur l'extension du théoreme limite du calcul des probabilités aux sommes de quantités dépendantes,* Math. Ann., 1 (1927).

ESSEEN, C. G.: *Fourier analysis of distribution functions. A mathematical study of the Laplace-Gaussian law,* Acta Math., vol. 77 (1945).

FELLER, W.: *Über den Zentralengrenzwertsatz der Wahrscheinlichkeitsrechnung,* Math. Zeit., vol. 40 (1935).

*GNEDENKO, B. V.: *Elements of the theory of distribution functions of random vectors,* Uspekhi Mat. Nauk, no. 10 (1944).

*——— *On the local limit theorem in the theory of probability,* Uspekhi Mat. Nauk, vol. 3, no. 3 (1948).

*——— *Local limit theorem for densities,* Dokl. Akad. Nauk SSSR, vol. 95, no. 1 (1954).

LIAPOUNOV, A. M.: *Sur une proposition de la théorie des probabilités,* Bull. Acad. Sci. Peter., 13 (1900).

——— *Nouvelle forme du théoreme sur la limite des probabilités,* Bull. Acad. Sci. Peter. (1901).

LINDEBERG, J. W.: *Eine neue Herleitung des Exponentialgesetz in der Wahrscheinlichkeitsrechnung,* Math. Zeit., vol. 15 (1922).

*LINNIK, YU. V.: *On the accuracy of the approach of sums of independent random variables to the Gaussian distribution,* Izv. Akad. Nauk SSSR, vol. 2 (1947).

*PROKHOROV, YU. V.: *Local theorem for densities,* Dokl. Akad. Nauk SSSR, vol. 83, no. 6 (1952).

*TCHEBYCHEV, P. L.: *On two theorems concerning probability,* Zar. Akad. Nauk (1887).

——— *Sur deux théorèmes relatifs aux probabilités.* Acta Math., vol. 14 (1890-91) [French version of above article]. *Oeuvres,* vol. 2. New York: Chelsea, 1962.

CHAPTER IX

*BAVLI, G. M.: *Über einige Verallgemeinerungen der Grenzwertsatz der Wahrscheinlichkeitsrechnung,* Mat. Sbornik, vol. 1 (43), no. 6 (1936).

*GNEDENKO, B. V.: *On a characteristic property of infinitely divisible distribution laws,* Bull. MGU, vol. 1, no. 5 (1937).

*——— *Limit theorems for sums of independent random variables,* Uspekhi Mat. Nauk, no. 10 (1944).

*KHINTCHINE, A. YA.: *A new derivation of a formula of P. Lévy,* Bull. MGU, vol. 1, no. 1 (1937).

CHAPTER X

CRAMÉR, H.: *On harmonic analysis in certain continuous functional spaces*, Ark. Mat. Astr. Fys., 28B, no. 12 (1942).

*DUBROVSKIĬ, V. M.: *Generalization of the theory of purely discontinuous stochastic processes*, Dokl. Akad. Nauk SSSR, vol. 19 (1938).

*——— *Investigation of purely discontinuous stochastic processes by the methods of integro-differential equations*, Izv. Akad. Nauk SSSR, vol. 8 (1944).

FELLER, W.: *Zur Theorie stochastische Prozesse (Existenz und Eindeutigkeitssätze)*, Math. Ann. 113 (1936).

KARHUNEN, K.: *Über lineare Methoden in der Wahrscheinlichkeitsrechnung*, Ann. Acad. Sci. Fennicae, ser. A, I, Math. Phys. 37 (1947).

KHINTCHINE, A. YA.: *Korrelationtheorie der stationäre stochastischen Prozesse*, Math. Ann. 109 (1934).

*KOLMOGOROV, A. N.: *Simplified proof of the ergodic theorem of Birkhoff-Khintchine*, Uspekhi Mat. Nauk, no. 5 (1938).

——— *Über die analytischen Methoden in der Wahrscheinlichkeitsrechnung*, Math. Ann., 104 (1931).

——— *Interpolation und Extrapolation von stationären zufälligen Folgen*, Bull. Acad. Sci. U.R.S.S., Ser. Math., 5 (1941).

*——— *Statistical theory of vibrations with a continuous spectrum*, Jubilee Sbornik Akad. Nauk SSSR, part 1 (1947).

KOLMOGOROV, A. N. and N. A. DMITRIEV: *Branching stochastic processes*, Compt. Rend. Acad. Sci. U.R.S.S., n.s., vol. 56 (1947).

*KOLMOGOROV, A. N. and B. A. SEVAST'YANOV: *The computation of final probabilities for branching stochastic processes*, Dokl. Akad. Nauk SSSR, vol. 56, no. 8 (1947).

LOÈVE, M.: *Sur les fonctions aléatoires stationnaires de second ordre*, Rev. Sci., 83, no. 5 (1945).

——— *Fonctions aléatoires a decomposition orthogonale exponentielle*, Rev. Sci., 84, no. 3 (1946).

MARUYAMA, G.: *The harmonic analysis of stationary stochastic processes*, Member Fac. Sci. Kyushu Univ., A, 4, no. 1 (1949).

*ROZANOV, YU. A.,: *Spectral theory of multi-dimensional stationary processes with discrete time*, Uspekhi Mat. Nauk, vol. 13, no. 2 (1958).

*SEVAST'YANOV, B. A.: *Branching stochastic processes*, Vestnik MGU, no. 3 (1948).

*YAGLOM, A. M.: *On the question of linear interpolation of stationary stochastic sequences and processes*, Uspekhi Mat. Nauk, vol. 4, no. 4 (1949).

*——— *Introduction to the theory of stationary stochastic functions*, Uspekhi Mat. Nauk, vol. 7, no. 5 (1952).

CHAPTER XI

*BERNSTEIN, S. N.: *On the fiduciary probabilities of Fisher*, Izv. Akad. Nauk SSSR, vol. 5 (1941).

DARLING, D. A.: *The Kolmogorov-Smirnov and Cramér-von Mises tests*, Ann. Math. Stat., 28, no. 4 (1957).

*DINKIN, E. B.: *Necessary and sufficient statistics for a family of probability distributions*, Uspekhi Mat. Nauk, vol. 6, no, 1 (1951).

FELLER, W.: *On the Kolmogorov-Smirnov limit theorems for empirical distributions,* Ann. Math. Stat., vol. 19, 2 (1948).

GLIVENKO, V. I.: *Sulla determinazione empirica di una legge di probabilita,* Giorn. dell'Inst. Itai. degli Attuari, vol. 4 (1933).

*GNEDENKO, B. V. and V. S. KOROLYUK:. *On the maximum deviation between two empirical distributions,* Dokl. Akad. Nauk SSSR, vol. 80, no. 4 (1951).

*GNEDENKO, B. V. and E. L. RVACHËVA: *On a problem in the comparison of two empirical distributions,* Dokl. Akad. Nauk SSSR, vol. 82, no. 4 (1952).

KOLMOGOROV, A. N.: *Sulla determinazione empirica di una legge di distribuzione,* Giorn. dell'Inst. Ital. degli Attuari, vol. 4 (1933).

*———— *Determination of the center of dispersion and measure of skewness on the basis of a limited number of observations,* Izv. Akad. Nauk SSSR, vol. 6 (1942).

*———— *Unbiased estimates,* Izv. Akad. Nauk SSSR, vol. 14 (1950).

*KOLMOGOROV, A. N. and YU. V. PROKHOROV: *On sums of a random number of terms,* Uspekhi Mat. Nauk, vol. 4, no. 4 (1949).

NEYMAN, J.: *L'estimation traitée comme une probleme classique de probabilité* (Actualités Scientifiques et Industrielles, no. 739). Paris: Gauthier-Villars, 1938.

*SMIRNOV, N. V.: *Estimation of the deviation between empirical distribution curves for two independent random samples,* Bull. MGU, vol. 2, no. 2 (1939).

*———— *Approximation of distribution laws of random variables on the basis of empirical data,* Uspekhi Mat. Nauk, vol. 10 (1944).

WALD, A.: *Sequential tests of statistical hypotheses,* Ann. Math. Stat., vol. 16 (1945).

INDEX

INDEX

ABSORBING BARRIERS, 119f., 127
Addition, axiom of, 54
　of probabilities, theorem on, 28, 55
Aitchison, J., 220
Axiom(s), of addition, 54
　　extended, 57
　of continuity, 57
　of Kolmogorov, 52, 54
　of probability theory, 54ff.
　　consistency of, 56
　　incompleteness of, 56

BACHELIER, L., 11
Banach's matchbox problem, 123
Bayes' formula, 65, 66, 342
Bernoulli, James, 9, 14, 17, 79, 130, 233
Bernoulli scheme of trials, 79, 100, 146f.
　See also Bernoulli distribution
Bernoulli's Theorem, 107, 108, 112, 227, 237
Bernstein, S. N., 10, 52, 111, 445, 447, 448, 449, 450
Bernstein's Theorem, 249
Bertrand, J., 39
Bertrand's paradox, 40ff.
Binomial distribution law, 81
Birkhoff, G., 378
Birkhoff-Khintchine Theorem, 378, 383
Blanc-Lapierre, A., 378, 445
Bochner, S., 273
Bochner-Khintchine Theorem, 274
Boltzmann statistics, 33
Borel, E., 10, 237, 445
Borel field of events, 53
Borel-Cantelli Lemma, 246, 247
Borel's Theorem, 237, 240f.
Bose-Einstein statistics, 33, 34
Brown, J. A., 220
Brownian motion, 118, 121

Buffon, J., 229
Buffon's needle problem, 41, 43
Bunyakovsky, V., 10
Bunyakovsky-Cauchy inequality.
　See Schwarz inequality

CANTELLI, F., 246, 391
Cases (outcomes) of trial. *See* trial
Cauchy distribution (or law), 176, 191
Central Limit Theorem, 288ff.
Certain events, 15, 24, 48, 54
Chain, simple Markov, 125
　homogeneous, 126
Characteristic function, 250ff.
　of n-dimensional random variable, 277
Chebyshev. *See* Tchebychev
Coefficient of correlation. *See* correlation coefficient
Compatibility condition, 367
Complementary events, 24, 54
Complete group, of events, 25
　of pairwise mutually exclusive events, 25
　of possible outcomes of trial, 27
Compound probability, theorem on, 61
Conditional probability, 59
Confidence limits, 387, 420
Convergence of sequence of random variables, 237
　almost certain, 238
　in probability, 224, 238
　of sequence of distribution functions, weak, 263
　with unit probability, 238
Converse limit theorem, 269
Correlation coefficient, 200, 201
Correlation function, 369f.
Correlation theory, 369
Covariance, 200

455

Covariance matrix, 200
Cramér, H., 46, 259, 378, 445, 448, 450
Cramér's Theorem, 171
Critical region, 403
   optimum, 404
Cumulants, 218, 253

DE MÉRÉ, problem of, 77
Decomposability of event into mutually exclusive events, 25
DeMoivre, A., 9, 84
DeMoivre-Laplace Limit Theorems, 84ff., 85, 88, 94, 106, 107, 286, 297
   generalization to chain-dependent trials, 136ff.
Density. *See* function, probability density
Deviation, standard, 195, 389
   normalized, 207
Die, 23
Difference of events, 23
Direct limit theorem, 269
Discrete random variables, 148, 153
Distribution, Bernoulli, 146f., 295
   Binomial, 81
   Cauchy, 176, 177, 191
   estimation of unknown parameters in, 385ff., 409ff.
      classical procedure for, 408ff.
   χ^2, 172, 174, 285
   Gaussian. *See* distribution, normal
   Laplace, 219
   lattice, 295
   logarithmically normal, 220
   Maxwell, 166, 173, 219
   normal (or Gaussian), 147, 154f., 159ff., 162, 164, 170, 171, 175, 180, 191, 192, 194, 199, 214, 254, 304, 310, 311
      n-dimensional, non-degenerate, 180, 282
         proper (non-degenerate), 180, 282
      two-dimensional, 159ff., 162, 170, 175, 179, 180, 194
         reduction to system of independent variables, 180
   Pascal, 219
   Poisson, 114, 147, 192, 254, 295, 304, 310
   Polya, 219
   Simpson (or triangular), 169
   Student's (or t), 176f.
   triangular (or Simpson), 169
   uniform, 154, 159, 168, 170
Distribution function. *See* function
Distribution law, 152
Distribution span, 296
   maximal, 297
Doeblin, W., 130, 448

ELEMENTARY EVENTS, 26, 53
Elementary system of random variables, 316
   conditions for, 319
Ellipse of equal probabilities, 163
Empirical distribution function.
   See function
Empty set, 53
Encounter problem, 38, 54
Equal likelihood, 22
Equivalence of events, 22
Ergodic Theorem, of Birkhoff-Khintchine, 378, 383
   of Markov, 132, 135
Errors of Type I and Type II, 403f.
Essential states, 130f. *See also* states
Estimation of distribution parameters, 385, 386, 401ff., 409ff.
Event(s), certain, 15, 24, 48, 54
   complementary, 24, 54
   complete group of, 25
   complete group of pairwise mutually exclusive, 25
   decomposable into mutually exclusive events, 25
   difference of, 23
   elementary, 26, 53
   equivalence of, 22
   field of, 25, 53
      Borel, 53
   implication of, 22
   impossible, 15, 24, 54
   (stochastically) independent, 61, 62
      collectively (or mutually), 62
   intersection of, 22, 24
   mutually (or collectively) independent, 62

Events (cont'd)
  mutually exclusive, 24, 29, 54
    events decomposable into sum of, 25
  pairwise mutually exclusive, 25
    complete group of, 25
  product of, 22, 24
  random, 15, 48, 53
    probability of, 47
  set of conditions for, 15
  sum of, 22, 24
  sure (or certain), 15
  union of, 22, 24
Expectation (or mathematical expectation), 190, 194, 199, 202ff.
  conditional, 192, 204
Expected value. *See* expectation
Extended axiom of addition, 57

FAVORABLE CASE (or outcome) of trial, 27
Feller, W., 10, 130, 322, 445, 447, 449, 450, 451
Fermat, P., 9, 17
Fermi-Dirac statistics, 33, 35
Field of events. *See* events
Fisher, R., 419
Fokker, A. D., 11, 347
Fokker-Planck equation, 330
Fortet, R., 378, 446
Fourier transforms, 250
Fry, T. C., 362
Function, characteristic, 250ff., 277
  distribution, 146, 151f., 259
    conditional, 157, 339
    empirical, 388
    estimation of unknown, 385ff.
    jump of, 151
    multi-dimensional, 157
    properties of, 150, 151, 152
    of quotient, 174, 185
    for sums, 167, 185, 317ff.
    two-dimensional, 194, 202
    uniqueness theorem for, 259, 281
  jump, 365
  non-negative definite. *See* positive-semidefinite
  positive-definite. *See* positive-semidefinite

  positive-semidefinite, 273ff.
  probability density, 154, 162
    conditional, 339
    of random vector, 162

GAUSS, K., 9
Gaussian distribution. *See* distribution, normal
Geiger-Müller counter, 74
Glivenko, V. I., 389, 446, 451
Glivenko's Theorem, 391
Gnedenko, B. V., 284, 297, 445, 446, 447, 449, 451
Gnedenko's Theorem, 297
Gosset, W. L. ("Student"), 177
Group. *See* complete group

HÁJEK-RÉNYI INEQUALITY, 242, 448
Helly's theorems, 264, 266f.
Huyghens, C., 9, 17
Hypotheses, composite, simple, etc.
  See statistical hypotheses
  formula for probabilities of.
  See Bayes' formula

IMPLICATION OF ONE EVENT BY ANOTHER, 22
Impossible event. *See* event
Independence, 61, 62, 164
  mutual (or collective) of several events, 62
    and pairwise independence, 62
  of random variables, 164
  of two events, 61
Independent trials, repeated, 79
Inequality, Hájek-Rényi, 242, 448
  Kolmogorov's, 241, 244
  Schwarz, 197
Infinitely divisible distribution law, 304, 307
Integral, Stieltjes, 181, 184
  stochastic, 374
Integral theorem of DeMoivre-Laplace.
  See DeMoivre-Laplace
Intersection of events, 22, 24
Inversion formula for characteristic functions, 256, 258, 281

458   INDEX

JUMP OF DISTRIBUTION FUNCTION, 151
Jump function, 365

KARHUNEN, K., 378, 450
Khinchin. *See* Khintchine
Khintchine, A., 10, 50, 233, 273, 367, 445, 446, 447, 448, 449, 450
Khintchine's Theorem, 230, 371
Kolmogorov, A., 10, 11, 52, 210, 211, 220, 233, 241, 244, 307, 326, 344, 354, 363, 378, 393, 446, 447, 450, 451
Kolmogorov, axioms of, 52, 54
Kolmogorov-Feller equations, 354
Kolmogorov's compatibility criterion, 394
Kolmogorov's equations, 344, 347, 352
Kolmogorov's inequality, 241, 244
Kolmogorov's Theorem, 241, 307, 393
Krein, M. G., 284, 449

LAPLACE, P., 9, 14, 45, 84
Laplace distribution, 219
Laplace Theorem. *See* DeMoivre-Laplace Theorem
Large numbers. *See* Law of Large Numbers
Lattice distribution, 295
Lattice points, 99
Law, Cauchy, 176, 177
  distribution, 152
  infinitely divisible, 304, 307
  of Large Numbers (Bernoulli Theorem), 107, 112, 223, 225, 237
    for dependent variables, 232, 234
    necessary and sufficient condition for, 233f.
    strong, 237, 241
    in Tchebychev's form, 225
  Poisson, 114, 304, 310
  probabilistic (or stochastic), 16
  stochastic (or probabilistic), 16
  Student's, 177
Lévy, P., 10, 446
Liapounov, A. M., 10, 287, 303, 449
Liapounov condition, 290
Liapounov's Theorem, 294
Likelihood, equal, 22
Limit theorem, central, 288

converse, 269
direct, 269
for infinitely divisible laws, 312
integral, 94ff., 143, 286
local, 85, 93, 95, 136ff., 295
for sums, 317
of Smirnov, 396, 400
See also DeMoivre-Laplace Limit Theorems
Limiting probabilities, theorem on, 133
Limits, in von Mises' theory, 50
See also convergence
Lindeberg, J., 11, 449
Lindeberg condition, 289, 290, 322
Linnik, Yu. V., 275, 449
Lobachevsky, N. I., 9, 14, 170
Local Laplace Theorem, 84ff. *See also* DeMoivre-Laplace
Local Limit Theorem, 85
Loève, M., 378, 446, 450
Logarithmic normal law, 220
Lomnitsky, A., 10
Lyapunov. *See* Liapounov

MAKEHAM'S FORMULA, 74
Markov, A. A., 10, 14, 125, 135, 224, 232, 233, 287, 289, 446, 448
Markov, generalized equation of, 343
Markov chains, 125ff.
  simple, 125
  homogeneous, 126
Markov's Theorem, 132, 135, 230, 232, 249
Markov processes, 326
Mass phenomena, 13, 16
Mathematical expectation.
  See expectation
Mathematical probability, 17. *See also* probability
Mathematical statistics, 11, 384ff.
  problems of, 385ff.
Matrix, covariance, 200
  of transition probabilities (transition matrix), 126
Maxwell distribution, 166, 173, 219
Method of truncation, 230
Mises, R. von, 14, 49, 50, 446
  theory of, 49ff.

Moments, 213
    absolute, 214ff.
    central, 213
        mixed, of second order.
            See covariances
    k-th order, 213, 253
    about origin, 213
    problem of, 218
Mortality tables, construction of, 73, 74
Most probable value, 83
Multiplication Theorem, 61
Mutually exclusive random events, 54

NEKRASOV, P., 14
Neyman, J., 419, 451
Neyman-Pearson Theorem, 404
Non-hereditary processes. See stochastic processes without aftereffect
Normalized deviation, 207
Normalized random variables, 207

ORDINARINESS, condition of, 331
Outcomes of trial. See trial

PASCAL, B., 9, 17
Pascal distribution, 219
Pearson, K., 14, 47, 229
Period of state, 131
Planck, M., 11, 347. See also Fokker-Planck
Poincaré, H., 11
Poisson, S. D., 9, 113, 233, 253
Poisson distribution (or law), 114, 147
Poisson integral, 155
Poisson process, 330ff., 335
Poisson's Theorem, 113ff., 229, 323
Polya distribution, 219
Probabilistic (or stochastic) law, 16
Probabilistic (or stochastic) process, 326ff., 342
Probabilities, theorem on addition of, 28
    compound, theorem on, 61
    theorem on limiting, 132ff.
Probability, 16, 17, 19, 28, 37, 47, 54, 58
    axiomatic definition of, 54ff., 58
    of causes (or hypotheses), 65
        formula for, 65

classical definition of, 19, 22, 26, 29, 37, 40, 41, 44, 52, 54
conditional, 59, 339
convergence in, 224, 238
as degree of certainty, 19ff.
existence of, as meaningful statement, 18
geometrical, 37ff., 38, 39
measure theory and, 55, 58, 148, 210ff.
numerical examples of, 30ff.
as objective property of phenomena, 17
properties of, 28, 29
statistical definition of, 19, 22, 44ff., 47, 48
total, formula of, 64
of transition, 125, 126
unconditional, 58
von Mises' definition of, 51, 52
Probability density function. See function
Problem, of Buffon, 41
    of de Méré, 77
    encounter, 38, 54
    of gambler's ruin, 67ff.
    (Banach's) matchbox, 123
    of moments, 218
Process, 325. See also stochastic processes
Product of events, 22, 24
Prokhorov, Yu. V., 211, 447, 448, 449, 451

RAIKOV, D. A., 260, 448, 449
Random events, 15, 48, 53
Random numbers, tables of, 46f.
Random process. See Stochastic process
Random variable(s), 145, 146, 148, 153, 155, 157, 180
    continuous, 154
    convergence of sequence of, 237
    discrete, 148, 153
        possible values of, 153
    distribution law of, 152
    functions of, 166ff., 180ff.
    independent, 164
    multi-dimensional, 157
    normally distributed, 147, 154
    two-dimensional, 159f., 170, 177ff.
    uncorrelated, 372
    See also distribution

Random vector (or multi-dimensional random variable), 157f.
  probability density function of, 162
  uniformly distributed, 159
Random walk, 118ff., 125, 126, 127
  as reformulation of ''gambler's ruin,'' 67
Randomness, von Mises' principle of, 50
Reflecting barriers, 119f., 125
Region, critical, 403
  optimum, 404

SCHWARZ INEQUALITY, 197
Semi-invariant (or cumulant), 218, 253
  of k-th order, 253
Sequence, of independent trials, 79ff.
  of random variables, convergence of, 238
  stationary, 373
    in wide sense, 373
σ-algebra of events, 53
Simpson distribution law, 169
Singular interval, 379
Slutsky, E. E., 10, 373, 448
Slutsky's Theorem, 373
Smirnov, N. V., 396, 445, 447, 451
Smirnov's Limit Theorem, 396, 400
Spectral decomposition of stationary processes, 378
s-Singular interval, 380
Standard deviation, 195
State(s), 125, 130
  classification of, 130
  essential, 130ff.
    equivalence classes of, 130
    communicating, 130
    period of, 131
  transient (or unessential), 130
  unessential (or transient), 130
Stationary process, 330
Stationary sequence. See sequence
Statistical hypotheses, comparison of, 401ff. 409
  composite, 387, 402
  simple, 387, 402
  testing of, 387, 394, 403f.

Statistics, mathematical, 384
  vital (or demographic), 45f.
  See also Boltzmann, Bose-Einstein, etc.
Stieltjes integral, 181, 184
  improper, 181, 184
Stirling's formula, 32
Stochastic independence of events.
  See independence
Stochastic integral, 374
  Stieltjes, 376
Stochastic (or probabilistic) law, 16
Stochastic (or probabilistic) process, 326ff., 342
  continuous, 344
  homogeneous, with independent increments, 360, 363
  of Markov type, 326, 327
  purely discontinuous, 353
  realization of, 328
  stationary, 367
    in wide sense, 369
    continuous, 370
  without aftereffect, 327
  with discrete spectrum, 373
Strong law of large numbers, 237, 241, 245
''Student'' (W. L. Gosset), 177
Student's distribution (or law), 176f.
Sum of events, 22, 24
Sure events. See Certain events
Symmetry condition, 367

TCHEBYCHEV, P. L., 10, 14, 224, 225, 231, 287, 449
Tchebychev's inequality, 225, 229, 240
Tchebychev's Theorem, 226, 227, 229, 230, 232
t-distribution. See distribution, Student
Total probability, formula of, 64
Trajectory, 397
Transient state, 130
Transition probabilities, 126ff.
Transition matrix, 126
Trial(s), 27
  Bernoulli (scheme of), 79
  complete group of possible outcomes of, 27
  favorable case (or outcome) of, 27

Trial(s), repeated independent, 79
    extension of Bernoulli Theorem to, 112
Triangular distribution law, 169
Truncation, method of, 230

UNCONDITIONAL PROBABILITY, 58
Unessential state, 130
Uniqueness theorem for distribution function, 259, 281
Union of events, 22, 24

VALUE, most probable, 83
Variables. *See* random variables
Variance, 195ff., 200, 206ff.
Variational series, 388
von Mises. *See* Mises

WEAK CONVERGENCE, 263

YAGLOM, A. M., 378, 445, 450

ANSWERS TO THE EXERCISES

ANSWERS TO THE EXERCISES

Chapter I

2. a) $B + AC$; b) A; c) AB.

4. $1/12$.   5. $2/5$.   6. $\binom{M}{m}\binom{N-M}{n-m} \Big/ \binom{N}{n}$.

7. $(n-b)(2m+n-b-1)/(m+n-b)(m+n-b-1)$.

9. $\sum_{j=1}^{n} \binom{m}{j}\cdot\binom{n}{j} \Big/ \binom{m+n}{2j}$.

10. $\sum_{j=1}^{n} (-1)^{j+1}/j!$.

11. Answer as in Ex. 10.

12. $\sum_{j=1}^{n}\binom{n}{j}^2 \Big/ \sum_{j=1}^{n}\binom{2n}{2j} = [(2n)!/(n!)^2 - 1]/(2^{2n-1} - 1)$.

13. P{at least one ace} $= 1 - (5/6)^4 = 0.518$,
    P{at least one double ace} $= 1 - (35/36)^{24} = 0.491$.

14. $1/2$.   15. $1/4$.   16. $a/2b$ $(a \leq b)$.

17. a) $(1 - r^3/R^3)^N$; b) $\exp(-4\pi\lambda r^3/3)$.

18. a) $1 - (1-p_1)(1-p_2)\ldots(1-p_n)$;
    b) $(1-p_1)(1-p_2)\ldots(1-p_n)$;
    c) $\sum_{i=1}^{n} p_i(1-p_1)\ldots(1-p_{i-1})(1-p_{i+1})\ldots(1-p_n)$.

20. $\exp(-\lambda t)$.

21. First show that $p_i(t)$ satisfies the system of differential equations
    $$p_i'(t) = -aip_i(t) + a(i-1)p_{i-1}(t) \qquad (i = 1, 2, \ldots, n).$$
    Using the initial conditions $p_1(0) = 1$ and $p_i(0) = 0$ for $i > 1$, conclude that $p_i(t) = e^{-at}(1 - e^{-at})^{i-1}$.

Chapter II

1. a) 0.238; b) 0.752.   2. a) 63/256; b) 957/1024.
3. a) $1/12^4$; b) $11/12^3$.   4. $1 - 2\Phi(3.8) = 0.00014$.

5. a) $\sum_{m=0}^{N}\binom{n}{m}p^m q^{n-m}$, where $N = [np]$ or $np - 1$, according as np is not or is an integer;

   b) For $r \geq 1/p$, the required probability is zero. For $r < 1/p$ and np large, it is given by
   $$1 - \Phi(np(r-1)/\sqrt{npq}) - \Phi(np/\sqrt{npq}).$$

6. a) 2; b) $1 - 730!/2^{365}365^{730}$.

7. a) 0.133; b) 0.858;

   c) 104. Let x be the number of bits over 100 in the box. Then one can write:

   $\mathsf{P}\{x \text{ or fewer defective bits}\} =$
   $$= \sum_{m=0}^{x}\binom{100+x}{m}(0.02)^m(0.98)^{100+x-m} \sim \sum_{m=0}^{x} 2^m e^{-2}/m! \geq 0.9.$$

8. a) $\mathsf{P}\{121 \text{ or more deaths}\} \sim 1 - 2\Phi(7.77)$;

   b) $\mathsf{P}\{80 \text{ or fewer deaths}\} \sim \Phi(2.59) + \Phi(7.77) = 0.995$,
   $\mathsf{P}\{60 \text{ or fewer deaths}\} \sim \Phi(7.77) = 0.50$,
   $\mathsf{P}\{40 \text{ or fewer deaths}\} \sim \Phi(7.77) - \Phi(2.59) = 0.005$.

9. $\binom{2n-r}{n}2^{-2n-r}$.

10. For $r = 1, 2, \ldots, n-1$,
    $$P_r'(t) = -[\beta(n-r) + \alpha r]P_r(t) + \alpha(r+1)P_{r+1}(t) + \beta(n-r+1)P_{r-1}(t).$$
    For $r = 0$, $P_0'(t) = -\beta n P_0(t) + \alpha P_1(t)$.
    For $r = n$, $P_n'(t) = -\alpha n P_n(t) + \beta P_{n-1}(t)$.

11. For $r = 1, 2, \ldots, n-1$,
    $$P_r'(t) = -[\alpha(n-r) + \beta]P_r(t) + \beta P_{r+1}(t) + \alpha(n-r+1)P_{r-1}(t).$$
    For $r = 0$, $P_0'(t) = -\alpha n P_0(t) + \beta P_1(t)$.
    For $r = n$, $P_n'(t) = -\beta P_n(t) + \alpha P_{n-1}(t)$.

13. a) Use mathematical induction;
    b) Apply the Local Limit Theorem.

Answers to Exercises

Chapter III

1. $\pi_n = \pi_1$.

2. a) $P_{ij}(2) = c_i \sum_{k=1}^{\infty} c_k \exp\{-a(|i-k|+|k-j|)\}$.

   b) $(e^a - 1)/(e^a + 1 - e^{(1-i)a})$.

3. $2P_{ij}(n) = 1 - P_{ij}(n-1)$ for all i and j; $p_j = 1/3$ for all j.

Chapter IV

2. a) $F[(x-b)/a]$ $(a>0)$, $1 - F[(x-b)/a]$ $(a<0)$; $|a|^{-1}p[(x-b)/a]$;

   b) $F(0) - F(1/x)$ $(x<0)$, $F(0) + 1 - F(1/x)$ $(x>0)$; $x^{-2}p(1/x)$;

   c) $F(\arctan x)$, $p(\arctan x)/(1+x^2)$;

   d) 0 $(x \leq -1)$, 1 $(x>1)$, $1 - F(\arccos x)$ $(-1 < x \leq 1)$; $p(\arccos x)/\sqrt{1-x^2}$ $(|x|<1)$, 0 (elsewhere);

   e) Let $a \leq f(x) \leq b$. Then $F_\eta(x) = 0$ $(x \leq a)$, 1 $(x>b)$, and $F(f^{-1}(x))$ or $1 - F(f^{-1}(x))$ $(a < x \leq b)$, depending on whether $f(x)$ is increasing or decreasing; in both cases, $p_\eta(x) = p(f^{-1})|df^{-1}/dx|$ $(a < x < b)$, and 0 elsewhere.

3. a) $2\pi^{-1}\arctan(x/a)$ $(0 \leq x)$; b) $1/2 + \pi^{-1}\arctan(x/a)$ $(|x| < \infty)$.

4. a) $1/\pi\sqrt{R^2-y^2}$ $(|y|<R)$, 0 (elsewhere);

   b) $2/\pi\sqrt{4R^2-y^2}$ $(0<y<2R)$, 0 (elsewhere).

5. 0 $(z \leq 0)$, $1 - \sqrt{4R^2-z^2}/2R$ $(0 \leq z \leq 2R)$, 1 $(z \geq 2R)$.

6. 0 $(x \leq \pi a^2/4)$, $(\sqrt{4x/\pi}-a)/(b-a)$ $(\pi a^2/4 \leq x \leq \pi b^2/4)$, and 1 $(x \geq \pi b^2/4)$.

7. a) $2/\pi$; b) $(2 \arctan e)^2/\pi^2$.

8. a) $F_3 = -F_2(-\infty)F_1(x)$. ξ and η are independent if $F_1(x)$ and $F_2(y) - F_2(-\infty)$ are their respective distribution functions;

   b) $F_4(y) = -F_3(-\infty) - F_1(-\infty)F_2(y)$, $F_3(x) = -F_2(-\infty)[F_1(x) - F_1(-\infty)] + F_3(-\infty)$, and so $F(x,y) = [F_1(x) - F_1(-\infty)][F_2(y) - F_2(-\infty)]$.

9. 0 $(x \leq 0)$, $x(2a-x)/a^2$ $(0 \leq x \leq a)$, 1 $(x \geq a)$.

10. a)
$$p(x_k) = \begin{cases} \dfrac{n!}{a^n(k-1)!(n-k)!} x_k^{k-1}(a-x_k)^{n-k} & (0 \leq x_k \leq a), \\ 0 & \text{(elsewhere)}; \end{cases}$$

b)
$$p(x_k, x_m) = \begin{cases} \dfrac{n!}{a^n(k-1)!(m-k-1)!(n-m)!} x_k^{k-1}(x_m-x_k)^{m-k-1}(a-x_m)^{n-m} \\ \hspace{4cm} (0 \leq x_k < x_m < a), \\ 0 \hspace{3cm} \text{(elsewhere)}. \end{cases}$$

11. a) $(F(x))^n$; b) $1-(1-F(x))^n$;

c) $F_{\xi_k}(x) = B(F(x); n-k+1, k)/B(n-k+1)$, the numerator and denominator being incomplete and complete Beta functions; ξ_1 is the largest observed value.

d) $F_{(\xi_m, \xi_k)}(x, y) = [B(k, m-k)B(m, n-m+1)]^{-1} \cdot$
$$\cdot \int_0^{F(y)} v^{n-k}(1-v)^{k-1} B(F(x)/v; n-m+1, m-k) dv.$$
$$(-\infty < x < y < +\infty)$$

12. a) $F(x, x, \ldots, x)$;

b) $\sum\limits_{i=1}^{n} p_i - \sum\limits_{1 \leq i < j \leq n} p_{ij} + \ldots + (-1)^{n+1} F(x, x, \ldots, x)$, with $p_{ij\ldots k}$ defined as in

(1), § 23, with $b_s = \infty$ and $a_s = x$.

13. Uniformly in $(0, 1)$.

14. a) $c_1 = \beta^{a+1}/\Gamma(a+1)$;

b) $(\beta^{a+\gamma+2} e^{-\beta x} x^{a+\beta+1})/\Gamma(a+\gamma+2)$ for $x > 0$, and 0 elsewhere.

15. $(2h)^{-1} \int\limits_{x-h}^{x+h} F(z) dz$.

16. $x^3/(1+x)^3$ $(x \geq 0)$, 0 $(x \leq 0)$.

17. a) $1/2 + \pi^{-1} \arctan(x/2)$;

b) 0 $(x \leq -4)$, $(x+4)^2/48$ $(-4 \leq x \leq 0)$, $(x+2)/6$ $(0 \leq x \leq 2)$, $1-(6-x)^2/48$ $(2 \leq x \leq 6)$, 1 $(x \geq 6)$.

c) $(1/2 - x/4a) \exp(x/a)$, for $x \leq 0$;
$1 - (1/2 + x/4a) \exp(-x/a)$, for $x \geq 0$.

18. a) $(x+1)^{-2}$ $(x > 0)$, 0 $(x < 0)$;

b) 0 $(x < 0)$, $1/2$ $(0 < x \leq 1)$, $1/2x^2$ $(x \geq 1)$.

19. $\int\limits_0^{\infty} F_1(x/z) dF_2(z) + \int\limits_{-\infty}^{0} \{1 - F_1(x/z)\} dF_2(z)$.

20. a) 0 $(x \leq -a^2)$, $(2a^2)^{-1}\{x \log(a^2/|x|) + a^2 + x\}$ $(|x| \leq a^2)$, 1 $(x \geq a^2)$;

b) $2\pi^{-1} \int_0^{\pi/4} \exp(x/\sin 2\varphi) d\varphi$ $(x \leq 0)$;

$1 - 2\pi^{-1} \int_0^{\pi/4} \exp(-x/\sin 2\varphi) d\varphi$ $(x \geq 0)$.

21. For $a \leq \pi/2$,

$$F_\xi(x) = \begin{cases} 0 & (x \leq 0), \\ \int_0^x F_\xi(A+B) dF_\eta(u) + \int_x^{x \csc a} [F_\xi(A+B) - F_\xi(A-B)] dF_\eta(u) & (x \geq 0), \end{cases}$$

where $B = \sqrt{x^2 - u^2 \sin^2 a}$, $A = u \cos a$. The second integral drops out when $a > \pi/2$.

25. $[1/2\pi\sigma^2 \sqrt{m(n-m)}] \cdot$
   $\cdot \exp[-(x-ma)^2/2m\sigma^2 - (x-y-(n-m)a)^2/2(n-m)\sigma^2]$.

27. $\binom{m+\mu}{\mu} \cdot \binom{M-m+N-\mu-1}{N-\mu} / \binom{M+N}{N}$.

28. $(n-2)x^{n-3}$ $(0 < x < 1)$, 0 (elsewhere). *Hint*: Use the procedure of Ex. 11d and the result of Ex. 13 to first derive the joint distribution of the variables $F(x_1)$, $F(x_2)$, and $F(x_3)$.

Chapter V

1. a) Show, for example, that $f(x) = (1 + a - ae^x)^{-1}$ is a moment-generating function, i.e., $\mathsf{E}\xi^n = f^{(n)}(0)$; $\mathsf{E}\xi = f'(0) = a$, $\mathsf{E}\xi^2 = f''(0) = a + 2a^2$, etc.; $\mathsf{D}\xi = a(a+1)$.

   b) $\mathsf{E}\xi = a$; $\mathsf{D}\xi = a(a\beta + 1)$.

2. a) $n(n-1)(n-2)p^3 + 3n(n-1)p^2 + np$;

   b) $n(n-1)(n-2)(n-3)p^4 + 6n(n-1)(n-2)p^3 + 7n(n-1)p^2 + np$;

   c) $2 \sum_{k=0}^{[np]} (np - k) \binom{n}{k} p^k q^{n-k}$.

3. c) $\sum_{i=1}^n p_i q_i (q_i - p_i)$;

   d) $\sum_{i=1}^n (p_i q_i - 3p_i^2 q_i^2) + 6 \sum_{1 \leq i < j \leq n} p_i q_i p_j q_j$.

5. $\sum_{k=0}^{N} \binom{n}{2k+1} p^{2k+1} q^{n-2k-1}$, where $N = [(n-1)/2]$.

6. $\mathsf{E}\xi = a$; $\mathsf{D}\xi = 2a^2$.

7. $\mathsf{E}v = 2a/\sqrt{\pi}$; $\mathsf{E}(\text{k.e.}) = 3ma^2/4$; $\mathsf{D}v = 3a^2/2 - 4a^2/\pi$; $\mathsf{D}(\text{k.e.}) = 3m^2a^4/8$.

8. $\mathsf{E}s = (2x_0\Phi(x_0/2\sqrt{Dt}) + 2\sqrt{Dt}\exp(-x_0^2/4Dt))/\sqrt{\pi}$; $\mathsf{E}s^2 = x_0^2 + 2Dt$.

12. $\mathsf{E}\xi = l/3$; $\mathsf{D}\xi = l^2/18$; $\mathsf{E}\xi^n = 2l^n/(n+1)(n+2)$.

13. $\mathsf{E}\xi = \exp(\alpha + \beta^2/2)$; $\mathsf{D}\xi = \{\exp(\beta^2) - 1\}\exp(2\alpha + \beta^2)$.

14. $(k+1)^{-1}(b/2)^k$.

15. $\mathsf{E}s = mn/2$; $\mathsf{D}s = mn(2^n - m)/(2^{n+2} - 4)$.

16. $(n-m)/n$.

17. $r = (\alpha^2 - \beta^2)/(\alpha^2 + \beta^2)$;
$$(4\pi\alpha\beta\sigma^2)^{-1}\exp[-Q(x,y)(\alpha^2+\beta^2)/8\alpha^2\beta^2\sigma^2],$$
where
$$Q(x,y) = [x-(\alpha+\beta)a]^2 - 2r[x-(\alpha+\beta)a][y-(\alpha-\beta)a] + \\ + [y-(\alpha-\beta)a]^2.$$

21. a) $\mathsf{E}|\xi - a| = a(\lambda - 1) + 2aF_\xi(a)$; b) $a\lambda = \sigma\sqrt{2/\pi}$.

Chapter VI

3. Use Khintchine's Theorem.

4. Use Markov's Theorem.

Chapter VII

1. a)
$$F(x) = \begin{cases} \sum_{j=k+1}^{\infty} a_j/2 & (-k-1 < x \leq -k) \\ & (k = 0, 1, \ldots) \\ (1 + \sum_{j=0}^{k} a_j)/2 & (k < x \leq k+1). \end{cases}$$

b) $0\ (x \leq \lambda_0)$, $\sum_{j=0}^{k} a_j\ (\lambda_k < x \leq \lambda_{k+1})$ $(k = 0, 1, \ldots)$.

2. a) $a^2/(a^2 + t^2)$; b) $e^{-a|t|}$; c) $(4/a^2t^2)\sin^2(at/2)$.
   d) $0\ (|t| \geq a)$, $(a-|t|)/a\ (|t| \leq a)$.

4. a) $0\ (x \leqq -1)$, $1/2\ (-1 < x \leqq 1)$, $1\ (x > 1)$;
   b) $0\ (x \leqq -2)$, $1/4\ (-2 < x \leqq 0)$, $3/4\ (0 < x \leqq 2)$, $1\ (x > 2)$.
   c) $e^x\ (x \leqq 0)$, $1\ (x \geqq 0)$;
   d) $0\ (x \leqq -a)$, $(x+a)/2a\ (|x| \leqq a)$, $1\ (x \geqq a)$.

Chapter IX

1. a) $\varphi_n(t) = (1 + a\beta - a\beta e^{it})^{-1/\beta}$, with $a = a/n$, $\beta = n$ [see b)];
   b) $\varphi_n(t) = (1 + a'\beta' - a'\beta' e^{it})^{-1/\beta'}$, with $\beta' = \beta n$, $a' = a/n$;
   c) $\varphi_n(t) = e^{-a|t|/n}$.

2. $\varphi_n(t) = \beta^{a'}/(\beta - it)^{a'}$, with $a' = a/n$.

3. Use the result of Ex. 2 and Theorem 2 of § 44. The characteristic function of the Laplace distribution is $e^{iat}/(1 + a^2 t^2)$.

4. a) $\gamma = a/\beta$, $G(x) = a[1 - e^{-\beta x}(\beta x + 1)/\beta^2]$;
   b) $\gamma = a$, $G(x) = \begin{cases} -axe^{-x/a} - a^2 e^{-x/a} + a^2 & (x \geqq 0), \\ -axe^{x/a} + a^2 e^{x/a} - a^2 & (x \leqq 0). \end{cases}$

CPSIA information can be obtained
at www.ICGtesting.com
Printed in the USA
BVHW030403180321
602778BV00007B/523